AF763335

NUMERICAL MODELS FOR SUBMERGED BREAKWATERS

NUMERICAL MODELS FOR SUBMERGED BREAKWATERS

Coastal Hydrodynamics and Morphodynamics

AMIR SHARIF AHMADIAN

ELSEVIER

AMSTERDAM • BOSTON • HEIDELBERG • LONDON
NEW YORK • OXFORD • PARIS • SAN DIEGO
SAN FRANCISCO • SINGAPORE • SYDNEY • TOKYO
Butterworth-Heinemann is an imprint of Elsevier

B H

Butterworth-Heinemann is an imprint of Elsevier
The Boulevard, Langford Lane, Kidlington, Oxford OX5 1GB, UK
225 Wyman Street, Waltham, MA 02451, USA

© 2016 Elsevier Ltd. All rights reserved.

No part of this publication may be reproduced or transmitted in any form or by any means, electronic or mechanical, including photocopying, recording, or any information storage and retrieval system, without permission in writing from the publisher. Details on how to seek permission, further information about the Publisher's permissions policies and our arrangements with organizations such as the Copyright Clearance Center and the Copyright Licensing Agency, can be found at our website: www.elsevier.com/permissions.

This book and the individual contributions contained in it are protected under copyright by the Publisher (other than as may be noted herein).

Notices
Knowledge and best practice in this field are constantly changing. As new research and experience broaden our understanding, changes in research methods, professional practices, or medical treatment may become necessary.

Practitioners and researchers must always rely on their own experience and knowledge in evaluating and using any information, methods, compounds, or experiments described herein. In using such information or methods they should be mindful of their own safety and the safety of others, including parties for whom they have a professional responsibility.

To the fullest extent of the law, neither the Publisher nor the authors, contributors, or editors, assume any liability for any injury and/or damage to persons or property as a matter of products liability, negligence or otherwise, or from any use or operation of any methods, products, instructions, or ideas contained in the material herein.

ISBN: 978-0-12-802413-3

Library of Congress Cataloging-in-Publication Data
A catalog record for this book is available from the Library of Congress.

British Library Cataloguing-in-Publication Data
A catalogue record for this book is available from the British Library.

For Information on all Butterworth-Heinemann publications
visit our website at http://store.elsevier.com/

ELSEVIER
Book Aid International
Working together to grow libraries in developing countries
www.elsevier.com • www.bookaid.org

CONTENTS

CHAPTER 1

Introduction

1.1 COASTAL EROSION AND DEFENSE

Coastal zones are commonly defined as the interface between land and sea. About 71% of the total surface of planet Earth is covered by water (361.13 million km^2), creating 1,634,701 km of coastline (Burke et al., 2001) either with the open oceans, inland seas or both, for 84% of the countries of the world (Martinez et al., 2007). Coasts around the world have been the most favored locations to live permanently, or to utilize for leisure, recreational activities, tourism or commerce and other human activities (Culliton et al., 1990; Miller and Hadley, 2005). The prominent importance of the coasts of the world, considering social, environmental and economic aspects, has been broadly distinguished and a majority of the world's population inhabits coastal zones. According to the United Nations Atlas of the Oceans, 44% of people live within 150 kilometers (93 miles) of the sea (UN Atlas, 2013).

Shorelines are naturally dynamic and continually changing because of the interaction of the sea-level changes, tides, currents, winds, waves, storms and extreme events with seacoasts (Prasetya, 2007). As waves approach the shore, high turbulence, wave-generated surges and currents created by wave breaking cause sediment transport and consequently changes in shoreline through processes of accretion and erosion. Wave-induced erosion and deposition occur on a continuous basis along the coasts. During storms the energy reaching the coast becomes high, resulting in natural hazards with high vulnerability (Martinez et al., 2007). In addition to natural processes, the coastal erosion is further aggravated by human interventions along the coasts, within river catchments and offshore, raising social, environmental and economic concerns in the long term. This range and variety of natural and human causes of erosion, especially in areas with rapidly rising coastal land value, have led to uncertainty on how to treat shoreline erosion (Prasetya, 2007).

One popular engineering approach is use of coastal structures to control longshore currents and offshore energy reaching the coast in order to reduce or even stop the rate of coastal erosion and trap longshore

© 2016 Elsevier Ltd.
All rights reserved.

sediment transport. Various coastal structures can be designed and constructed to control and decrease wave-induced coastal erosion. These structures include groins, seawalls, revetments, dikes, artificial headlands and breakwaters, etc. (Prasetya, 2007).

A groin is a long, narrow structure built perpendicular to the coastline, extending from shore into sea. These structures induce local scour at their toes and cause erosion downdrift. Typically multiple groins should be constructed to successfully contribute to beach stability and often regular maintenance is required. Seawalls are shore-parallel structures that are generally massive structures emplaced along a considerable stretch of shoreline. Scour generally occurs at the base of these structures. Seawalls may also accelerate erosion of adjacent coastline. Artificial headlands are another form of coastal defense that are relatively large structures and can also cause erosion downdrift. Breakwaters, either emerged, semi-submerged or fully submerged, are a form of coastal defense that is designed and constructed for shoreline protection purposes. Emerged breakwaters are large structures and relatively difficult to build and need special design. They are vulnerable to strong wave action. The costs of installing these structures for coastal protection are very high. In addition, strong negative public reaction to rock emplacements along the coast often aggravates the problem (Prasetya, 2007).

1.2 SUBMERGED BREAKWATERS FOR COAST PROTECTION

Recently, submerged breakwaters have become particularly attractive as coastal protection for recreational and residential coastal areas due to their reduced environmental and visual impact. Since they are underwater, they are less subjected to wave action and consequently not exposed to severe wave breaking. A successful design of submerged breakwaters may also cause beach restoration by trapping natural sediments. Lower construction cost compared with other kinds of detached breakwaters is another advantage. The advantages of submerged breakwaters over conventional structures make them more attractive for protecting natural and developed beaches.

Submerged breakwaters are appropriate for all coastlines. They are often constructed for beach protection or to restore eroded beaches, being applied as a preliminary defense system to protect the principal coastal structures, redistribute sediment transport patterns, create desirable beach features, create calmer zones in harbors, prevent siltation or alter the sediment deposition area in port access ways and navigation channel entrances. Therefore, they are one of the major engineering priorities at the moment, playing an important role in beach protection, and use of this kind of structure is continuously increasing. However, on the other hand, their design/project is very complicated.

Basically, a successful application of submerged breakwaters strongly depends on its accurate and effective design. For instance, some unsuccessful applications of these structures due to bad design can be seen around the world. A submerged breakwater placed alongshore in Palm Beach, Florida, is one example (Browder et al., 1996). The main problem in this area has been storm waves and erosion of the beach, as well as very little sediment supplying. Although field measurements and monitoring showed that the reef slightly reduced the incident wave height, erosion in the breakwater lee side was detected. It was observed that the breakwater prevented the overtopped flow to be returned offshore normally and it was instead redirected alongshore and consequently increased the longshore currents and pumping out of sediments (Browder et al., 1996).

Emerged breakwaters cannot be constructed in the form of long continuous structures without gaps (Pinto and Neves, 2003). The gaps between the barriers are necessary in the emergent breakwaters for continuous water exchange between the protected area and shoreside, but often produce rip currents, bed irregularities and tombolos (Pilarczyk, 1996). However, with submerged breakwaters, while they reduce the intensity of wave action, some overtopping is permitted, allowing circulation along the shoreline zone. The sufficient water exchange results in a nature-friendly beach. These structures contribute to dissipation of incident wave energy and provide a calm, sheltered area behind the structure. In some cases, the submerged structures dissipated wave energy more efficiently than the emerged ones. Therefore, submerged breakwaters do not have the disadvantageous features of other structures and can offer significant benefits, making them very suitable for shoreline stabilization (Pinto and Neves, 2003).

Significant change of the nearshore wave field and circulation is caused by construction of submerged breakwaters. These are actually driven by several coastal phenomena such as wave overtopping and breaking over the crest, permeability through the body and wave diffraction around the head of the breakwater. Some of these physical processes have been well understood and widely researched either through numerical simulations or experimental models in the laboratory. However, wave transformations over or around submerged breakwaters, such as wave breaking and wave diffraction and their influence on circulation patterns behind the breakwater, are still not very clear and require further investigation. A review of the literature shows that most work has focused on the two-dimensional effects of submerged breakwaters while, surprisingly, three-dimensional effects have not been studied comprehensively and in detail. This might be because of the higher expense of 3D models and the complex physical processes of these models. There is therefore a need to improve our understanding of flow around submerged breakwaters and thereby to produce better design methods that take 3D effects into account.

1.3 COASTAL PROCESSES AND SUBMERGED BREAKWATERS

Water waves are characterized by a number of physical parameters such as wave height, wave length and period. The wave height is the vertical distance between wave trough and crest, while the wave length is the horizontal distance between consecutive wave crests. The wave period is defined as the time needed for two consecutive crests to pass a stationary point. Waves are commonly created by wind and carry significant amounts of energy. The magnitude of the energy is related to the square of the wave height. However, these characteristics are usually subject to change by different coastal processes when the waves enter shallow water. They are also altered by wave-structure interaction. The main physical processes involved in wave transformation over and around submerged breakwaters are wave shoaling, reflection, refraction, diffraction and breaking, each of which affects transmitted wave height and pattern behind the structure in a very complex way. Understanding the influence of these processes on the wave field around a submerged breakwater is essential in the design process.

Wave transformation by shoaling occurs as the waves approach shallower water perpendicularly, where wave speed and wave length decrease. Therefore, assuming the energy flux is conserved, the energy per unit area of the wave changes, and in shallow water the wave height increases while the wave period is constant. By decreasing the wave length and increasing the wave height to approximately the same as the water depth and consequently increasing the wave steepness, wave breaking might happen. This usually causes the wave to become unstable, curling forward and breaking. However, if waves approach the shallow waters at an angle to the sea floor contours, since wave celerity is dependent on water depth, the part of the wave crest in shallow water moves slower than the part in deeper water and the wave crest bends to align with the bottom contours. This process is called refraction. Wave reflection also occurs when a wave strikes a reflective surface. Wave diffraction is concerned with the transmission of wave energy, in this case across wave rays or along the wave crest. This occurs when waves pass through a gap between two segments of breakwater or around the head of a single breakwater.

In reality, diffraction, refraction and shoaling all occur simultaneously. When waves encounter an obstacle or a sudden change in bathymetry, some of the wave energy will be forced to move across the wave ray or along the wave crest. Although shoaling, refraction and diffraction theory may predict a wave of a certain height, there is a physical limit to the steepness of a wave. Beyond this steepness, the wave can no longer retain its form and will break, dissipating a large portion of its energy (see McCowan, 1891; Miche, 1944; Munk, 1949).

Some physical parameters involved in wave transformation over a submerged breakwater are incident wave height (H_i), offshore wave length (L_o), water depth (h), breakwater crest width (B), submergence depth (h_s), breakwater seaward slope (m) and transmitted wave height H_t. These parameters are required when two-dimensional (hereafter 2D) transformation of waves over a breakwater crest is being studied. However, when three-dimensional (hereafter 3D) processes are being considered, some additional parameters depending on the breakwater geometry and its location might be needed. These parameters include breakwater length, gap size, distance to the beach, etc. A 3D coordinate system will also be necessary to describe the wave field and its spatial variation in three dimensions properly.

Previous research has been published for two-dimensional phenomena such as overtopping (Bruce et al., 2006), reflection (Zanuttigh and van der Meer, 2006), set-up (Calabrese et al., 2008), and wave-induced current (Tajziehchi and Sharif Ahmadian, 2009). The impact of structures has often been expressed in terms of a wave transmission coefficient K_t, because it represents a dominant variable in the shoreline response to structure placement (Hanson and Kraus, 1991). K_t is defined as the ratio between the wave height transmitted behind the submerged breakwater and the incident wave height. Wave transmission is often an important criterion in the design of a breakwater structure and influences early decisions on the type of structure and the choice of construction material.

At design stage classical formulae are used to predict the wave transmission coefficient. Recent results from tests on wave transmission over submerged breakwaters are very encouraging and have led to several design tools including empirical formulae and and neural networks (Buccino and Calabrese, 2007; Goda and Ahrens, 2008; Panizzo and Briganti, 2007; van der Meer et al., 2005).

1.4 NUMERICAL MODELING FOR SUBMERGED BREAKWATERS

The rapid advancement of computers has increased the application of numerical models employed in coastal engineering problems. Various numerical models and modeling techniques have been introduced and applied in different fields of coastal engineering and obviously also in the case of modeling, analysis and design of submerged breakwaters (Chau, 2010).

The modeling can be categorized into different spatial dimensions including one-dimensional, two-dimensional, quasi three-dimensional, and fully three-dimensional models. Some of the discretization techniques are: finite difference method, finite element method, finite volume method, boundary element method, spectral element method, high-resolution discretization schemes, etc. Numerical wave models can also be classified into two main categories of phase-resolving models and phase-averaged models. The application of phase-resolving models is limited to relatively small regions, while phase-averaged models may be applied for larger areas (Liu, 1994; Liu and Wu, 2000).

In addition, there are a large number of other types of computational fluid dynamics or numerical models and techniques which have been successfully used in different coastal engineering problems, including submerged breakwaters or other kind of coastal structures. Some examples are meshless models like smoothed-particle hydrodynamics, statistical models, computational intelligence models such as artificial neural networks, genetic algorithms and programming, evolutionary computations, etc. Further information and description of these models will be presented in the next chapters.

However, the most commonly used wave models in coastal engineering problems as well as coastal structures, and particularly submerged breakwaters, include spectral models, mild-slope equation models, Boussinesq equation models, shallow-water equation models, and quasi and fully three-dimensional models. Disregarding the theoretical concepts behind these models, a brief description of their applicability is presented here. Detailed information about the models will be presented in the next chapters.

The wave spectral model can be used for large scales. These models only provide far-field wave information and not the detailed wave pattern around coastal structures. The mild-slope equation model can be used in both offshore and nearshore areas where wave nonlinearity is not very strong (Lin, 2008).

However, to model nearshore waves with strong wave nonlinearity, a Boussinesq equation wave model is more likely to be adopted. It should be noted, however, that the Boussinesq equations are valid only from intermediate water depth to shallow water before the surf zone. Unlike the wave spectral model and the mild-slope equation model, the Boussinesq model can be used for solitary waves (Lin, 2008).

However, to model tsunami or tides, a shallow-water equation model could be a better option. The limitation of the shallow-water equation model is that it is applicable only when the vertical scale is very small in comparison with the horizontal scale (Lin, 2008).

However, besides two-dimensional models as mentioned here, to model 3D problems, quasi or fully three-dimensional models are necessary, although the computation costs may be more expensive than the two-dimensional models.

The performance of numerical models or predictive tools to a great extent depends on many parameters such as model parameters, the numerical scheme, the numerical techniques, dimensions, boundary conditions, etc. These simulation and design tools necessarily involve certain assumptions and limitations. Therefore, to have a proper numerical model for a practical coastal problem, a comprehensive understanding of the problem domain is essential. An appropriate theoretical representation of the specific physical problem is essential. Then a numerical solution of the mathematical equations should be well performed to obtain accurate and reasonable outcomes.

1.5 PURPOSES AND SIGNIFICANCES

Past research has shown that submerged breakwaters have the potential to be effective in protecting shorelines from destructive water waves. Submerged breakwaters are used to dissipate wave energy, preventing waves reaching the beach, on which they might otherwise cause sediment transport and significant changes in the shoreline through processes of accretion and erosion. The significant role of submerged breakwaters in reducing incident wave height is well understood; however, the design knowledge of submerged breakwaters, including their impacts on wave transmission, currents, sediment processes and shoreline response, is still developing. The effectiveness of submerged breakwaters depends on many physical parameters such as the incident wave climate, breakwater geometry, distance from the shoreline, total water depth and water depth over the structure. The pattern and magnitude of wave energy and the current behind a submerged breakwater have a significant influence on shoreline change behind the structure. Therefore, to obtain a better understanding of the impact of submerged breakwaters, it is necessary to consider both 2D and 3D effects. In general, the hydrodynamics induced by the submerged breakwater is driven by a number of phenomena related to wave transmission behind the structure, namely wave overtopping, percolation and diffraction (Vicinanza et al., 2009).

2D effects of submerged breakwaters have been widely studied, both theoretically and experimentally. Wave energy dissipation over 2D submerged breakwaters is caused by several mechanisms including wave breaking, bottom friction and percolation through the porous structure. Of these mechanisms, wave breaking is the most dominant. However, existing numerical models for energy dissipation due to wave breaking have been developed for beaches with gentle slopes and are not directly applicable to breaking over submerged breakwaters. Many empirical approaches for estimating 2D wave transmission have also been developed (Allsop, 1983; Bleck and Oumeraci, 2002; Buccino and Calabrese, 2007; d'Angremond et al., 1996; Goda, 1969; Goda and Ahrens, 2008; Panizzo and Briganti, 2007; Seabrook and Hall, 1998; van der Meer et al., 2005; Wamsley and Ahrens, 2003). The empirical equations are very useful as an initial assessment of the protection level. However, the only information provided by these equations is the average K_t and not any information on the spatial distribution of wave height behind the structure. This does not guarantee reliable and accurate design tools for engineers (Vicinanza et al., 2009). These approaches assume that wave breaking occurs continuously over the entire structure, which is incorrect and leads to an incorrect gradient in wave radiation stresses and wave-generated flow over the breakwater, particularly near the ends of the breakwater where 3D effects of diffraction are important (Vicinanza et al., 2009).

This may emerge as a critical problem in working with these models, when users need to consider the rate of wave diffraction around the head of the structure as well as the wave overtopping rate. However, very few studies have focused on the 3D effects of structures such as breakwater-induced circulation and even fewer on combined diffraction-overtopping effects on wave transmission (Sharif Ahmadian and Simons, 2014 and Vicinanza et al., 2009). Using the limited experimental data available on 3D effects of submerged breakwaters, simplified methods have been proposed for calculating near-shore flows and rip currents through the gaps of segmented breakwaters (Bellotti, 2004; Zanuttigh et al., 2008). However, wave diffraction around submerged breakwaters has not yet been studied sufficiently (Sharif Ahmadian and Simons, 2012; Sharif Ahmadian and Simons, 2014; Vicinanza et al., 2009); this is an important omission, as it is clear that a proper tool for predicting wave height behind such breakwaters must include diffraction effects as well as the 2D wave transmission (Vicinanza et al., 2009). It is also important to note that the wave height variation due to diffraction affects the forcing of the circulation pattern in the leeside of the breakwater (Hanson and Kraus, 1991).

As indicated previously, 2D models have been found to be ineffective in predicting the full spatial distribution of wave height behind submerged breakwaters. Therefore, more investigations are needed to extend the 2D studies to 3D and to quantify their improved efficiency. The present book is part of an effort to improve nearshore hydrodynamic and morphodynamic modeling in the vicinity of submerged breakwaters. A first step towards this is an accurate prediction of the rate of 2D energy dissipation over the submerged structure through an improved understanding of the wave-breaking process, as this is a key requirement for simulating the wave field around the structure. The next step is a model of the 3D effects of submerged breakwaters on wave transmission by diffraction. Combined overtopping and diffraction effects will then be considered in 3D simulations to get a comprehensive model of the wave field around submerged breakwaters. To properly evaluate the contribution of diffraction, impermeable breakwater models have been used to prevent percolation. This clearly limits the 2D transmission process to the overtopping contribution (Sharif Ahmadian and Simons, 2014 and Vicinanza et al., 2009).

This book leads to numerical modeling of submerged structures, considering different physical processes such as the wave energy dissipation and transmission, wave-induced current, sediment transportation, etc., with the long-term aim of using such modeling methods in coastal area hydrodynamic and morphodynamic models and design of submerged breakwaters. The results of this study will be of use for future research into the forces driving flow circulation, sediment transport and morphodynamics. The approaches are based on various types of models and in some cases the obtained results have been evaluated against laboratory measurements for submerged breakwaters. A comparison is also made between the proposed models and predictions from existing models.

For this book, hydraulic model tests have been conducted using different wave and current flumes or basins, including 2D tests and 3D tests. Tests adopted a range of breakwater crest widths and heights, water depths, submergence depths, distances from beach and wave climate. Collected wave data is used to calibrate and evaluate numerical models based on different methods for predicting the wave height distribution, flow field and sediment transport pattern around the submerged breakwaters, and particularly in the lee of the structures. Diffraction, refraction and shoaling effects on wave transmission around the submerged breakwaters for a wide range of submergence depths and wave conditions are specifically accounted for.

1.6 MAIN OBJECTIVES OF BOOK

The main goal of this book is to investigate wave transformation, flow field and sediment transport pattern over and around submerged breakwaters. Comprehensive 2D-3D experimental laboratory test programs have been conducted to investigate the effect of different parameters on the wave and flow field. Numerical modeling of these processes has also been carried out to provide a functional design tool to predict the spatial wave and flow around submerged breakwaters. The specific steps towards the achievement of these goals are given as follows:

1. Understanding different physical processes in coastal areas
2. Understanding hydrodynamic and morphodynamic impact of submerged breakwaters
3. Familiarity with mathematical models and their theories
4. Familiarity with methods of numerical modeling and developing models for submerged breakwaters in coastal area
5. Setting up the models for simulation of wave and flow fields and sediment transport patterns over and around submerged breakwaters for a broad range of geometric, water level and wave conditions
6. Calibration of the models, testing and evaluations

7. Analysis of the results to investigate different phenomena related to submerged breakwaters
8. Analysis of the results to investigate the importance of various parameters (dimensional and nondimensional) for submerged breakwater design
9. Accuracy analysis and comparison of model performance using different statistical and graphical methods to select the most accurate and reliable models as appropriate design tools
10. Validation of the results from the proposed model against the experimental values at different scales and test conditions
11. Analysis of predicted results to investigate the sensitivity of the models under different conditions
12. Evaluation of the proposed model against predictions from existing models
13. Case studies and practical examples

1.7 LAYOUT OF BOOK

This book deals with the numerical modeling of the wave fields and flow and sediment transport patterns around submerged breakwaters. The main body of the book is divided into 11 chapters including this introduction. The introduction provides an overview of the background, topics, purposes and significance of the study, motivation for the book and objectives.

Chapter 2 introduces the most important physical processes in the coastal area and discusses the hydrodynamic and morphodynamic impacts of submerged breakwaters. Wave transmission, flow field and shore response to these structures are investigated in this chapter. Wave transformation and wave-induced current over the submerged breakwater are analyzed and the effectiveness of several parameters (dimensional and dimensionless) is examined, giving an improved understanding of the hydrodynamic and morphodynamic processes.

Past research on submerged breakwaters focusing on numerical modeling of wave transmission and flow pattern around submerged breakwaters as well as sediment transport and shore response is reviewed in Chapter 3. Previously published numerical studies for submerged breakwaters are discussed. Recent numerical investigations simulating wave transmission behind submerged breakwaters are also reviewed briefly. A full literature review and background to the study are also presented.

Chapters 4 and 5 present a brief description of mathematical models and their basic concepts including theories, methodologies, algorithm development and an explanation of the specific models used in this book in detail.

Chapter 6 presents a short description of numerical modeling techniques, followed by predictive modeling for the submerged breakwater. The methods of simulation throughout the modeling are also defined in this chapter.

Chapter 7 details the numerical methods, modeling set-up and procedure utilized in the simulation. Approaches for numerical simulations for analysis of the physical processes for submerged breakwaters are described. This chapter presents numerical modeling applicable in modeling wave field, flow and sediment transport patterns for submerged breakwaters. Wave propagation around 3D submerged breakwaters is discussed numerically using the shallow water equations and machine learning algorithms. A description of the model is briefly presented in the chapter. The main processes consists of model set up, calibration, and simulations are introduced and described in detail.

Chapter 8 deals with approaches for development of design models for submerged breakwaters. This chapter discusses a numerical model based on radial basis function (RBF) methods for predicting the spatial wave transmission coefficient in the shadow region of a 3D submerged breakwater. In this chapter, a discussion on the shallow water equations are also presented for comparison with the results of the three dimensional RBF model, with particular focus on the 3D wave field behind the submerged breakwater in various scales. This chapter describes model development and analysis of results, essential parts of model construction. The process is outlined in detail in the chapter. The analysis approaches of the results are also presented, followed by a short discussion of the analysis of simulation results. These analyses are essential to ensure the capability of the proposed model to predict real applications at larger scale and a broad range of input parameters. The proposed models should be tested in the numerical simulations and the results are required to be compared to the laboratory experimental data as will be discussed further in the next chapters.

Chapter 9 introduces existing simulation and design tools with a short description of the models. The results of the available numerical models are compared with those from other methods. The analysis of the results is presented and the capability of the selected numerical model is also investigated. This chapter is completed with the main conclusions and recommendations for future application of the models and further works to improve their capabilities to deal with the design of submerged breakwaters.

Step-by-step procedures for practical applications are outlined in Chapter 10 with the intention of providing initial prediction schemes according to the proposed models described in the book. The main aim is to introduce simplified tools for preliminary design purposes which can be used in submerged breakwater design. 2D and 3D calculators are described in the chapter, based on the proposed models. In this chapter, the simplified models and examples are presented with a step-by-step procedure including details about the model mathematics. Guidelines for application of the models are also provided in this chapter. Applying the proposed models for predictions, the guidance provided in this chapter can be used in choosing the most appropriate models to predict different design parameters for submerged breakwaters. Application of the numerical methods particularly to solve partial differential equations (PDEs) is discussed in detail and with a large number of solved examples. The Navier-Stokes equations are also investigated. For the solution of these equations, the most popular methods by engineers and scientists in the field of fluid mechanics are used; such as the finite difference, finite volume methods, etc. Thus in this chapter we intend to study these partial differential equations and different governing equations and methods for solving them, with examples from simple problems to complex cases.

And finally, Chapter 11 completes the book with case studies and worked-out examples for the submerged breakwaters including several examples of numerical simulation, analysis and design of submerged breakwaters. A simplified 3D tool with a step-by-step procedure and prediction scheme for practical applications and preliminary design purposes of the submerged breakwaters in an easy to use and user-friendly form is presented in the chapter.

REFERENCES

Allsop, N., 1983. Low-crested breakwaters, studies in random waves. Proc. Coastal Structure 83. ASCE.

Bellotti, G., 2004. A simplifed model of rip currents systems around discontinuous submerged barriers. Coast. Eng. 51 (4), 323–335.

Bleck, M., Oumeraci, H., 2002. Hydraulic performance of articial reefs: global and local description. Proc. 28th Int. Conf. on Coastal Engineering. ASCE.

Browder, A.E., Dean, R.G., Chen, R., 1996. Performance of a Submerged Breakwater for Shore Protection, Proceeding of International Conference on Coastal Engineering, 2312–2323.

Bruce, T., van der Meer, J., Franco, L., Pearson, J., 2006. A comparison of overtopping performance of different rubble mound breakwater armour. Proceedings of the International Conference on Coastal Engineering. ASCE.
Buccino, M., Calabrese, M., 2007. Conceptual approach for prediction of wave transmission at low-crested breakwaters. J. Waterway, Port, Coast. Ocean Eng. 133, 213–224.
Burke, L., Kura, Y., Kasem, K., Revenga, C., Spalding, M., McAllister, D., 2001. Coastal Ecosystems. World Resources Institute, Washington DC.
Calabrese, M., Vicinanza, D., Buccino, M., 2008. 2d wave setup behind submerged breakwaters. Ocean Eng. 35, 1015–1028.
Chau, K.W., 2010. Modelling for Coastal Hydraulics and Engineering. Taylor and Francis Ltd..
Culliton, T.J., Warren, M.A., Goodspeed, T.R., Remer, D.G., Blackwell, C.M., MacDonough, J.J., 1990. 50 Years of Population Change along the Nation's Coasts, 1960–2010. National Oceanic and Atmospheric Administration, Rockville, MD.
d'Angremond, K., van der Meer, J., De Jong, R., 1996. Wave transmis-sion at lowcrested structures. In: Proc. 25th Int. Conf. on Coastal Engineering, vol. 24. ASCE, pp. 18–27.
Goda, Y., 1969. Re-analysis of laboratory data on wave transmission over breakwaters. Report of the port and Harbour Research Institute 8 (3), 3–18.
Goda, Y., Ahrens, J., 2008. New formulation for wave transmission over and through low crested structures. Proceedings of the International Conference on Coastal Engineering. ASCE.
Hanson, H., Kraus, N., 1991. Numerical simulation of shoreline change at lorain, Ohio. J. Waterway, Port, Coast. Ocean Eng. 117 (1), 1–18.
Lin, P., 2008. Numerical Modeling of Water Waves. Taylor & Francis Routledge.
Liu, P.L.-F., 1994. Model equations for wave propagation from deep to shallow water. In: Advances in Coastal and Ocean Engineering, vol. 1, pp. 125–158.
Liu, P.L.-F., Wu, C.-S., 2000. Recent Advancement in Modeling of Wave Propogation and Breaking Waves in Surf Zone. The Meteorological Research and Development.
Martínez, M.L., Intralawan, A., Vázquez, G., Pérez-Maqueo, O., Sutton, P., Landgrave, R., 2007. The coasts of our world: ecological, economic and social importance. Ecol. Econ. 63, 254–272.
McCowan, J., 1891. On the solitary wave. Philos. Mag., J. Sci., London 32.
Miche, R., 1944. Mouvements ondulatoires de la mer en profoundeur constante ou decroissante (Breaking wave motion in water of constant water depth). Ann. Ponts et Chaussees 121, 285–318 (in French).
Miller, M.L., Hadley, N.P., 2005. Tourism and coastal development. In: Schwartz, M.L. (Ed.), Encyclopedia of Coastal Science. Springer-Verlag, The Netherlands, pp. 1002–1008.
Munk, W.H., 1949. The solitary wave theory and its applications to surf problems. N. Y. Acad. Sci. Ann. 51, 376.
Panizzo, A., Briganti, R., 2007. Analysis of wave transmission behind low-crested breakwaters using neural networks. Coast. Eng. 54, 643–656.
Pilarczyk, W.K., Zeidler, R.B., 1996. Offshore Breakwaters and Shore Evolution Control. A.A. Balkema, Rotterdam, Netherlands.
Pinto, F.T., Neves A.C.V., 2003. Environmental aspects of using detached breakwaters for coastal protection purposes, Situation and Perspectives for the European Union, Porto, Portugal.
Prasetya, G., 2007. Chapter 4 Protection from coastal erosion: thematic paper: the role of coastal forests and trees in protecting against coastal erosion. Proceedings of the

Regional Technical Workshop. Regional Office for Asia and the Pacific, Rap Publication, Khao Lak, Thailand.

Seabrook, S., Hall, K., 1998. Wave transmission at submerged rubblemound breakwaters. Proc. 26th Int. Conf. on Coastal Engineering 2, 20002013.

Sharif Ahmadian, A., Simons, R.R., 2012. 3-D Wave Field Around Submerged Breakwater. ASCE, New York. Available from: http://dx.doi.org/10.9753/icce.v33.structures.13.

Sharif Ahmadian, A., Simons, R.R., 2014. A 3D numerical model of nearshore wave field behind submerged breakwaters. Coast. Eng. 83, 190–204. Available from: http://dx.doi.org/10.1016/j.coastaleng.2013.10.016.

Tajziehchi, M., Sharif Ahmadian, A., 2009. Artificial Neural Network Modeling of Wave-Induced Current Over Submerged Breakwaters, Coastal Dynamics 09, vol. 14. World Scientific, Tokyo, pp. 1–10. Available from: http://dx.doi.org/10.1142/97898142 82475_0017.

United Nations Atlas of the Oceans, 2013. UN-OCEANS.

van der Meer, J.W., Briganti, R., Zanuttigh, B., Wang, B., 2005. Wave transmission at low-crested structures, including oblique wave attack. J. Coast. Eng. 52, 915–929.

Vicinanza, D., Caceres, I., Buccino, M., Gironella, X., Calabrese, M., 2009. Wave disturbance behind low-crested structures: diraction and overtopping eects. Coast. Eng. 56, 1173–1185.

Wamsley, T., Ahrens, J., 2003. Computation of wave transmission coe-cients at detached breakwaters for shoreline response modeling. ASCE Conf. Proc. 147, 593–605.

Zanuttigh, B., van der Meer, J., 2006. Wave reflection from coastal structures. Proc. Int. Conf. Coast. Eng. 5, 4337–4349.

Zanuttigh, B., Martinelli, L., Lamberti, A., 2008. Wave overtopping and piling-up at permeable low crested structures. Coast. Eng. 55, 484498.

CHAPTER 2

Fundamental Concepts

2.1 INTRODUCTION

Past research and studies on submerged breakwaters indicate that these structures cause various physical processes. Some aspects have indubitable importance, such as the structural design, the stability of the rubble mound, scour, wave transmission and flow pattern.

The wave transmission plays an important role in the design of submerged breakwaters. Sufficient breakwater height and width reduce waves transmitted onshore. Sufficient distance of the structure from the shoreline also decreases the turbulence in the sheltered region. Roughness of the breakwater body also helps to significantly dissipate incident wave energy and consequently transmitted wave height at leeside. Bottom friction exerts a major influence on the wave energy dissipation and decrease of the incident wave height when water depth is low enough and the breakwater crest width is sufficiently large.

This reduction of transmitted wave energy and turbulence in the shadow zone entraps and accumulates suspended sediment traveling with the long shore current and prevents sediment from being transported offshore with the return flow (Pilarczyk and Zeidler, 1996). Therefore, it is very important to make sure that, based on the calculated parameters during the design process, the residual wave energy is not able to erode further beach.

2.2 PHYSICAL PARAMETERS RELATED TO SUBMERGED BREAKWATERS

A brief description is presented here of some of the physical parameters that can be characterized as the most important factors for the wave transformation process. The incident wave height is defined as the vertical distance between the crest and the trough of the wave measured at the toe of the breakwater. The incident wave period is also defined as the time interval between the start and the end of the wave, which is related to the wave length and water depth. Incident wave length in deep water corresponding

© 2016 Elsevier Ltd.
All rights reserved.

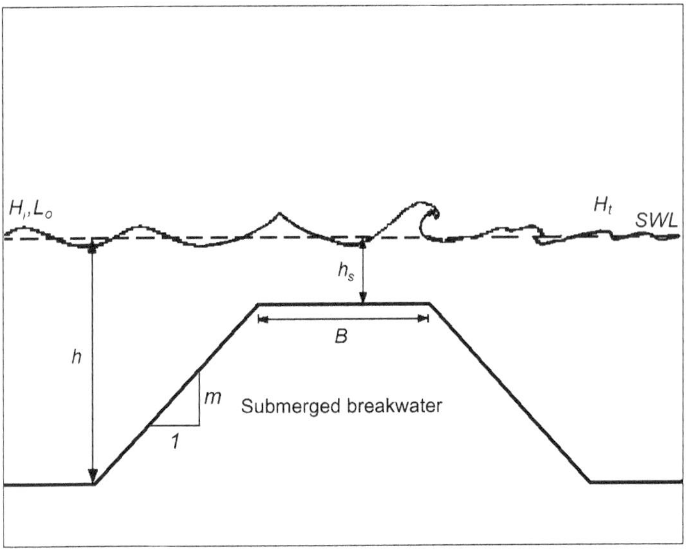

Figure 2.1 Sketch showing physical parameters relevant to the wave transmission over a breakwater.

to the wave period is defined as the horizontal distance between the start and the end of the wave. Breakwater crest height is the vertical distance between the seabed and the structure's crest. Water depth is the vertical distance between the seabed and the still water level at the seaward toe of the breakwater, while the submergence depth is the vertical distance between water surface and the crest of the breakwater. Crest width is defined as the distance between the seaside edge of the crest and the leeside edge. The slope of the seaward face of the breakwater and the offshore slope of the bottom seaward face of the submerged breakwater also influence the wave-structure interaction. The diameter of armor rock is defined as the diameter of the stone comprising the breakwater and the porosity is the ratio of the volume of voids to the total volume of the breakwater.

Some physical parameters involved in wave transformation over a submerged breakwater are illustrated in Figure 2.1. These parameters include incident wave height (H_i), onshore wave length (L_o), water depth (h), breakwater crest width (B), submergence depth (h_s), breakwater seaward slope (m) and transmitted wave height (H_t). Figure 2.2 shows an experiment with a breakwater in the flume where wave breaking occurs over the breakwater crest. These parameters are required when two-dimensional (hereafter 2D) transformation of waves over a

Figure 2.2 Wave transformation over a 2D small-scale submerged breakwater in the wave-current flume.

breakwater crest is being studied. However, when three-dimensional (hereafter 3D) processes are being considered, some additional parameters depending on the breakwater geometry and its location might be needed. These parameters include breakwater length, gap size, distance to the beach, etc. A 3D coordinate system will also be necessary to describe the wave field and its spatial variation in three dimensions properly (Figure 2.3).

A small-scale 3D submerged breakwater in a wave and current flume is shown in Figure 2.4. The breakwater roundhead and an adjacent trapezoidal trunk section are constructed with smooth marine plywood material with horizontal top and side slopes of 1V:1H in all directions.

2.3 PHYSICAL PROCESSES IN THE PRESENCE OF SUBMERGED BREAKWATERS

There are numerous types of physical processes involved in the wave transformation in the presence of submerged breakwaters, including wave refraction, wave diffraction, wave reflection, wave breaking, wave overtopping, creating waves with higher harmonic frequencies, etc.

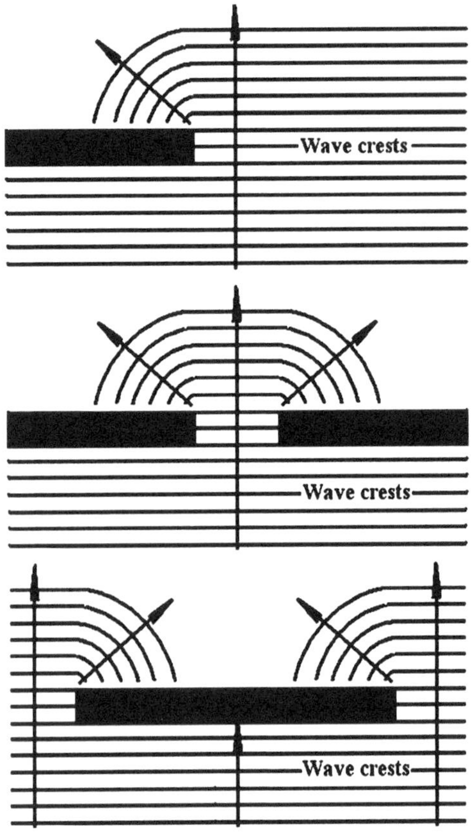

Figure 2.3 Plan view of diffraction pattern for a single semi-infinite obstacle, a double semi-infinite obstacle (an opening in the obstacle) and a single obstacle with infinite length.

The breakwater geometry and the local bathymetry affect the wave transformation. The equation of energy conservation is written as follows:

$$E_{\text{inc}} = E_{\text{ref}} + E_{\text{diss}} + E_{\text{gen}} + E_{\text{trans}} \tag{2.1}$$

where E_{inc}, E_{ref}, E_{diss}, E_{gen} and E_{trans} are energies related to the incident wave, reflected wave, and dissipated wave; energy generated in the form of other types of motion; and energy related to the transmitted wave, respectively.

Wave transformation phenomena are very complex and depend on the wave characteristics and the geometric properties of the structure. Understanding of the physical wave transformations is, therefore, very important for design purposes. Some main wave transformation

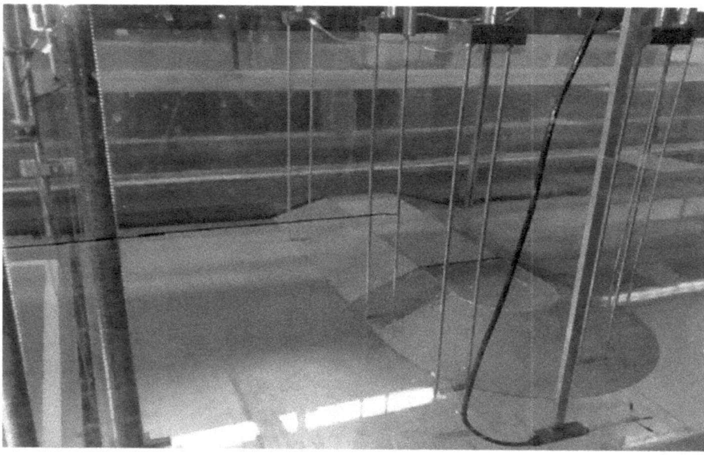

Figure 2.4 The 3D small-scale submerged breakwater in the 20 × 1.2 × 1.3 m wave flume.

phenomena are discussed in more detail following. However, for more information, see Dean and Dalrymple (1991).

Considering the wave motion for nonbreaking waves, the energy flux per unit crest length (Φ_e) is:

$$\Phi_e = c_g u_e \tag{2.2}$$

where c_g is the wave group celerity and u_e is the wave energy density:

$$u_e = \frac{1}{8}\rho g H^2 \tag{2.3}$$

where ρ is the density of water, g is the gravitational acceleration and H is wave height.

Under stationary conditions the total energy transport must be constant along the wave ray (Burnside, 1915):

$$\frac{d\Phi_e}{ds} = 0 \tag{2.4}$$

where s is the coordinate along the wave ray.

Considering wave–submerged breakwater interaction, as incident waves perpendicularly encounter the rising face of the submerged breakwater and approach shallower water over the crest, they slow down. Over the seaside slope of a submerged breakwater, the wave field may be irregular and complex with low-frequency wave heights and some degrees of nonlinearities increasing (Liberatore and Petti, 1993;

Petti and Ruol, 1991, 1992). It can be noted from the law of conservation of energy that the energy flux must remain constant. Wave celerity is dependent on water depth. Therefore, entering the shallower water, as group speed decreases (and also the wave length is reduced), energy per unit area of the wave and consequently the wave height have to increase. Therefore, when the waves approach the seaside slope of the submerged breakwater, they begin to shoal. As a result, bound harmonic waves are generated (Beji and Batjes, 1993, 1994). This kind of wave transformation is described by the phenomenon known as shoaling (Dean and Dalrymple,1991; Goda, 2000). Shoaling by a submerged breakwater may increase the nonlinearity of the wave field.

As a result of shoaling, an additional wave-induced flux of horizontal momentum is generated, which must be balanced. This, therefore, needs a decrease in the mean water level. This wave-induced decrease of the surface level is called set down and takes place before the wave breaks. On the other hand, the increase in wave height due to the shoaling causes the wave crest to become too steep and lose energy through a breaking process. Basically, when wave height reaches a critical level greater than a certain fraction of the water depth, the wave crest steepens and the wave can no longer retain its form; therefore, the crest of the wave overturns, wave breaking occurs and a considerable amount of wave energy is transformed into turbulent kinetic energy and dissipated (for further details see McCowan, 1891; Miche, 1944; Munk, 1949). Wave breaking at submerged breakwaters is influenced by the seaside slope and the submergence depth of the structure (Smith and Kraus, 1990, 1991). The slope of the sea bottom also affects wave breaking and its type. It should be noted, as will be discussed later, that wave breaking models need to calculate wave energy dissipation over the submerged breakwaters (Zanuttigh and Lamberti, 2006). The wave energy dissipation at the place of the breakwater defines the forces on the structure. These forces influence the stability of the breakwater. Breaking waves passing over a submerged breakwater also cause water to flow into the onshore of the breakwater due to the difference in mean water level.

After the break point, the wave energy flux and the radiation stress decrease due to energy dissipation and the reduction in wave height through the surf zone. The horizontal momentum equations of the mean flow again need to be balanced. This requires an increase in the mean water level. Therefore, a radiation stress gradient is created that drives wave set-up. This is another physical process which takes place after the wave breaking. The gradient of the wave-induced set-up behind the submerged breakwaters also causes water to flow along the shoreline. As a consequence of the wave set-up created because of the wave breaking on a submerged breakwater, wave-induced currents are also generated. This phenomenon plays an important role in the hydrodynamics of submerged breakwaters. The wave-generated set-up and current depend on mass flux over the breakwater contributed by mean flow, breaking wave rollers and percolation of water through the breakwater.

However, if waves approach the submerged structure at an angle to the breakwater contours, the section of wave crest in deeper water offshore of the breakwater travels faster than the section in shallow water over the seaside slope and crest. Therefore, a different physical process, known as refraction, will take place, which is the change in direction of the incident wave. The conservation of energy flux can be used to analyze refraction effects. Refraction by a submerged breakwater may also increase the nonlinearity of the wave field.

Reflection is also another phenomenon which involves a change in direction of incident waves, when waves bounce from the breakwater back toward the ocean. Therefore, a fraction of the incident wave energy is reflected offshore. The amount of the wave reflection somewhat depends on the forward face slope and the submergence depth of the breakwater (Ahrens, 1987, 1988; Dattari et al., 1979; van der Meer, 1991). Diffraction also happens when waves encounter the submerged barrier. The wave energy is carried along the crest into the sheltered region due to the changes in wave height and consequently the waves turn into the shadow zone. In reality, diffraction, refraction and shoaling all occur simultaneously. When waves encounter an obstruction or a sudden change in bathymetry, some of the wave energy will be forced to dissipate across the wave ray (or along the wave crest).

Wave overtopping takes place generally when waves meet fully submerged breakwaters. It might also happen when waves meet semi-submerged breakwaters or low-crested structures, particularly those with a crest lower than the incident wave height. During overtopping, three processes important to the submerged breakwaters take place: the wave transmission, the passing of water flow over the breakwater and percolation. When the waves encounter permeable structure, percolation of the incident waves through the porous submerged breakwater also occurs.

Over the crest, energy from the fundamental wave frequency is transferred to higher harmonic frequencies (Driscoll et al., 1993). This may be affected to some degree by the wave reflection and breaking, as well as the wave energy dissipation by frictional resistance, permeability and flow through the structure.

After the breakwater crest, the developed higher frequency wave components are separated from the fundamental components and travel with their own celerity (Beji and Battjes, 1993, 1994; Cornett, Mansard and Funke, 1994; Petti and Ruol, 1992). Flow separation and eddies are also developed in the shore side of the breakwater crest (Ting and Kim, 1994).

After breaking in the surf zone, the waves continue to move in a new zone where the broken waves are propagated onto the beach, a region extending from the limit of run down to the limit of run up, called the swash zone. Investigation of hydrodynamic and morphodynamic processes in this zone is also required to be considered since in this zone usually intense erosion occurs. The swash process consists of two phases: uprush and backwash. Uprush is an onshore phase and is known as movement of a turbulent layer of sea water up the beach, which is basically an upward flow of energy back to the sea as the waves lose their momentum. But downrush is the offshore phase where a downward flow is movement of water down the beach. It could also be important for rock-armored slopes and rubble mound structures.

The sum of the wave set-up and the uprush is called wave runup, which is defined as the vertical difference between the highest extent the wave reaches on a beach or structure and the still water level (Sorensen, 1997). Wave run-up is particularly important for emergent breakwaters as their crest height is defined based on allowable overtopping. It might be even important for low-crested structures or submerged breakwaters where wave transmission over the crest is mostly allowed to have a prediction of velocity, depth and volume of overtopping flow. An interaction between incident waves and return currents either over the crest or in the gaps may also be observed in the vicinity of the submerged breakwaters (Fulford, 1985; Petti and Ruol, 1991, 1992).

Morphodynamic response of submerged breakwaters including the seabed topography and the beach profile is also required to be considered appropriately in simulations and analysis because of its significant role in performance of these types of coastal structures. Prediction of the changes in the morphology of the coastal area in the presence of the submerged breakwaters is a major challenge in coastal engineering. One needs to investigate the fundamental processes coupling hydrodynamics and morphodynamics that drive coastal morphology in the presence of the submerged breakwaters. Morphodynamic response of submerged breakwaters still includes many uncertainties and should be considered carefully during the process of designing such structures.

The presence of submerged structures alters the pattern of water flux and sediment transport along and across a coast due to the existence of the various hydrodynamic phenomena (as mentioned in previous section). The transport of sediment in a coastal area basically takes place when the seabed is exposed to large enough shear stresses caused by external forces due to the water movements. The coastal area is exposed to a very complex environmental forcing such as water-level changes, tides, winds, waves, currents, storm surges and extreme events, wave set-ups, the depth variations, the sources and sinks of sediment and human interventions. Therefore, the prediction of bed and suspended loads with very unknown and highly nonlinear sediment transport processes together with turbulence and many other complicated hydrodynamic processes becomes extremely difficult for these specific structures. The most important factors involved in the process of sediment transport in a coastal area are: wave climate, current condition, sea-level changes, bathymetry, sediment characteristics, existing rivers, eroding coasts or tidal inlets, and human interventions.

All the hydrodynamic effects of submerged breakwaters should be considered for an accurate and comprehensive morphodynamic analysis and simulation. In practice, different morphodynamic behaviors might be observed depending on the effective environmental and design factors, including the local wave climate, water level condition, geometry and location of the breakwater, beach state, characteristics and availability of sediment, actual hydrodynamic patterns, etc. Both longshore and cross-shore water fluxes and associated sediment transport should be analyzed. Longshore currents and their associated sediment fluxes are basically generated due to the oblique wave incidence, wave transforming over the structure or coastal profile, generating wave set-up, overtopping or percolation, eddy return fluxes and overall mass balances. Depending on the wave conditions, cross-shore fluxes (also referred to as onshore and offshore currents) over the coastal profile may also take place and, in terms of short-term coastal stability, extreme storm surges and waves may result in significant coastal erosion. Cross-shore currents are derived from the undertow, wave asymmetry and reflection fluxes. All these wave-driven currents and their associated sediment transport fluxes for a beach with a submerged breakwater should be considered carefully in order to have an accurate analysis of the morphodynamic responses of the structure.

2.4 PERFORMANCE OF THE SUBMERGED BREAKWATERS

Submerged breakwaters are designed to reduce the incident wave energy in order to protect the coasts against erosion and port channels from sand deposition. They can also be designed to redistribute the sediment transport pattern and trap the traveling sediment for the purpose of beaches restoration. These structures sometimes are used together with other coastal defense techniques such as beach nourishment to protect the coasts. They are constructed in different forms, such as single breakwaters, a series of segmented breakwaters, or continuously. Performance of a submerged breakwater particularly depends on its geometrical and structural characteristics, the location of the breakwater (the water depth at the toe and its distance from the coastline), wave-structure interaction, sediment transport and local morphodynamics.

The wave transmission coefficient increases with submergence depth and wave period and decreases with raising the incident wave height and the breakwater crest width. Basically, a large submergence depth relative to the incoming wave height results in high wave transmission coefficients. The energy of longer waves to a lesser extent is dissipated by wave breaking or turbulent losses. Broader breakwaters also result in a higher degree of wave energy dissipation due to turbulent losses or wave breaking. In addition, the seaside slope of the submerged breakwater influences the energy dissipation. Porosity, together with the rock armor diameter, can also play an important role in the dissipation of wave energy. The wave-induced set-up and discharge flux over the submerged breakwater increases with incident wave height and breakwater crest width due to higher energy dissipation during the wave breaking.

REFERENCES

Ahrens, J.P., 1987. Characteristics of reef breakwaters. Technical Report, CERC, pp. 87–17, 45.

Ahrens, J.P., 1988. Reef breakwater response to wave attack. Berm Breakwaters: Inconventional Rubble-Mound. American Society of Civil Engineers.

Beji, S., Battjes, J.K., 1993. Experimental investigation of wave propagation over a bar. Coast. Eng. 19 (1–2), 151–162.

Beji, S., Battjes, J.A., 1994. Numerical simulation of nonlinear wave propagation over a bar. Coast. Eng. 23 (1–2), 1–16.
Burnside, W., 1915. On the modification of a train of waves as it advances into shallow water. Proceedings of the London Mathematical Society. Series 214: 131–133.
Cornett, A., Mansard, E., Funke, E., 1994. Wave transformation and load reduction using a small tandem reef rreakwater - physical model tests. Ocean Wave Measurement and Analysis. American Society of Civil Engineers, New York.
Dattari, J., Raman, H., Shankar, N.J., 1979. Performance characteristics of submerged breakwaters. 16 International Conference on Coastal Engineering. American Society of Civil Engineers, New York.
Dean, R.G., Dalrymple, R.A., 1991. Water wave mechanics for engineers and scientists. Advanced Series on Ocean Engineering 2. World Scientific, Singapore, ISBN 978-981-02-0420-4.
Driscoll, A.M., Dalrymple, R.A., Grilli, S.T., 1993. Harmonic generation and transmission past a submerged rectangular obstacle, 23 International Conference in Coastal Engineering, vol. 1. American Society of Civil Engineers, New York.
Fulford, E.T., 1985. Reef type breakwaters for shore stabilization. Coastal Zone 85. American Society of Civil Engineers, New York.
Goda, Y., 2000. Random seas and design of maritime structures, Advanced Series on Ocean Engineering 15, second ed. World Scientific, Singapore, ISBN 978-981-02-3256-6.
Liberatore, G., Petti, M., 1993. Wave transformations over a submerged bar: experiments and theoretical interpretations, 23 International Conference on Coastal Engineering, vol. 1. American Society of Civil Engineers, New York.
McCowan, J., 1891. On the solitary wave. Philos. Mag., J. Sci., London 32.
Miche, R., 1944. Mouvements ondulatoires de la mer en profoundeur constante ou decroissante (Breaking wave motion in water of constant water depth). Ann. Ponts et Chaussees 121, 285–318 (in French).
Munk, W.H., 1949. The solitary wave theory and its applications to surf problems. N. Y. Acad. Sci. Ann. 51, 376.
Petti, M., Ruol, P., 1991. Experimental study on the behaviour of submerged detached breakwaters. 3rd International Conference on Coastal and Port Engineering in Developing Countries, vol. 1, pp. 167–179.
Petti, M., Ruol, P., 1992. Laboratory tests on the interaction between nonlinear long waves and submerged breakwaters. Proc. 23th Int. Conf. on Coastal Engineering. ASCE.
Pilarczyk, W.K., Zeidler, R.B., 1996. Offshore Breakwaters and Shore Evolution Control. A.A. Balkema, Rotterdam, Netherlands.
Smith, E.R., Kraus, N.C., 1990. Laboratory Study on Macro-Features of Wave Breaking Over Bars and Artificial Reefs, CERC Technical Report 90-12.
Smith, E.R., Kraus, N.C., 1991. Laboratory study of wave-breaking over bars and artificial reefs. J. Waterway, Port, Coast. Ocean Eng. 117 (4), 307–325.
Sorensen, J., 1997. National and international efforts at integrated coastal management: definitions, achievement, and lessons. Coast. Manage. 25, 3–41.
Ting, F.C.K., Kim, Y.K., 1994. Vortex generation in water waves propagating over a submerged obstacle. J. Coast. Eng. 24 (1–2), 23–49, American Society of Civil Engineers. New York.
van der Meer, J.W., 1991. Stability and transmission at low-crested structures. Technical report, Delft Hydraulic, Report No. H 453. 33p.
Zanuttigh, B., Lamberti, A., 2006. Experimental analysis and numerical simulations of waves and current flows around low-crested rubble-mound structures. J. Waterway, Port, Coast. Ocean Eng. 132 (1), 10–27.

CHAPTER 3

Literature Review and Background

The functional design knowledge of submerged breakwaters, including their impacts on wave transformation, wave and current fields around the structure, sediment transport and shoreline response, is still developing.

Various experimental studies and numerical models have been developed for predicting different physical phenomena related to submerged breakwaters. Previous laboratory experiments, numerical models and field measurements have been performed in an effort to investigate the hydrodynamic and morphodynamic impacts of submerged breakwaters on the coastal zone and studies still continue to be developed. This chapter looks into the historical studies and research on submerged breakwaters.

Many researchers have simulated different wave transformation processes in coastal areas in the presence of submerged breakwaters applying nonlinear wave theory for shallow waters. These processes include wave transformation, wave-induced set-up and current across submerged breakwaters, etc. For example, Boussinesq equations with some simplifications have been applied by Cruz et al. (1992, 1997) to model the transformation of long waves of moderate amplitude over a submerged breakwater with a slowly varying slope. Although they obtained reasonable results for waves calculated over the breakwater's crest, the accuracy of the simulated wave field behind the structure was not encouraging. Another type of equation is the mild-slope equation. However, again, one may need many simplifying assumptions during the simulation of submerged breakwaters, as these equations are less accurate for steep slopes, which is the case for the most of the submerged breakwaters.

Previous research has been published for two dimensional phenomena on the submerged breakwaters, for instance: wave transmission/filtration (Goda and Ahrens, 2008), wave overtopping (Bruce et al., 2006), wave reflection (Zanuttigh and van der Meer, 2006), wave set-up (Calabrese et al., 2008). The wave transmission either over the crest or through the body plays an important role in environmental effect of submerged

© 2016 Elsevier Ltd.
All rights reserved.

breakwaters and shoreline response to these breakwaters. Wave transmission is basically a very important criterion in the design of a submerged breakwater particularly in preliminary stages of design. Very encouraging 2D design tools have been presented by many researchers such as van der Meer et al., 2005, Panizzo and Briganti, 2007, Buccino and Calabrese, 2007 and Goda and Ahrens, 2008, etc. However, very few studies have been focused on the 3D effects of the submerged breakwaters and even fewer on wave transmission in particular. The need to include 3D effects in the wave transmission process behind low-crested breakwaters (or submerged breakwaters) comes from evidence that formulas for 2D wave transmission tend to underestimate 3D data (Vicinanza et al., 2009). Examples of this behavior are Adams and Sonu (1986), Seabrook and Hall (1998), Calabrese et al. (2005) and Buccino and Calabrese (2007). The primary consequence of underestimating wave transmission is that the average degree of protection provided by the structure would be overestimated. Therefore, an accurate and reliable calculation of the 3D spatial variation of the wave height across the sheltered area is of engineering interest in the initial design of submerged breakwaters (Vicinanza et al., 2009). However, the studies carried out for the submerged breakwaters in the past are not yet sufficient for an appropriate understanding of their behavior.

The impact of submerged breakwaters on seabed and shoreline changes depends on the wave and current field behind the structure. Although physical and mathematical modeling approaches and field studies continue, the hydrodynamics and morphological effects of submerged breakwaters have not been understood appropriately. This chapter discusses previous research on submerged breakwaters in detail.

A precise calculation of the hydrodynamic effects in the vicinity of submerged breakwaters plays a major part in an accurate and reliable morphological modeling of seabed and shoreline response for such structures. Existing numerical models mostly employ the radiation stress formula presented by Longuet-Higgins and Stewart (1964). Mass transport induced by breaking wave rollers is not considered in most of the hydrodynamic models for submerged breakwaters.

Many theoretical studies, experimental works, numerical models and empirical approaches have been developed for prediction of wave transmission over and behind submerged breakwaters since very long time ago in the past and over the years (Adams and Sonu, 1986; d'Angremond et al., 1996; Dick and Brebner, 1968; Johnson et al., 1951; Losada et al.,

1996; Petti and Ruol 1992; Schlurmann et al., 2002; Seabrook and Hall, 1998; van der Meer et al., 2005).

The first reported physical experiment on wave transmission behind submerged breakwaters was done by Stucky and Bonnard (1937), who used a 2D type of structure with trapezoidal section. Their study was followed in more comprehensive 2D geometrical detail by the Beach Erosion Board (1940). Morison (1949) and Johnson et al. (1951) examined rectangular breakwaters. Johnson et al. (1951) showed better energy dissipation of steeper waves with a wider barrier, while longer period waves were generally less affected by breakwater crest width and height.

Laboratory experiments on vertical and composite overtopping breakwaters with regular waves conducted by Goda et al. (1967) showed that the transmission coefficient K_t is mostly governed by the submergence ratio (h_s/H_i). They found higher harmonic waves in the transmitted waves in their experiments. They proposed an empirical equation for calculating K_t over mounded vertical wall and composite breakwaters based on the incident wave height and the submergence depth, as well as some variables related to the type of structure. Later, Goda (1969) proposed revised values of these parameters.

Longuet-Higgins (1967) presented an analytical formula to calculate the set-up for the submerged breakwaters exposed to nonbreaking waves. The mean sea level difference between the two sides of the breakwater was calculated based on the wave height and the wave number of incident, reflected and transmitted waves as well as water depth corresponding to seaside and shoreside of the breakwater.

Experimental results by Dick (1968) illustrated that the analytical formula presented by Longuet-Higgins (1967) significantly underestimates the set-up.

Diskin et al. (1970) conducted a series of 2D laboratory experiments on submerged breakwaters exposed to regular waves. They presented a formula for prediction of the relative set-up ratio δ/H_i behind submerged breakwaters, based on the incident wave height and submergence depth for a specific range.

Laboratory tests with a smooth and impermeable breakwater by Seelig (1980) verified the empirical model of Goda (1969) for regular and irregular waves proposing the variables $\alpha = 2.6$ and $\beta = 0.15$. 2D experiments on smooth and impermeable submerged breakwaters with various geometries by Abdul Khader and Rai (1980) revealed that the dissipation of wave energy is dependent on relative crest height (h_c/h), where h_c is the breakwater crest height and h is total water depth.

Massel and Butowski (1980) developed a model for prediction of two-dimensional wave transmission and reflection in the presence of porous breakwaters. However, the model does not have the capability of predicting three-dimensional wave transmission.

Allsop (1983) carried out a series of experimental tests to study interaction of low-crested, rock-armored structures with random waves under different wave conditions. Allsop (1983) changed the Goda's empirical formula (Goda, 1969) considering H_s, significant wave height, and s, the wave steepness (H_i/L_o).

Two years later, Powell and Allsop (1985) provided empirical curves as design guidelines for prediction of wave transmission coefficient over low-crested permeable breakwaters with 40% porosity.

Ahrens (1987) investigated low-crested rubble-mound breakwaters experimentally and proposed an expression considering a large number of the physical variables including the following: H_s, significant wave height; A_t, area of breakwater cross-section; D_{50}, dimension of stone; L_p, wave length corresponding to peak wave period; h_c, breakwater crest height; h, still water level, and h_s, the submergence depth. Ahrens (1989) discussed the stability of a kind of rubble mounds breakwaters with a homogeneous pile of stone known as reef breakwaters. A large number of experimental tests was conducted under irregular wave attacks for a wide range of wave conditions from very minor to very severe levels. They measured stability of reef breakwater according to reduction in breakwater crest height rather than number of displaced stones. The model was shown to have accurate predictions of damage to the reef breakwater. They illustrated that the model is also suitable for the submerged rubble mound breakwaters even if the degree of overtopping is very high (Ahrens, 1989).

Several studies have been conducted to analyze the shoreline response to submerged breakwaters (Watanabe et al., 1986; Hanson and Kraus, 1990; O'Connor and Nicholson, 1995). Hanson and Kraus (1990) applied the 1-line shoreline change numerical model to predict shoreline response to the detached barriers. They compared the results of simulations with field data (Hanson and Kraus, 1990). They showed that shoreline advances with increasing wave period and decreasing wave height. They also discussed the different forms of shoreline response to the offshore structures. However, their model cannot predict the amount of piling up and the overtopping and morphodynamic effects of longshore current behind the structure. Sawaragi (1992) used field data measured in the Niagata beach where shoreline was subject to erosion and two submerged breakwaters were constructed in the gap between two existing emergent breakwaters in order to protect the beach. The field measurements by Sawaragi (1992) showed a moderate degree of shoreline response in a steady manner. Zyserman et al. (1999) and Jiménez and Sánchez-Arcilla (2002) analyzed the morphodynamic effects of a detached breakwater considering its geometry and location at the beach in short term and long term, respectively. A short-term analysis of the morphodynamic impact of submerged breakwaters on the shoreline was also made in the DELOS project numerically. Numerical models under different hydrodynamic conditions were employed for simulation of the beach response to submerged breakwaters in a Catalonian beach, Altafulla Beach. Aminty et al. (1983) conducted a series of laboratory tests in a wave flume to investigate the hydrodynamic and morphodynamic effects of breakwater parameters. Two-dimensional morphodynamic effects of submerged breakwaters were also investigated experimentally by Chiaia et al. (1992).

Kobayashi and Wurjanto (1988) presented a numerical model based on the finite-amplitude shallow water equations to predict monochromatic wave transmission over an impermeable submerged breakwater. Relatively good agreement with 2D laboratory data was obtained. Rojanakamthorn et al. (1990) developed a mathematical model based on the equation of waves on a porous layer for the computation of wave transformation over a permeable submerged breakwater. However, the model presented lower-estimated results in comparison with measured laboratory data. Davies and Kriebel (1992) carried out a series of experimental tests to study the interaction of the solid and rubble-mound submerged breakwaters with regular and irregular wave attacks. It was shown that the transmission coefficient does not considerably differ between these two type of breakwaters in the case of crest height below still-water level.

Battjes (1994) developed a time domain Boussinesq model to simulate interaction of long, unidirectional waves and submerged breakwaters. The model showed good agreement with experimental data for non-breaking waves.

Van Gent (1994) developed a general two-dimensional model for a number of popular coastal structures. The model, therefore, was also capable of being applied for wave-structure interaction analysis in the case of permeable and impermeable submerged breakwaters. However, the model was not designed for predicting the spatial variation of the wave heights behind the structure. Losada et al. (1995) developed a model for prediction of the wave transformation of regular waves with oblique incidence over the porous submerged breakwaters. The model had the possibility of being applied in two- or three-dimensional modes.

2D and 3D laboratory experiments were conducted (Gourlay, 1993, 1996a, 1996b) to measure the wave transformation, wave induced set-up and current over the reefs. They found that both current and set-up increases with incident wave height, while submergence depth has respectively direct and inverse effects on current and set-up. The results also revealed that set-up is sensitive to wave period.

Gourlay (1993) proposed two dimensionless representations for set-up and discharge over the reefs. Gourlay (1993) indicated that for low submergence ($S = (h_s + \delta)/H_i < 0.7$) waves break at the reef edge, while for high submergence ($S > 0.7$) waves break over the reef. Gourlay (1996a) observed that set-up occurs over the reef when S is less than 2.5.

Symonds (1994) and Symonds et al. (1995) solved the momentum and continuity equations analytically, assuming linear friction, and presented an equation for current velocity over the reef top considering parameters U, $\gamma_r = H_r/h_s$, H_r and h_s which are respectively integrated current velocity, wave deformation index, the maximum wave height and water depth over the reef. B, reef length, h_b, wave breaking depth over the slope, α, reef face slope and r, the friction term in the momentum equation were other parameters included in the equation (see Nelson, 1996 and Longuet-Higgins, 1970). Symonds et al. (1995) calibrated the model using field data collected from John Brewer reef (reef width $B = 300$ m) in Townsville, Australia (Hardy, 1993); $\gamma_r = 0.35$ and $r = 0.25$.

D'Angremond et al. (1996) reanalyzed data sets collected from other experimental tests (Allsop, 1983; Daemen, 1991; Daemrich and Kahle, 1985; Powell and Allsop, 1985; Seelig, 1980; van der Meer, 1988) revealing that the formula provided by Ahrens (1987) is not suitable for calculating K_t particularly when crest width is large. They proposed a more precise expression for transmitted wave height passing over permeable and impermeable submerged breakwaters with limits $0.075 < K_t < 0.80$ for both conditions. The incident wave height, the submergence depth and the breakwater crest width were included in the formula. ξ_o, which is the Iribarren number, was also considered. ξ_o can be calculated by knowing the offshore face slope of the breakwater and the incident wave steepness:

$$\xi_o = \frac{tan\alpha}{\sqrt{\frac{H_i}{L_o}}} \tag{3.1}$$

in which H_i is incident wave height and L_o is incident wave length in deep water.

The results of 3-year monitoring of a 594 m shoreline parallel precast concrete submerged breakwater in Palm Beach, Florida, USA by Browder and Dean (1996) showed that the erosion of the beach behind the structure was 2.3 times higher than before breakwater construction. Tomasicchio (1996) investigated long rubble mound submerged breakwaters. At Osita, Roma, Italy, the field measurements indicate the low environmental impact as well as the low construction costs of the rubble mound submerged breakwaters and their ability to dissipate wave energy, decrease sediment transport and protect the shoreline. 3D experiments conducted by Groenewoud et al. (1996) in a wave basin illustrated that the influence of the lengths of gaps between the segmented submerged breakwaters on shoreline response and sediment transport offshore is significantly important.

Analyses of an extensive series of laboratory measurements carried out in six European universities by van der Biezen et al. (1997) showed significant sediment loss through gaps. Therefore, van der Biezen et al. (1997) indicated that the segmented breakwaters have a negative effect on the beach profile development. Van der Biezen et al. (1998) also conducted numerical simulations applying nonlinear shallow water equations. Mass transport of water due to wave breaking over the submerged breakwater was not taken into account in the model and unrealistic values of bottom friction were also applied for the calibration of the model. The results were compared with the measured current behind the breakwaters (De Later, 1996). The analysis showed lower values of current calculated in the numerical model.

Seabrook and Hall (1997) found that the armor stone size has little effect on wave transmission. Seabrook and Hall (1998) also developed a predictive expression for K_t at submerged rubble mound breakwaters including the effect of crest width, the wave conditions, rock diameter and submergence depth. The proposed equation is valid in some specific ranges.

Debski and Loveless (1997) and Loveless et al. (1998) conducted an extensive series of 2D and 3D laboratory tests in wave flume and wave basin for permeable and impermeable breakwaters. They observed that the wave-induced set-up is significantly smaller in 3D tests as the water level gradient behind the breakwater generates longshore flow. Loveless et al. (1997) presented an empirical formula for calculating the set-up behind submerged breakwaters based on the incident wave height, the wave period, the wave length in deepwater, the water depth at the offshore toe of the breakwater and the breakwater crest height. Loveless et al. (1998) also provided an improved expression of the formula to predict the dimensionless set-up $\frac{\delta}{B}$ behind a permeable submerged breakwater. The wave conditions, the water depth, rock diameter and the submergence depth parameters were included in the equation. They found that set-up increases with breakwater crest width B and it is significantly affected by the rock size d_{50} or porosity. They observed that the average nondimensional set-up δ/H_i is approximately the value 0.3.

Lin and Liu (1998) developed a Reynolds Average Navier-Stokes model based on a previous developed model by Kothe et al. (1991), capable of simulating wave-porous structure interactions considering turbulent generation and dissipation mechanisms. The model was extensively tested and validated under various conditions. However, it requires high computation times to be used for simulations in real life.

Gironella and Sanchez-Arcilla (1999) developed a design equation for the transmission coefficient in the presence of a submerged breakwater using multilinear regression to fit an equation to a wide range of data sets (the variables include the submergence depth and the offshore wave length; a set of the constant parameters were also considered and proposed after the calibration). The equation was proposed to be valid in a specific range of parameters.

Drei and Lamberti (1999) carried out laboratory experiments under regular and irregular wave attack in wave flume in the presence of narrow-crested impermeable submerged breakwaters. They demonstrated that wave set-up at submerged breakwaters is higher than provided by the set-up theory for beaches. They mentioned that this is because of a higher breaker index for the submerged breakwater in contrast to the beaches. They found the average nondimensional set-up δ/H_i for impermeable breakwaters to be about 0.8 for zero submergence and about 0.49 for all other submergence depths. Drei and Lamberti (1999) provided a very simple formula for the overtopping discharge over a semi-submerged breakwater. The only variable in the formula is the incident wave height.

Hearn (1999) developed an analytical hydrodynamic model using the depth-averaged steady-state momentum equation and both linear and quadratic friction theory to calculate wave-driven flow across a coral reef. Good agreement was obtained with a linear friction law. Massel and Brinkman (2001) provided an analytical approach to model wave-induced set-up and current over shoals and coral reefs without steep sea faces. The model has not been verified with laboratory or field data.

Gonzalez et al. (1999) modified the Dean (1977) model and provided a model for submerged breakwaters. Good agreements were observed between the model and 2D experimental model data of Chatham (1972) and Sorensen and Beil (1988). A similar model for reef-protected beaches was also proposed and validated against field data from Spain and Australia by Muooz-Pérez et al. (1999).

Lynett et al. (2000) also presented a nonlinear shallow water model using the modified Boussinesq equations. Dalrymple et al. (2001) developed a model for tracking deformations of free surface flows based on a smoothed particle hydrodynamics approach. Both Lynett et al. (2000) and Dalrymple et al. (2001) require very high computation times to be used for simulations in real life.

Two-dimensional experimental modeling was conducted by Bleck and Oumeraci (2001) for a rectangular submerged breakwater in a wave flume with both regular and irregular waves. Bleck and Oumeraci (2001) proposed an empirical equation for K_t based on the relative water depth h_s/H_i; however, reanalyzing the data, Bleck and Oumeraci (2002) modified the equation for a rectangular impermeable breakwater. Later an analytical model for wave transmission at artificial reefs was also provided by Bleck and Oumeraci (2004).

Mendez, Losada, and Losada (2001) presented analytical expressions for second-order mean quantities such as mass flux, energy flux, radiation stress, and mean water level when wave reflection and energy dissipation occurred due to wave breaking and porous flow over and through a permeable submerged structure. These expressions were presented in the form of shape functions of the incident wave characteristics, structure geometry, and permeable material characteristics. A 2D model with submerged trapezoidal breakwaters and a 3D model for submerged permeable rectangular breakwaters were presented. Regular waves were considered in both models. Good agreement with experimental data in prediction of wave height transformation and mean water level variations along the wave flume was obtained by the two models (Mendez, Losada, and Losada, 2001).

Roul and Faedo (2002) applied dimensionless free board $(h_s + R_u)/H_i$ as proposed by Davies and Kriebel (1992) and developed an empirical expression for K_t. The parameter R_u is the potential run-up as defined by Ahrens and McCartney (1975) as a function of Iribarren number ξ_o.

Calabrese et al. (2002) found that the d'Angremond et al. (1996) equation is more reliable than other previous methods for prediction of K_t. They provided a model for predicting the wave transmission coefficient using data obtained from the large-scale experimental tests with the parameters: the incident wave height, the breakwater crest width, the water depth and the Irribaren number. The proposed equation by Calabrese et al. (2002) is valid in the range of experimental test values.

Melito and Melby (2002) proposed some equations for estimating the transmission coefficient past a CORE-LOC armor layer low-crested breakwater based on the variation of the submergence ratio h_s/H_i.

Calabrese et al. (2003) assumed that wave set-up is dominated by the amount of momentum and continuity set-up. The effects of wave transmission coefficient, the breakwater crest height, the submergence depth, the wave breaker depth over the offshore slope, the effective crest width of the breakwater and the friction factor were considered.

Reniers and Battjes (1997) studied random wave transformation, set-up and longshore current velocity experimentally and numerically. They indicated that, in comparison with measured data from a large wave basin, the numerical model underestimated set-up and longshore current without the concept of a surface roller model being included in the equations. The unrealistic friction parameters need to be adopted as calibration parameters in the model for fitting to laboratory or field data without the concept of a surface roller model being included in the equations (Lesser et al., 2003). Therefore, based on the study done by Lesser et al. (2003), the wave roller effects should be included in the numerical models to make them more reliable in predicting hydrodynamic and morphological effects of submerged breakwaters. Svendsen (1984a, 1984b) provided an equation for radiation stress in the surf zone including wave roller, considering the effects of the wave period, the wave height and the water surface profile. A roller model was also proposed and applied by Dally and Brown (1995) to calculate wave-induced longshore current over plane beaches.

Losada et al. (1996) presented two models to study the impact of breakwater geometry, porous material properties and random wave characteristics including oblique incidence, on the kinematics and dynamics over and inside a submerged permeable breakwater. The model included various engineering tools such as breakwater efficiency, stability of armor units, bottom scour and determination of harbor tranquility. Losada et al. (1996) mentioned that the limitations of the model are the use of potential, linear theory and the lack of considering wave breaking and their consequent implications on engineering applications. They showed that random waves (unidirectional or multidirectional) induce lower reflection and transmission coefficients than regular waves. They also illustrated that reflection is more sensitive to incident wave characteristics than transmission. It was shown by Losada et al. (1996) that the wave randomness significantly decreases friction coefficients. Their model reproduced the experimental results well.

Hayakawa et al. (1998) conducted both two- and three-dimensional numerical simulation of wave fields around the submerged breakwater. The results of the calculation were compared with the laboratory experimental data. Laboratory experiments were carried out with a wave tank, 15 m long and 0.6 m wide, with a permeable submerged breakwater. The 3D submerged breakwater occupied half width of the wave tank (Hayakawa et al., 1998). They illustrated that to have a good agreement with the measured data in the laboratory, the side wall boundary condition for the 3D simulation of the wave flume should be nonviscous. (Hayakawa et al., 1998).

Following previous works (Schaap, 1997; Torrini, 1997; van der Biezen et al., 1997), van der Biezen et al. (1998) presented the results of 2DH numerical modeling of morphological effects of submerged breakwaters. The calculated results by their numerical model were compared with measured experimental data in the wave basin (length = 28 m, width = 14 m, depth = 0.60 m). Experiments were carried out with and without segmented submerged breakwaters under various configurations with regular waves normally incident on a beach (van der Biezen et al., 1998). All tests in their study were conducted with a rubble mound breakwater with an impermeable core. The breakwater crest width was 0.16 m and the water depth at the toe of the breakwater was 0.4 m. The submergence depth was 1.5 m and the initial beach slope for all experiments was 1:15. Regular waves with heights of 0.08 m, 0.10 m and 0.12 m and period of 1.55 s were generated (For further details see: van der Biezen et al., 1998). Van der Biezen et al. (1998) used the Delft3D package, developed by Delft Hydraulics, for their numerical modeling. The Delft3D wave model used in their study was a phase-resolving module, based on the nonlinear shallow water (NSW) equations, which calculated a stationary wave field including wave heights, set-up, energy dissipation, etc. They concluded that the numerical modeling of segmented submerged breakwaters based on the nonlinear shallow water equations yields promising results (van der Biezen et al., 1998).

Considering different design parameters, the morphodynamic effects of submerged breakwaters were investigated numerically by Lesser et al. (2003). They indicated that an inappropriate design of submerged breakwaters may result in significant erosion around the structure and at the shoreline.

Shen et al. (2004) simulated the propagation of Cnoidal waves over a submerged breakwater using a two-equation k-ε turbulence model based on the volume of fluid (VOF) method. They compared their results with two-dimensional laboratory data reported by Ohyama et al. (1995). The comparisons showed relatively good agreement; however, noting some discrepancy in the results, Shen et al. (2004) indicated that a more refined turbulence model might be needed to address the differences.

Friebel and Harris (2004) developed a 2D engineering aid for preliminary design of submerged breakwaters by improvement of existing two-dimensional K_t models. A total of five data sets were combined and analyzed (Daemen, 1991; Daemrich and Kahle, 1985; Seabrook and Hall, 1997; Seelig, 1980; van der Meer, 1988). Their model provided the required design criteria for use in determining wave attenuation in the presence of submerged breakwaters. It was shown that their model reproduces better results compared with other models. Their results confirm that K_t is significantly dependent upon the dimensionless ratio of submergence (h_s/H_{moi}). They also mentioned in their paper that the importance of other nondimensional parameters in order of decreasing importance were the ratios of breakwater crest width to water depth (B/h), breakwater crest height to water depth (h_c/h), breakwater crest width to offshore wave length (B/L_o), and submergence depth to breakwater crest width (h_s/B) (Friebel and Harris, 2004). It was suggested by Friebel and Harris (2004) that the proposed model is valid for specific variable ranges and is recommended to be employed as an engineering tool for preliminary design purposes (For further details please see: Friebel and Harris, 2004).

Garcia et al. (2004) developed a numerical model based on the decomposition of the instant velocity and pressure fields into mean and turbulent portions to calculate water surface elevation and flow in the presence of permeable low-crested breakwaters for regular breaking waves. The model was named COrnell BReaking waves And Structures (COBRAS) and used in the simulation of wave propagation in the surf zone, wave breaking and wave-structure interactions (Garcia et al., 2004). COBRAS solves the Reynolds Averaged Navier-Stokes Equations (RANS) numerically for the 2D flow field. It uses a modified k-ε turbulence model. COBRAS applies the volume of fluid (VOF) method for free surface tracking (Garcia et al., 2004; Lin and Liu, 1998). The comparison with 2D experimental tests carried out by Vidal et al. (2001) showed that the model reproduces the measured quantities with good agreement. The model was extended by Lara et al. (2006) for random waves. Good results were obtained compared to the measured data (Vidal et al., 2001).

Roul et al. (2004) investigated the overtopping discharge and set-up behind the rubble mound breakwater. They observed that the net transmitted discharge has an inverse effect on the set-up.

Van der Meer et al. (2004), van der Meer et al. (2005) improved the 2D empirical formula for rubble mound low-crested breakwaters provided by van der Meer (1991) and d'Angremond et al. (1996) using data collected from more than 2300 tests from numerous studies. Van der Meer et al. (2004), van der Meer et al. (2005) found that for rubble mound low-crested structures with $B/H_i < 10$ the equation provided by d'Angremond et al. (1996) is still applicable, but for $B/H_i > 10$, they proposed a modified equation considering both effects of and $\frac{h_s}{H_i}$.

Van der Meer et al. (2004), van der Meer et al. (2005) also proposed that the lower limit of K_t may be kept constant as suggested by d'Angremond et al. (1996) while the upper limit of K_t needs to be modified. Two equations were provided by van der Meer et al. (2004), van der Meer et al. (2005) for smooth and impermeable low-crested structures based on the variation of the Irribaren number. The values of 0.075 and 0.8 were proposed for the lower and higher limit of K_t, respectively.

Van der Meer et al. (2004), van der Meer et al. (2005) realized that a wave transmission coefficient is influenced by wave angle in smooth breakwaters. They indicated that K_t decreases with increasing incident wave angle. Thus a prediction formula was proposed by van der Meer et al. (2004), van der Meer et al. (2005) for smooth structures, including obliquity. They provided the valid range of variables for the equation. Van der Meer et al. (2004), van der Meer et al. (2005) also indicated that K_t is not sensitive to θ in the case of rubble mound structures.

As mentioned before, Van der Meer et al. (2005) gave a summary of a large number of collected data on the low-crested structures. The data for wave transmission over LCS in the DELOS project was used. The DELOS database (Kramer et al., 2005) consists of more than 2300 tests. These data sets were applied and analysed to provide new empirical formulas on wave transmission coefficient for rubble-mound low-crested structures. Within the DELOS project, the interaction of low-crested structures and oblique wave attack was also considered (Kramer et al., 2005). Analysis and discussion for spectral form changes, reflection and 3D effects were also presented (van der Meer et al., 2005).

Kramer et al. (2003) and Kramer et al. (2005) also conducted a series of 3D laboratory experiments including wave obliquity in the wave basin at Aalborg University, Denmark (9.7 × 12.5 m) on low-crested structures (LCS) as part of the DELOS project. One of the layouts tested was a symmetric arrangement of two detached breakwaters with a narrow/wide berm, separated by a gap and subjected to experiments at zero free board, positive freeboard (emerged) and negative freeboard (submerged) with regular unidirectional waves and random directional waves (Kramer et al., 2003; Kramer et al., 2005). The JONSWAP spectral shape was used for the random waves (Kramer et al., 2003).

Johnson et al. (2005) applied numerical models to model wave and current fields around submerged breakwaters. They used both phase-averaged and phase resolving methods in their research. Johnson et al. (2005) used the experimental data collected during the DELOS project to validate the models. They realised that wave heights predicted by models were comparable to that observed in the experiments if the wave breaking sub-model was properly calibrated for wave energy dissipation over the submerged breakwater. The outcomes of the numerical models were compared with three dimensional experimental data sets collected from wave basin (Lamberti et al., 2003). The calculated transmitted wave height using both models was observed to be in good agreement with the measured data. Furthermore, simulated instant water surface elevations obtained from the phase-resolving model was compared satisfactorily with the laboratory measured data in the wave flume tests. It was illustrated by Johnson et al. (2005) that the water flow pattern obtained by both models was qualitatively close to the measurements. The phase-averaged model shows qualitatively good agreement with the measured data for the water flow pattern around the submerged breakwaters (for more details, see Johnson et al., 2005).

Van Oosten and Peixo (2005) and van Oosten et al. (2006) developed an artificial neural network model for prediction of two-dimensional wave transmission behind submerged structured based on the DELOS database. Panizzo and Briganti (2007) also similarly developed an artificial neural network model for prediction of the wave transmission coefficient behind two-dimensional submerged structures. Both methods produced much better predictions in comparison with existing prediction formulas for wave transmission coefficient in two-dimensional conditions; however, both models were not capable of being used for three-dimensional situations and prediction of the spatial variation of the wave field around the submerged breakwaters (Panizzo and Briganti, 2007; van Oosten et al., 2006).

Gourlay and Colleter (2005) applied the energy balance equation for reefs with low submergence ($S < 0.7$) and found the equation of discharge over the reef based on a discharge coefficient of broad-crested weir and a friction factor of flow in open channel. The effects of the width of the reef and the incident wave height were also considered.

Gourlay and Colleter (2005) also used the energy equation between the point of maximum set-up over the reef and onshore reef edge for higher submergence ($0.7 < S$). Conditions in which flow over the reef is subcritical and flow is open channel were assumed. A reef profile shape factor (see Gourlay, 1996b) and a wave deformation index over the reef were considered (1996a).

Hydrodynamic and morphodynamic effects of submerged structures were investigated by Ranasinghe et al. (2006) experimentally and numerically. They conducted a series of 3D experimental tests in wave basin in the presence of submerged structures. An empirical formula as a preliminary engineering tool for shoreline response to submerged structures was also proposed. The results illustrated that the influence of submerged breakwaters on erosion or accretion of shoreline significantly depends on the predominant incident wave angle, the submergence depth and the distance to the beach. Turner (2006) applied an odd-even function analysis to evaluate the erosion/accretion in the vicinity of detached structures. The results were verified by the field data from a submerged reef structure at the northern Gold Coast, Queensland, Australia (Turner, 2006).

2D experimental and numerical modeling of wave transmission and wave-induced current over submerged breakwaters was carried out by Tajziehchi and Cox (2007). Laboratory 2D experiments were carried out in a 3 m wide wave flume to investigate wave transmission, wave-induced set-up and current over submerged breakwaters with different crest widths under different submergences. They also examined and reported the capability of the Delft3D model in the case of modeling and calculating wave transformation and wave-generated current over the submerged breakwaters (see Cox and Tajziehchi, 2006; Tajziehchi, 2006; Tajziehchi and Cox, 2007).

Kobayashi et al. (2007) developed a numerical model based on an existing time-averaged model of continuity, momentum, and energy equations to predict the mean and standard deviation of the free surface elevation and horizontal fluid velocities above and inside a porous submerged breakwater. The original model was designed to predict irregular breaking wave transformation on impermeable beaches. A modified formula for intense wave breaking on the steep seaward slope of the breakwater was used to estimate the wave energy dissipation rate due to irregular breaking waves (Kobayashi et al., 2007). The model predicts the measured cross-shore variations of the water surface level. The measured experimental data were collected in a laboratory. A submerged porous breakwater on a gentle impermeable beach slope was constructed in the laboratory (Kobayashi et al., 2007).

Buccino and Calabrese (2007) presented semi-empirical models and predictive equations for K_t behind both submerged and emergent breakwaters based on a simplified modelling of wave breaking and overtopping as well as seepage through the body. The results show good agreement with a large set of measured data. The proposed formula for submerged breakwaters was suggested to be valid in a specific range of data.

An artificial neural network model to predict K_t behind low-crested structures was developed by Panizzo and Briganti (2007). The datasets gathered within the European Union research project DELOS were used to train and validate their model. Their model covers a set of non-dimensional physical parameters including h_s/H_i, H_i/D_{n50}, B/H_i, B/L_{op}, ξ_{op} and H_i/h. In their model, the parameter h_s is negative for submerged breakwaters. B is the crest width, α is the breakwater seaward slope angle and D_{n50} is the nominal diameter of the armor units defined as:

$$D_{n50} = \left(\frac{M_{50}}{\rho_r}\right)^{\frac{1}{3}} \tag{3.2}$$

where M_{50} is the median mass of unit given by 50% on the mass distribution curve and ρ_r is the mass density of the rock. As for the hydraulic parameters, h is the water depth at the toe of the structure, L_{op} is the offshore wave length associated to the peak period T_p by means of the relationship $L_{op} = \frac{gT_p^2}{2\pi}$, where g is the earth gravity acceleration.

In order to represent the combined effect of the wave steepness $\left(S_{op} = \frac{H_i}{L_{op}}\right)$ and the seaward slope of the structure α, the surf similarity parameter (Iribarren number) associated with the peak period T_p was used, which is calculated as previously explained. The calibrated and optimized model was found to perform more accurately compared with van der Meer et al. (2005).

Christou et al. (2008) compared the outcomes of a numerical model based on the boundary element method (BEM) with measured data. Their research studied the interaction between nonlinear regular waves and rectangular submerged breakwaters. The numerical model was illustrated to predict the spatial and temporal water surface profiles and spectra in good agreements with the experimental data. Christou et al. (2008) also showed that water wave reflection from rectangular submerged breakwaters is linear.

As for waves transmitted over low-crested structures (LCS), Goda and Ahrens (2008) proposed a formula by empirical fitting to the curves in the design diagram of the wave transmission coefficient derived by Tanaka (1976). An effective width of structure and an approximate limit of dimensionless wave run-up were introduced (Goda and Ahrens, 2008). An effective diameter of the materials composing LCS was also considered equal to the median diameter D_{50} for rubble stone and $\left(\frac{M_{50}}{\rho_r}\right)^{1/3}$ for concrete blocks with M_{50} being the mass and ρ_r being the density. A parameter for effective width B_{eff} of the low-crested structures was introduced. This parameter can be calculated based upon the structure crest height relative to the still water level (Goda and Ahrens, 2008).

Developing a 3D design tool for prediction of global wave transmission coefficient $K_{d,t}$ behind the low-crested structures, Vicinanza et al.

(2009) proposed a method based on a combined diffraction-overtopping scheme for the first time. Vicinanza et al. (2009) indicated that the significance of the 3D studies increases with the importance of diffraction and with reductions either in the overtopping rate or the permeability of the breakwater or the length of the structure compared to the incident wavelength.

Vicinanza et al. (2009) did not consider permeability effects as they indicated that wave diffraction, wave overtopping and permeability are often significantly correlated; when permeability increases, the diffracted waves are expected to reduce, as can be derived from a quick application of the Penny and Price (1952) theory for cushion breakwaters. They mentioned that the same holds for wave overtopping, because the higher the permeability, the lower the run-up height. So their study was restricted only to wave diffraction and overtopping. They proposed to employ the available wave diffraction theories combined with a 2D wave transmission formula. They assumed that wave diffraction occurred in a same way as for nonovertopped structures with no statistical correlation with 2D wave overtopping. This gives, by summing energies (Buccino et al., 2009; Vicinanza et al., 2009):

$$K_{d,t} = \sqrt{K_d^2 + K_t^2} \tag{3.3}$$

where K_d is the wave diffraction coefficient and can be estimated through the simplified approach, to be discussed later, or by a more sophisticated numerical model; K_t is the 2D wave transmission coefficient and can be evaluated by selecting the predictive equation (previously discussed in the two-dimensional wave transmission section) that best fits the breakwater layout one.

The two-dimensional wave transmission coefficient ($K_{t,\mathrm{2D}}$) is calculated by the Goda and Ahrens (2008) method and the wave diffraction coefficient (K_d) is calculated by the Penny and Price (1952) method. Then according to the computed values $K_{t,\mathrm{2D}}$ and K_d, the three-dimensional wave transmission coefficient ($K_{t,\mathrm{3D}}$) will be calculated. In order to perform a meaningful comparison with other models, the selected models are required to be calibrated on the same range of data.

K_d is calculated by Penny and Price (1952) based on penetration of waves through a gap smaller than one wave length. When the gap width is small compared with wave length, an approximate solution may be obtained by assuming that the streamlines for the motion of the water through the gap are the same as for a simple uniform streaming of water through the gap (Penny and Price, 1952). Then the problem of water waves becomes similar to the sound waves; which was treated by Lamb (1924). Using Lamb's solution, an expression for $F(x,y)$ can be deduced for points not too near the gap (Penny and Price, 1952) based on the distance from the center of the gap, the gap width and Euler's constant. This approach is consistent with waves propagating in all directions behind the breakwater in a uniform manner. Thus the gap between two segments of breakwater is considered as a point source. The heights of the diffracted wave at any point to incident wave height were proposed to be proportional to $r^{-1/2}$ (Penny and Price, 1952).

The hypothesis that assumes wave diffraction and 2D transmission are uncorrelated seems quite reasonable, as they are totally different physical processes (Vicinanza et al., 2009). Vicinanza et al. (2009) employed the results of hydraulic model tests conducted at Delft University of Technology (Caceres et al., 2008) to derive and discuss prediction tool for wave and current field behind a low-crested (not fully submerged) breakwater.

Considering the wave diffraction models at a single breakwater of finite length, Morse and Rubenstein (1938), Goda et al. (1971), Stiassnie and Dagan (1972) and Filianoti (2000) developed exact theories, where the diffraction coefficient at an isolated detached barrier is given in terms of Mathieu functions (Vicinanza et al., 2009). McIver (2005) also proposed a solution as a series of integrals. However, two models of Penny and Price (1952) for monochromatic waves and of Goda et al. (1978) for multidirectional seas, in the case of semi-infinite breakwaters, were recommended by Vicinanza et al. (2009), as these methods are much easier to use, particularly for the initial stages of design. Nevertheless, these methods have not been properly supported by experimental evidence so far and this may lead to some uncertainties in practical applications (Vicinanza et al., 2009).

As mentioned previously, data on this matter are limited. Van der Meer et al. (2005) analyzed 3D random wave experiments conducted at Aalborg University (Kramer et al., 2005) and concluded that wave height behind the barrier was the same, or nearly the same, as in a 2D flume. Nevertheless, it is probable that the authors were referring to the wave height generated by overtopping and seepage only (2D transmission), since wave probes were located quite far from the breakwater heads (Buccino et al., 2009; Vicinanza et al., 2009). However, Calabrese et al. (2005) compared the experiments of van der Meer et al. (2005), and found that the wave height measured at the lee of the structure was larger than that predicted. This result agrees with the findings of Adams and Sonu (1986) and Seabrook and Hall (1998), who indicate the presence of an additional contribution to the global transmission process associated with the 3D nature of wave motion (Buccino et al., 2009; Vicinanza et al., 2009). The computed transmission coefficient, $K_{d,t}$, can be compared with the Buccino and Calabrese (2007), Goda et al. (2008), van der Meer et al. (2005) or other formula for 2D wave transmission to measure the importance of the 3D effects of submerged breakwaters.

Johnson (2006) applied MIKE 21 PMS to simulate wave transformation around submerged breakwaters. MIKE 21 PMS is basically a refraction-diffraction model (Battjes and Janssen, 1978). Johnson (2006) applied the model to calculate the wave transmission coefficient. The outcomes of the model were also compared with measured data (Zanuttigh et al., 2003). It was shown that the model reproduces higher energy dissipation over a submerged breakwater, although, the model have relatively good agreement with the experimental results.

Lin (2006) extended the model of Lin and Li (2003) for solving the Navier-Stokes equations and presented a three dimensional multiple-layer σ-coordinate model for interaction of water waves with submerged/emerged structures. The earlier model was originally designed for wave propagation over varying topography. Lin (2006) extended the old model to be able to calculate wave propagation over the submerged rectangular breakwaters with sudden topography changes. The results obtained by the extended model was compared with a VOF model (Lin and Liu, 1998) for simulation of 2D solitary wave propagation over a submerged rectangular structure. The computed values were also verified with the measured data (Zhuang and Lee, 1996). The model was shown to present promising results.

Zanuttigh and Lamberti (2006) analyzed fluxes around and through the rubble-mound low-crested structures experimentally and numerically. Their experiments were performed in a $12.5 \times 9.7\ m^2$ wave basin at Aalborg University, Denmark. 2D and 3D tests on low-crested structures were performed and data collected for wave set-up, wave overtopping discharge, returning flow and filtration flux within the framework of the DELOS project. Segmented detached breakwaters were constructed. Zanuttigh and Lamberti (2006) tested narrow and wide berm under regular, irregular and oblique wave attacks. Numerical simulations were carried out and the accuracy of the model MIKE 21 was evaluated (Zanuttigh and Lamberti, 2006).

Lykke Andersen and Burcharth (2008) published results of 3D experiments on wave overtopping, where the structure model had a core made of a very fine material, with a 2 mm median diameter. Based on those tests, they found a best-fit value of the roughness factor equal to 0.39 that was considered as the most suited to their experiments, mainly because it came from tests conducted in a wave basin and also because it took into account the results of research by Zanuttigh and Andersen (2010) on 3D wave reflection. Altogether, previous results supported the idea that wave diffraction may play a significant role in the global wave transmission process. They mentioned that its significance should increase with reducing wave overtopping and permeability, when barriers are short compared to the incoming wave length and, for a given structure length, when one moves from the central part of the shadow zone towards the head (Lykke Andersen and Burcharth, 2008).

Zanuttigh et al. (2008) presented a conceptual model for the prediction of wave transmission coefficients and wave spectra in the existence of a low-crested emerged structure based on the structure geometry and wave climate. Permeability of the structure was also considered. Thus both effects of filtration through the structure body and wave overtopping over the crest were studied. A relatively good agreement with the measured data in the wave flume and wave basin was observed (Zanuttigh et al., 2008).

Kriezi and Karambas (2010) studied wave transmission and reflection in presence of rectangular submerged breakwaters experimentally and numerically. Irregular and regular waves were tested with breaking and nonbreaking conditions. The experiments were conducted in large-scale facilities in Universitat Politecnica de Catalunya, Laboratory de Ingeneria Maritima, Spain. Improved Boussinesq type equations were solved numerically. A breaking model (Okamoto and Basco, 2006) was used. The results of the proposed model were shown to be in good agreement with measured values data for both breaking and nonbreaking wave conditions.

Vanlishout et al. (2010) studied the interaction of oblique waves with impermeable rubble-mound submerged breakwaters experimentally. The tests were carried out using three dimensional physical models. It was shown that the permeability has no significant impact on the wave transmission behind the rubble-mound submerged structures considering a oblique wave attack (Vanlishout et al., 2010).

Del Jesus (2011) studied the performance of an existing submerged breakwater in Santa Monica experimentally. 3D laboratory model was used (Adams and Sonu, 1986). The effect of the permeability of the beach were also investigated numerically by solving a Navier-Stokes model (IH-3VOF) (del Jesus, 2011).

Liu et al. (2013) developed an improved two-dimensional model based on incompressible smoothed particle hydrodynamics (ISPH) to simulate free surface flow in presence of various impermeable coastal structures. The performance of the model for predicting wave transmission and reflection behind and in front of the structures, respectively (Liu et al., 2013). It was shown that a relatively good agreement with existing analytical, experimental and numerical models was observed.

Sharif Ahmadian and Simons (2014) presented a numerical method and prediction tool to calculate the spatial wave transmission coefficient and wave field for regular waves around and behind three dimensional detached submerged breakwaters. Two distinct methods were developed based on machine learning algorithms; multi-layer perceptron (MLP) and radial-basis function (RBF). The models were designed using a number of new laboratory data sets Sharif Ahmadian and Simons (2014). The models were expressed in terms of both dimensional and non-dimensional parameters. Predictions obtained by the non-dimensional RBF model was observed to have best results as a wave transmission calculator in the case of the three dimensional wave field around the submerged breakwaters. It was shown that the proposed model has a good agreement with the measured data. The results of the model was verified for interpolation, and extrapolation cases. The predictions of model were also tested at larger scale conditions using different laboratory facilities (see: Sharif Ahmadian and Simons, 2014). The comparisons revealed sufficient agreement with the laboratory results to suggest that the model has potential to be employed as a 3D prediction tool in real applications (Sharif Ahmadian and Simons, 2012; Sharif Ahmadian and Simons, 2014).

REFERENCES

Abdul Khader, M.H., Rai, S.P., 1980. A study of submerged breakwaters. J. Hydraulic Res. 18 (2), 113–121.

Adams, C.B., Sonu, C.J., 1986. Wave transmission across submerged near-surface breakwater. Proc. 20th Int. Conf. on Coastal Engineering. ASCE.

Ahrens, J.P., 1987. Characteristics of reef breakwaters. Technical Report, CERC-87-17, 45p.
Ahrens, J.P., 1989. Stability of reef breakwaters. J. Waterway, Port, Coast. Ocean Eng. 115 (2), 221–234.
Ahrens, J.P., McCartney, B.L., 1975. Wave period effect on the stability of riprap. Proc. of Civil Engineering in the Ocean/III. ASCE.
Allsop, N.W.H., 1983. Low-crested breakwaters, studies in random waves. Proc. Coastal Structure '83. ASCE.
Amnity, P., Lamberti, A., Liberatore, G., 1983. Experimental studies on submerged barriers as shore protection structures. Proc. Int. Conf. on Coastal and Port Eng. in Developing Countries, Colombo, 201–218.
Battjes, J.A., 1994. Shallow water wave modelling. In: Isaacson, M., Quick, M. (Eds.), Proc. Waves-Physical and Numerical Modelling. University of British Columbia, Vancouver.
Battjes, J.A., Janssen, J.P.F.M., 1978. Energy loss and set-up due to breaking of random waves. Proc. 16th Int. Conf. on Coastal Engineering. ASCE.
Beach Erosion Board, 1940. A model study of the effect of submerged breakwaters on wave action. Technical memorandum No. 1, Chief of Engineers, U.S. War Dept.
Bleck, M., Oumeraci, H., 2001. Wave damping and spectral evolution at artificial reefs. Proc. 4th Int. Symp. on Ocean Wave Measurement and Analysis. ASCE.
Bleck, M., Oumeraci, H., 2002. Hydraulic performance of artificial reefs: global and 179 local description. Proc. 28th Int. Conf. on Coastal Engineering. ASCE.
Bleck, M., Oumeraci, H., 2004. Analytical model for wave transmission at artificial reefs. Proc. 29th Int. Conf. on Coastal Engineering. ASCE.
Browder, A., Dean, R.G., 1996. Performance of a submerged breakwater for shore protection. Proc. 25th Int. Conf. on Coastal Engineering. ASCE.
Bruce, T., van der Meer, J., Franco, L., Pearson, J., 2006. A comparison of overtopping performance of different rubble mound breakwater armour. Proc. of the Int. Conf. on Coastal Engineering. ASCE, 1034–1043.
Buccino, M., Calabrese, M., 2007. Conceptual approach for prediction of wave transmission at low-crested breakwaters. J. Waterway, Port, Coast. Ocean Eng. 133, 213–224.
Buccino, M., Vicinanza, D., Caceres, I., Calabrese, M., 2009. 3D wave field behind impermeable low crested structures. Proc. of the 10th Int. Coastal Symposium. J. Coast. Res., SI 56, Lisbon, Portugal, 477–481.
Caceres, I., Stive, M., Sanchez-Arcilla, A., Trung, L., 2008. Quantification of changes in current intensities induced by wave overtopping around low-crested structures. Coast. Eng. 55 (2), 113–124.
Calabrese, M., Vicinanza, D., Buccino, M., 2002. Large-scale experiments on the behaviour of low crested and submerged breakwaters in presence of broken waves. Proc. 28th Int. Conf. on Coastal Engineering. ASCE.
Calabrese, M., Vicinanza, D., Buccino, M., 2003. Low-crested and submerged breakwater in presence of broken waves. Proc. 13th ISOPE, 831–836.
Calabrese, M., Buccino, M., Zanuttigh, B., Cappietti, L., Paris, A., 2005. Riflessione e trasmissione di scogliere a cresta bassa. Studi Costieri 9, 93–106 (in Italian).
Calabrese, M., Vicinanza, D., Buccino, M., 2008. 2D wave setup behind submerged breakwaters. Ocean Eng. 35, 1015–1028.
Chatham, C.E., 1972. Movable-bed model studies of perched beach concept. Proc. 13th Int. Conf. on Coastal Engineering. ASCE.
Chiaia, G., Damiani, L., Petrillo, A., 1992. Evolution of a beach with and without a submerged breakwater: Experimental investigation. Proc. 23th Int. Conf. on Coastal Engineering. ASCE.

Christou, M., Swan, C., Gudmestad, O., 2008. The interaction of surface water waves with submerged breakwaters. Coast. Eng. 55 (12), 945958.
Cruz, E.C., Isobe, M., Watanabe, A., 1992. Nonlinear wave transformation over a submerged permeable breakwater. Proa, Int. Conf. Coastal Eng., pp. 1101−1114.
Cruz, E.C., Isobe, M., Watanabe, A., 1997. Boussinesq equations for wave transformation on porous beds. Coast. Eng. 30 (1), 125−156.
Daemen, I.F.R., 1991. Wave transmission at low crested structures. MSc. Thesis. Delft University of Technology, Faculty of Civil Engineering, Delft Hydraulics Report H 462.
Daemrich, K., Kahle, W., 1985. Schutzwirkung von Unterwasser Wellen brechern unter dem Einfluss unregelmassiger seegangswellen. Technical Report. Franzius Instituts fur Wasserbau und Kusteningenieurswesen, Report Heft 61 (in German).
Dalrymple, R.A., Knio, O., 2001. SPH Modeling of Water Waves, Proc. Coastal Dynamics. ACSE, Lund, Sweden.
Dally, W.R., Brown, C.A., 1995. A modeling investigation of the breaking wave roller with application to cross-shore current. J. Geophys. Res. 100 (C12), 24873−24883.
d'Angremond, K., Van der Meer, J.W., De Jong, R.J., 1996. Wave transmission at low-crested structures. Proc. 25th Int. Conf. on Coastal Engineering. ASCE.
Davies, B., Kriebel, D.L., 1992. Model testing on wave transmission past low crested breakwaters. Proc. 23th Int. Conf. on Coastal Engineering. ASCE.
Dean, R.G., 1977. Equilibrium beach profiles: U.S Atlantic and Gulf coasts. Department of Civil Engineering, Ocean Engineering Report No. 12. University of Delaware, Newark, DE.
Debski, D., Loveless, J.H., 1997. The design and performance of submerged breakwaters. Research Report. University of Bristol.
De Later, J., 1996. Effect of submerged breakwaters on a beach profile exposed to regular waves in a wave basin. Msc. Thesis. Delft University of Technology, Netherlands.
del Jesus, M., 2011. Three-dimensional interaction of water waves with coastal structures. Ph.D. thesis. Universidad de Cantabria.
Dick, T.M., Brebner, A., 1968. Solid and permeable submerged breakwaters. Proc. 11th Int. Conf. on Coastal Engineering. ASCE, 181.
Diskin, M.H., Vajda, M.L., Amir, I., 1970. Piling-up behind low and submerged permeable breakwaters. J. Waterways Harbours Div., WW2, ASCE, 359−372.
Drei, E., Lamberti, A., 1999. Wave pumping effect of a submerged barrier. Coastal Structures '99. ASCE.
Filianoti, P., 2000. Diffraction of random wind-generated waves by detached breakwater or breakwater gap. Ocean Eng. 27 (11), 1249−1263.
Friebel, H., Harris, L., 2004. A new wave transmission coefficient model for submerged breakwaters. In: Proceedings 29th International Conference on Coastal Engineering. Lisbon Portugal.
Garcia, N., Lara, J.L., Losada, I.J., 2004. 2-D numerical analysis of near-field flow at low-crested permeable breakwaters. J. Coast. Eng. 51, 991−1020.
Gironella, X., Sanchez-Arcilla, A., 1999. Hydrodynamic behaviour of submerged breakwaters. Some remarks based on experimental results. Proc. Coastal Structures '99. ASCE, 182.
Goda, Y., 1969. Re-analysis of laboratory data on wave transmission over breakwaters. Rep. Port Harbour Res. Inst. 18 (3), 3−18.
Goda, Y., Ahrens, J., 2008. New formulation for wave transmission over and through low crested structures. Proceedings of the International Conference on Coastal Engineering. ASCE.
Goda, Y., Takeda, H., Moriya, Y., 1967. Laboratory investigation on wave transmission over breakwaters. Rep. Port Harbour Res. Inst. 13, 38.

Goda, Y., Yoshimura, T., Ito, M., 1971. Reflection and diffraction of water waves by an insular breakwater. Rep. Harbour Res. Inst. 10, 4–52, Japan.
Goda, Y., Takaiama, T., Suzuki, Y., 1978. Diffraction diagrams for directional random waves. Proceedings of the International Conference on Coastal Engineering. ASCE.
González, M., Medina, R., Losada, M.A., 1999. Equilibrium beach profile for perched beaches. J. Coast. Eng. 36, 343–357.
Gourlay, M.R., 1993. Wave setup and wave generated currents on coral reefs. Proc. 11th Australian Conf. on Coastal and Ocean Engineering, Townsville, 479–484.
Gourlay, M.R., 1996a. Wave set-up on coral reefs. 1. Set-up and wave-generated flow on an idealized two dimensional horizontal reef. J. Coast. Eng. 27, 161–193.
Gourlay, M.R., 1996b. Wave set-up on coral reefs. 2. Set-up on reefs with various profiles. J. Coast. Eng. 27, 161–193.
Gourlay, M.R., Colleter, G., 2005. Wave-generated flow on coral reefs – an analysis for two-dimensional horizontal reef-tops with steep faces. J. Coast. Eng. 52, 353–387.
Groenewoud, M., van der Graaff, J., Claessen, E., van der Beizen, S., 1996. Effect of submerged breakwater on profile development. 25th Int. Conf. on Coastal Engineering. ASCE.
Hanson, H., Kraus, N.C., 1990. Shoreline response to a single transmissive breakwater. Proc. 22th Int. Conf. on Coastal Engineering. ASCE.
Hardy, T.A., 1993. The attenuation and spectral transformation of wind waves on a coral reef. PhD. Thesis. James Cook University, Townsville, Queensland.
Hayakawa, N., Hosoyamada, T., Yoshida, S., Tsujimoto, G., 1998. Numerical simulation of wave fields around the submerged breakwater with sola-surf method. International Conference on Coastal Engineering. ASCE.
Hearn, C.J., 1999. Wave-breaking hydrodynamics within coral reef systems and the effect of changing relative sea level. J. Geophys. Res. 104 (C12), 30007–30019.
Jiménez, J.A., Sánchez-Arcilla, A., 2002. Preliminary analysis of shoreline evolution in the leeward of low-crested breakwaters using one-line models. Proceedings of the 1st year meeting of Delos Project, Barcelona, Spain.
Johnson, H.K., 2006. Wave modeling in the vicinity of submerged breakwaters. J. Coast. Eng. 53, 39–48.
Johnson, H.K., Karambas, T.V., Avgeris, I., Zanuttigh, B., Gonzalez-Marco, D., Caceres, I., 2005. Modelling of waves and currents around submerged breakwaters. J. Coast. Eng. 52, 949–969.
Johnson, J.W., Fuchs, R.A., Morison, J.R., 1951. The damping action of submerged breakwaters. Trans., Am. Geophys. Union 32 (5), 704–718.
Kobayashi, N., Wurjanto, A., 1988. Wave transmission over submerged breakwaters. J. Waterway, Port, Coast. Ocean Eng. 115 (5), 662–680.
Kobayashi, N., Meigs, L., Ota, T., Melby, J., 2007. Irregular breaking wave transmission over submerged porous breakwater. J. Waterway, Port, Coast., Ocean Eng. 133 (2), 104–116.
Kothe, D.B., Mjolsness, R.C., Torrey, M.D., 1991. RIPPLE: A Computer Program for Incompressible Flows with Free Surfaces. Los Alamos National Laboratory.
Kramer, M., Zanuttigh, B., Baoxing, W., Van der Meer, J., Lamberti, A., Burcharth, H., 2003. Wave basin experiments. Tech. rep., Internal report, DELOS deliverable D31.
Kramer, M., Zanuttigh, B., Vandermeer, J., Vidal, C., Gironella, F., 2005. Laboratory experiments on low-crested breakwaters. Coast. Eng. 52, 867–885.
Kriezi, E., Karambas, T., 2010. Modelling wave deformation due to submerged breakwaters. Proceedings of the ICE - Maritime Engineering 163-1, 19–29.
Lamb, H., 1924. Hydrodynamics. Cambridge University Press.

Lamberti, A., Zanuttigh, B., Tirindelli, M., 2003. 3D hydrodynamic tests with low-crested structures: analysis of overtopping and velocity fields. Proc. ISOPE, 2003, EA (2003), pp. 562–569.
Lara, J.L., Garcia, N., Losada, I.J., 2006. RANS modelling applied to random wave interaction with submerged permeable structures. J. Coast. Eng. 53, 395–417.
Lesser, G.R., de Vroeg, J.H., Roelvink, J.A., de Gerloni, M., Ardone, V., 2003. Modeling the morphological impact of submerged offshore breakwaters. Coastal Sediment'2003, Annual Symposium on Coastal Engineering and Science of Coastal Sediment Processes, 5th, Clearwater Beach, Fla., May 18–23, 2003, Proceedings, 14p. [CD-ROM].
Lin, P., 2006. A multiple-layer Ï-coordinate model for simulation of wave-structure interaction. Comput. Fluids 35 (2), 147–167.
Lin, P., Li, C., 2003. Wave-current interaction with a vertical square cylinder. Ocean Eng. 30 (7), 55–76.
Lin, P., Liu, P.L.-F., 1998. A numerical study of breaking waves in the surf zone. J. Fluid Mech. 359, 239–264.
Liu, X., Xu, H., Shao, S., Lin, P., 2013. An improved incompressible SPH model for simulation of wave-structure interaction. Comput. Fluids 71, 113–123.
Longuet-Higgins, M.S., Stewart, R.W., 1964. Radiation stress in water waves, a physical discussion with applications. Deep Sea Res. 11, 529–562.
Longuet-Higgins, M.S., 1967. On the wave induced difference in mean sea level between two sides of a submerged breakwater. J. Mar. Res. 148–153.
Longuet-Higgins, M.S., 1970. Longshore currents generated by obliquely incident sea waves, 1. J. Geophys. Res. 75 (33), 6778–6789.
Losada, I.J., Silva, R., Losada, M.A., 1995. 3-D non-breaking regular wave interaction with submerged breakwaters. Coast. Eng. 28 (1), 229–248.
Losada, I.J., Silva, R., Losada, M.A., 1996. 3-D non-breaking regular wave interaction with submerged breakwaters. J. Coast. Eng. 28, 229–248.
Loveless, J.H., Debski, D., 1997. Wave transmission and set-up at detached breakwater. Coastal Dynamics'97. ASCE.
Loveless, J.H., Debski, D., MacLeod, A.B., 1998. Sea level set-up behind detached breakwaters. Proc. 26th Int. Conf. on Coastal Engineering. ASCE.
Lykke Andersen, T., Burcharth, H., 2008. Three dimensional investigations of wave overtopping on rubble mound structures. Coast. Eng. 56, 180–189.
Lynett, P.J., Liu, P.L.F., Losada, I.J., Vidal, C., 2000. Solitary wave interaction with porous breakwaters. J. Waterway, Port, Coast., Ocean Eng. 1266, 314–322.
Massel, S., Butowski, P., 1980. Wind waves transmission through porous breakwater. Coast. Eng. Proc. 1 (17).
Massel, S.R., Brinkman, R.M., 2001. Wave-induced set-up and flow over shoals and coral reefs. Part 1. A simplified bottom geometry case. Oceanologia 43 (4), 373–388.
Melito, I., Melby, J., 2002. Wave runup, transmission, and reflection for structures armored with core-loc. J. Coast. Eng. 42, 33–52.
Mendez, F.J., Losada, I.J., Losada, M.A., 2001. Wave-induced mean magnitudes in permeable submerged breakwaters. Journal of Waterway, Port, Coastal and Ocean Engineering, 127, 7–15.
Morison, J., 1949. Model study of wave action on underwater barriers. Tech. rep., Report HE-116-304, Inst. Eng. Res., Univ. Calif., Berkeley.
Morse, P., Rubenstein, P., 1938. The diffraction of waves by ribbons and by slits. Phys. Rev. 54, 895–898.
Muñóz-Pérez, J.J., Tejedor, L., Medina, R., 1999. Equilibrium beach model for reef protected beaches. J. Coast. Res. 15 (4), 950–957.
Nelson, R.C., 1996. Hydraulic roughness of coral reef platforms. Appl. Ocean Res. 18, 265–274.

Ohyama, T., Kioka, W., Tada, A., 1995. Applicability of numerical models to nonlinear dispersive waves. J. Coast. Eng. 24, 297–313.
Okamoto, T., Basco, D., 2006. The relative trough Froude number for initiation of wave breaking: theory, experiments and numerical model confirmation. Coast. Eng. 53 (8), 675–690.
Panizzo, A., Briganti, R., 2007. Analysis of wave transmission behind low-crested breakwaters using neural networks. Coast. Eng. 54, 643–656.
Penny, W., Price, A., 1952. The diffraction theory of sea waves and the shelter afforded by breakwaters. Philos. Trans. R. Soc. Lond. ÙA 244, 236–253.
Petti, M., Ruol, P., 1992. Laboratory tests on the interaction between nonlinear long waves and submerged breakwaters. Proc. 23th Int. Conf. on Coastal Engineering. ASCE.
Powell, K., Allsop, N., 1985. Low-crested breakwaters, hydraulic performance and stability. Tech. rep., Technical report SR 57, HR Wallingford.
Ranasinghe, R., Turner, I.L., Symonds, G., 2006. Shoreline response to multifunctional artificial surfing reefs: a numerical and physical modelling study. J. Coast. Eng. 53, 589–611.
Reniers, A.J.H.M., Battjes, J.A., 1997. A laboratory study of longshore currents over barred and non-barred beaches. J. Coast. Eng. 30, 1–22.
Rojanakamthorn, S., Isobe, M., Watanabe, A., 1990. Modeling of wave transformation on submerged breakwater. Proc. 22th Int. Conf. on Coastal Engineering. ASCE.
Roul, P., Faedo, A., 2002. Physical model study on low-crested structures under breaking wave conditions. Proc. Int. MEDCOAST Workshop on Beaches of the Mediterranean & the Black Sea, Kusadasi, Turkey, 83–96.
Roul, P., Faedo, A., Paris, A., 2004. Physical model study of water piling-up behind low-crested structures. Proc. 29th Int. Conf. on Coastal Engineering. ASCE.
Sawaragi, T., 1992. Detached breakwaters; Short Course on Design and Reliability of Coastal Structures, Venice, 1-3 October 1992.
Schaap, J., 1997. Modelling the effects of submerged breakwaters in a wave basin. Ph.D. thesis. Delft University of Technology, Msc.Thesis.
Schlurmann, T., Bleck, M., Oumeraci, H., 2002. Wave transformation at artificial reefs described by the Hilbert-Huang transformation (HHT). Proc. 28th Int. Conf. On Coastal Engineering. ASCE.
Seabrook, S.R., Hall, K.R., 1997. Effect of crest width and geometry on submerged breakwater performance. Proc. Canadian Coastal Conference, CCSEA, 58–72.
Seabrook, S.R., Hall, K.R., 1998. Wave transmission at submerged rubblemound breakwaters. Proc. 26th Int. Conf. on Coastal Engineering. ASCE.
Seelig, W.N., 1980. Two-dimensional tests of wave transmission and reflection characteristics of laboratory breakwaters. Technical report, CERC, Fort Belvoir, Report No. 80-1.
Sharif Ahmadian, A., Simons, R.R., 2012. 3-D Wave Field Around Submerged Breakwater. ASCE, New York. Available from: http://dx.doi.org/10.9753/icce.v33.structures.13.
Sharif Ahmadian, A., Simons, R.R., 2014. A 3D numerical model of nearshore wave field behind submerged breakwaters. Coast. Eng. 83, 190204. Available from: http://dx.doi.org/10.1016/j.coastaleng.2013.10.016.
Shen, Y.M., Ng, C.O., Zheng, Y.H., 2004. Simulation of wave propagation over a submerged bar using the VOF method with a two-equation k-e turbulence modeling. J. Coast. Eng. 31, 87–95.
Sorensen, R.M., Beil, N.J., 1988. Perched beach profile response to wave action. Proc. 21st Int. Conf. on Coastal Engineering. ASCE.

Stiassnie, M., Dagan, G., 1972. Wave diffraction by detached breakwaters. J. Waterways Harbours Coast. Div. 98 (2), 209–224.

Stucky, A., Bonnard, D., 1937. Contribution to the experimental study of marine rockfill dikesbull. Tech. rep., Technique de Suizze Romande, (also see summary in Tech. Memo. 1, Beach Erosion Board, 1940).

Svendsen, I.A., 1984a. Wave heights and set-up in a surf zone. J. Coast. Eng. 8, 303–329.

Svendsen, I.A., 1984b. Mass flux and undertow in a surf zone. J. Coast. Eng. 8, 347–365.

Symonds, G., 1994. Theory and observation of currents and setup over a shallow reef. Proc. Coastal Dynamics'98. ASCE.

Symonds, G., Black, K.P., Young, I.R., 1995. Wave-driven flow over shallow reefs. J. Geophys. Res. 100 (C2), 2639–2648.

Tajziehchi, M., 2006. Experimental and numerical modelling of wave-induced current and wave transformation in presence of submerged breakwaters. Ph.D. thesis. Civil & Environmental Engineering, Faculty of Engineering, UNSW.

Tajziehchi, M., Cox, R., 2007. Wave transmission and wave-induced current in presence of submerged breakwaters. 30th International Conference on Coastal Engineering 2006, San Diego USA. 5, 4970–4982.

Tanaka, N., 1976. Effects of submerged rubble-mound breakwater on wave attenuation and shoreline stabilization. Proc. 23th Japanese Coastal Engineering Conference. ASCE, 190.

Tomasicchio, U., 1996. Submerged breakwaters for the defence of the shoreline at Osita field experiences, comparison. Proc. 25th Int. Conf. on Coastal Engineering. ASCE.

Torrini, L., 1997. Nearshore effects of submerged breakwaters. Ph.D. thesis. Delft University of Technology, Msc.Thesis.

Turner, I.L., 2006. Discriminating modes of shoreline response to offshore-detached structures. J. Waterway, Port, Coast. Ocean Eng. 132 (3), 180–191.

van der Biezen, S.C., Van der Graaff, J., Schaap, J., Torrini, L., 1997. Small scale tests and numerical modeling of the hydrodynamic and morphological effects of submerged breakwaters. Proc. Combined Australian Coastal Engineering and Ports Conf., 219–224.

van der Biezen, S.C., Roelvik, J.A., van der Graaff, J., Schaap, J., Torrini, L., 1998. Wave transmission at low-crested structures. Proc. 26th Int. Conf. on Coastal Engineering. ASCE.

van der Meer, J.W., 1988. Rock slopes and gravel beaches under wave attack. PhD. Thesis. Delft University of Technology, Delft Hydraulics Report No. 396.

van der Meer, J.W., 1991. Stability and transmission at low-crested structures. Technical report, Delft Hydraulic, Report No. H 453. 33p.

van der Meer, J.W., Briganti, R., Wang, B., Zanuttigh, B., 2004. Wave transmission 191 at low-crested structures, including oblique wave attack. Proc. 29th Int. Conf. On Coastal Engineering. ASCE.

van der Meer, J.W., Briganti, R., Zanuttigh, B., Wang, B., 2005. Wave transmission at low-crested structures, including oblique wave attack. J. Coast. Eng. 52, 915–929.

van Oosten, R., Peixo Marco, J., van der Meer, J.W., Verhagen, H., 2006. Wave transmission at low-crested structures using neural networks. International Conference on Coastal Engineering. ASCE, San Diego.

Van Oosten, R.P., Peixo Marco, J., 2005. Wave transmission at various types of low-crested structures using neural networks, student thesis. Technische Universiteit Delft, Delft, The Netherlands.

Vanlishout, V., Verhagen, H., Troch, P., 2010. Oblique wave transmission through rough impermeable rubble mound submerged breakwaters. In: Proc. ICCE 2010-Shanghai.

Vicinanza, D., Caceres, I., Buccino, M., Gironella, X., Calabrese, M., 2009. Wave disturbance behind low-crested structures: Diffraction and overtopping effects. Coast. Eng. 56, 1173–1185.

Vidal, C., Lomónaco, P., Migoya, L., Archetti, R., Turchetti, M. Srci, M., et al., 2001. Laboratory experiments on flow around and inside LCS structures. Description of tests and data base. DELOS European Project, Technical Report.

Watanabe, A., Maruyama, K., Shimizu, T., Sakakiyama, T., 1986. Numerical prediction model of three-dimensional beach deformation around a structure. Coast. Eng. Jpn. 29, 179–194.

Zanuttigh, B., Andersen, T.L., 2010. Wave reflection in 3D conditions. Coast. Eng. 57 (5), 531–538.

Zanuttigh, B., Lamberti, A., 2006. Experimental analysis and numerical simulations of waves and current flows around low-crested rubble-mound structures. J. Waterway, Port, Coast. Ocean Eng. 132 (1), 10–27.

Zanuttigh, B., van der Meer, J., 2006. Wave reflection from coastal structures. Proc. Int. Conf. Coast. Eng. 5, 4337–4349.

Zanuttigh, B., Guerrero, M., Lamberti, A., 2003. 3D experimental analysis and numerical simulations of hydrodynamics around low crested structures. Proc. IAHR Conference, Thessaloniki, Greece, 369–376.

Zanuttigh, B., Martinelli, L., Lamberti, A., 2008. Wave overtopping and piling-up at permeable low crested structures. Coast. Eng. 55, 484–498.

Zhuang, F., Lee, J., 1996. A viscous rotational model for wave overtopping over marine structure, Proc 25th Int Conf Coast Eng, 25. ASCE.

Zyserman, J.A., Jorgensen, K., Christensen, E.D., 1999. Sediment transport and morphology in the vicinity of shore parallel breakwaters. Proceedings Coastal Structures 99, Santander, Spain, pp. 857–863.

CHAPTER 4

Theories and Methodologies

4.1 INTRODUCTION

In the analysis of hydrodynamic and morphodynamic processes as well as water quality changes in the presence of submerged breakwaters, numerical simulation is often employed to model the physical phenomena. Numerical modeling basically refers to the application of computer simulation, either alone or integrated with physical models. In this chapter a general brief outline is given with specific insight into the main characteristics, concepts and methodologies of the numerical models being used for the submerged breakwater analysis and design.

The hydrodynamic and morphodynamic analysis of the submerged breakwater under a wide range of environmental forces, such as currents, tides, waves, turbulence, storm surges, tsunamis and others, is required to be considered. Particularly, the depth variation at the site of the submerged breakwaters and the shoreline impacts play important roles in the performance of these structures. The structure geometry and water depth variation have significant influences on the waves breaking over the breakwater and the degree of turbulence flows and wave-induced currents. The nonlinearity of the different flows over and around the submerged breakwater as well as their complex interaction makes the problem even more complicated.

Nowadays, with the rapid development of numerical models and computational techniques, hydrodynamic and morphodynamic models are becoming more and more widespread in the coastal engineering community. A variety of numerical models are accessible, including one-dimensional, two-dimensional or three-dimensional models. These models, used in a wide range of coastal problems, solve the governing equations of conservation of mass and momentum numerically using computational techniques. These models are mostly based on the simplified Navier–Stokes equations, known as the shallow water equations, for coastal areas with considerably large scales in length and width in comparison with the depth scale. The most widely used numerical techniques are basically based on the finite difference method, the finite element method, finite volume method, boundary element method, etc.

© 2016 Elsevier Ltd.
All rights reserved.

A numerical model for wave modeling around submerged breakwaters may be either phase-averaged or phase-resolving. In a phase-resolving model, the sea surface is resolved and the model treats the wave field deterministically, while in a phase-averaged model, the statistics of the sea surface are computed, so this model treats the wave field stochastically. However, the accuracy of predictions by these models is to a great extent dependent on input data, such as bathymetry files and wind or wave data, as well as open boundary conditions, model parameters, simplifications, modeling techniques and the numerical scheme.

One of the main concerns in most of the numerical models, particularly in the case of submerged breakwaters with relatively steep slopes and sudden changes in water depth, is related to simulation of the wave breaking. Therefore, these models are generally provided along with some limitations. Depth-averaged equations, for example, include some simple expressions in the governing equations for wave energy dissipation when wave breaking occurs. These models cannot provide vertical variation of the velocity field and details of the turbulence and current field around the submerged breakwaters.

This chapter begins with an introduction to computational fluid dynamics and its application to the Navier–Stokes equations for submerged breakwaters. This is followed by an overview of a series of numerical models. A description of some of the most-used models, highlighting their main features, strengths, weaknesses and differences, is included in the next sections. Finally, a summary of the models, as well as further details of the models that are explained in this book, is presented.

Although, for most coastal structures, the effects of bottom friction and wave breaking play an important role in the design process, the mentioned wave equations do not include any term on the wave energy dissipation. This is a particularly important point for submerged breakwaters and should be noted carefully.

As discussed before, many different physical processes may occur when waves propagate over submerged breakwaters, which cannot be neglected in the structure design. These processes include wave reflection, shoaling, refraction, diffraction, etc. Therefore, developing an appropriate model to enable researchers, scientists and engineers to have a reliable and accurate simulation of all these forms of wave transformation is desired. A wide range of mathematical models have been developed by many people to include the effects of these processes and consequently to allow better predictions of wave transformation over and around the submerged breakwaters.

The outputs of these models can subsequently be used to determine wave effects on sediment transport and resulting geomorphology changes of the sea bed and coastline in the presence of submerged breakwaters.

4.2 TRADITIONAL MODELS FOR WATER WAVES

Numerical models basically approximate a set of the governing equations using numerical methods. The governing equations themselves represent a physical problem mathematically.

In the case of analysis of the interaction between the waves and coastal structures, and particularly submerged breakwaters, several models can be applied. The spectral wave models, Boussinesq equation, the mild-slope equation and shallow-water equation models are some of the most frequently used for modeling water wave propagation.

The spectral wave models are applied for a wide range of problems from deep to shallow waters and can simulate the propagation and transformation of random, short-crested, wind-generated waves in coastal waters. The spectral wave models consider an infinite number of linear waves for the sea waves, with each component then having its own parameters such as the wave amplitude, frequency and direction. The rate of change of wave energy flux is balanced by the wave energy transfer among different wave components in different directions and frequencies as well as energy input and dissipation. These models predict directional spectra, significant wave height, mean period, average wave direction, radiation stresses, and bottom motions over the model domain. These models may also have the ability to model different coastal processes including nonlinear wave interactions, refraction, shoaling, depth-induced breaking, white capping, etc. However, these models are mostly used for very large scales, deep waters and far-field regions. Therefore, they are not actually very appropriate for the near-field regions, for example near coastal structures, particularly with steep slopes and where the diffraction effects are important, such as with submerged breakwaters.

When dealing with nearshore wave field modeling, we may need to employ a more appropriate model. This is particularly due to the nonlinearity in waves in this zone. Therefore, for nearshore modeling, Boussinesq equation wave models, which are depth-averaged equations, are preferred. However, as a major limitation, application of the

Boussinesq equations are restricted to only intermediate and shallow water depths. In some cases we may also experience some minor limitations regarding the application of the Boussinesq equations resulting in less accurate results to some extent, for instance, those related to the rapidly varying bathymetries, fully nonlinearity and the coastal regions where the wave breaking may occur. An important difference of the Boussinesq model from the wave spectral and the mild-slope model (to be discussed later) is that the Boussinesq model can also be applied to nonperiodic flows. However, the shallow-water equation models, which are another type of wave model for coastal areas and which will be described later, are more suitable for such flows and particularly for long waves where the scales in width and length are very much larger than the depth scale. The Boussinesq equations for nonlinear and long waves include frequency dispersion, while the shallow water equations are not frequency dispersive.

The Boussinesq equations have been improved and extended by many researchers over the past years to overcome those restrictions or weaknesses. For example regarding the large nonlinearities, wave breaking or deep water depths. The Boussinesq equations can be modified by adding a new term to the depth-integrated momentum equation for the wave-breaking process with, for example, an eddy viscosity model (Karambas and Koutitas, 1992; Kennedy et al., 2000; Zelt, 1991), or a roller model (Brocchini et al., 1991; Schaffer et al., 1993). The modified Boussinesq equations are able to simulate surface wave propagation from intermediate water depth (water depth to wave-length ratio is about 0.5) to shallow water, including the wave-current interaction. Additionally, the Boussinesq equations can be modified to be applied for deeper water depths. Other restrictions can be also modified to some extent. The readers are referred to papers on these topics for further informations on modification approaches.

One of the widely known models in coastal engineering is the model based on the mild-slope equation. This model is basically used for linear waves in both shallow and deep waters, while the Boussinesq equation is used for nonlinear and long waves in shallow waters. The mild-slope equation (MSE) models, which are based on linear waves and slowly varying depths with lateral boundaries, are suitable to compute the wave field considering the wave diffraction effects from offshore to nearshore. The mild-slope equation was originally developed for shallow water waves (Eckart, 1952). The mild-slope equation (also known as the classical mild-slope equation) was further developed for arbitrary water depth by Berkhoff (1972). The mild-slope equation was primitively developed for wave propagation over mildly sloping sea beds and particularly combines the effects of the wave refraction, diffraction and shoaling into one consistent model. One of the good features of the mild-slope equation is that the vertical profile of the velocity is appropriately considered in the model, based on the linear wave theory. The models based on the mild-slope equation are today often used in order to approximate wave propagation and transformation as well as wave field changes over varying depths and bathymetries with lateral boundaries (i.e. coastal structures, seawalls, breakwaters, harbors, cliffs and coasts). The mild-slope equation calculates the wave and flow velocity amplitudes. Past studies show that the mild-slope equation provides relatively good outcomes with satisfactory accuracy for these problems.

However, the mild-slope equation can be employed only as long as the problem is linear and no interaction between waves exists. The model has been basically derived for irrotational linear harmonic waves. Furthermore, the mild-slope equation is not applicable to steep bottom slopes since the terms of the higher-order bottom effects, such as bottom curvature and square of the bottom slope, were neglected. The wave energy dissipation due to wave breaking and bottom friction, which are effectively two dominant dissipative mechanisms, has not been included in the model. Therefore, the applicability of the original theory in practical problems in the coastal area has been substantially restricted by these simplifying assumptions. Particularly in the case of submerged breakwaters, since the water depth varies substantially in the area of the structure and also due to the relatively high steepness of the slopes of the submerged breakwaters, the original mild-slope equation based on linear waves and slowly varying depths should be applied very carefully.

The mild-slope equation may be modified by including an energy dissipation function defined empirically based on the dissipative processes. For the particular case of submerged breakwaters, the wave breaking, the bottom friction and the steep sloping should be included. The mild-slope equation, therefore, has been modified and extended by other people in different ways in order to model random waves, steeply sloping and rapidly varying topographies, wave–current interaction, wave nonlinearity, bed friction and wave breaking, etc. (Chamberlain and Porter, 1995; Lee and Kirby, 1994; Massel, 1993; Massel, 1995; Radder and Dingemans, 1985; Smith and Sprinks, 1975).

The extended mild-slope equation by Massel (1993), and Chamberlain and Porter (1995) for steep slopes, and another extended equation by Chandrasekera and Cheung (1997) for random waves, are some examples. Lee et al. (1998) and Lee et al. (2002) also developed an extended mild-slope equation with the possibility of predicting the wave transformation of random waves over rapidly varying topography.

Massel and Gourlay (2000) developed a numerical model based on the extended refraction-diffraction equation (Massel, 1993) for rapidly varying slopes and arbitrary bottom shapes considering the wave breaking and bottom friction effects. Massel and Gourlay (2000) have presented promising results. A model based on the extended mild-slope equations would be preferable because of the wave-breaking process over the steep slope and the shallow crest of the submerged breakwaters.

However, all these models can be applied on a two-dimensional plane with depth integration. Therefore, in cases where the variations in depth are important, the previous models cannot be very useful. Therefore, the full three-dimensional Navier–Stokes equations should be solved. However, sometimes, due to the high complexity of the three-dimensional equations, a simplified scheme is applied. For example, the hydrostatic pressure is assumed and a quasi-3D model is obtained. The Navier–Stokes equations can also be solved completely in order to model fully 3D problems. In this case the full three-dimensional equations are solved without considering the simplified assumption of hydrostatic pressure. If a proper turbulence model is also included in the model, many complex physical processes can be simulated. However, working with the full three-dimensional Navier–Stokes equations with turbulence may be very time-consuming and expensive. In problems with low turbulence, the Navier–Stokes equations are reduced to the Laplace equation based on potential flow theory for irrotational flows. This type of model can be used for nonlinear wave transformation over varying depths.

The Reynolds Averaged Navier–Stokes (RANS) equations have also been employed in the studies of wave and submerged breakwater interaction particularly to describe turbulent flows. The RANS equations are basically time-averaged equations of flow. The Reynolds decomposition is used to derive the RANS equations from Navier-Stokes equations by separating the flow variable into the time-averaged and fluctuating components.

A number of numerical techniques are available for modeling hydrodynamics, morphology or water quality for submerged breakwaters in coastal area. These models have been developed based on different numerical schemes to solve the governing equations of fluid motion in the form of partial differential equations, such as the continuity and momentum equations.

Traditional methods such as finite difference, finite volume, finite element, boundary element and spectral methods based on the elements are some of the most often used methods. Models are mostly available for regular waves rather than irregular waves. Most of the models were developed by finite difference methods. A discretization of the problem domain in both time and space is needed for all types of these numerical models.

4.3 NEW APPROACHES

4.3.1 Meshless Methods

Traditional numerical methods such as the finite difference method, finite-volume method, and finite element method basically require a mesh of data points over the problem domain. The connections of nodes with their neighbors are used to define the derivatives in the governing equations like the Euler equations or the Navier–Stokes equations. In some cases where generating a mesh for problems with complex geometry is difficult, meshfree methods which do not require a mesh may be more applicable.

Recently, some new approaches based on meshless methods have been proposed. These models are based on particles rather than elements. Some examples are generalized finite difference, finite point, smoothed particle hydrodynamics, reproducing kernel particle, and element-free Galerkin methods. Smoothed-particle hydrodynamics (SPH) is one of the most-used mesh-free computational methods (Gingold and Monaghan, 1977; Lucy, 1977). The smoothed-particle hydrodynamics method divides the fluid into a set of particles with a spatial distance, so that the physical quantity of any particle depends on all the particles within a specific radial distance. It has been applied for various engineering problems in the field of fluid dynamics.

4.3.2 Artificial Intelligence Methods

In addition to the classic methods, in recent years methods based on artificial intelligence have been widely applied in the field of coastal engineering, and particularly in the case of submerged breakwaters.

Artificial intelligence is a branch of computer science concerned with developing computer-based methods for solving challenging problems. These methods basically store knowledge and use the learned knowledge

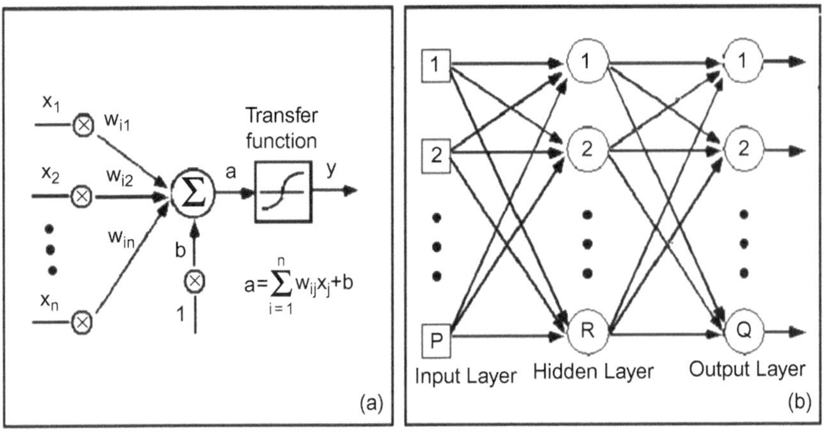

Figure 4.1 (a) A single neuron and its connections; (b) A feed-forward neural network model.

to solve problems. A wide range of artificially intelligent tools has been developed to solve an increasing range of problems in various fields.

Artificial neural networks (ANNs) are multidimensionally nonlinear parametric algorithms. Artificial neural networks imitate the human brain. They are capable of learning relationships between parameters of an unknown process where no mathematical analysis is available. They basically learn the knowledge with repeated examples. The input neurons receive information and pass it to the next layer. Artificial neural networks have found applications in many engineering fields including coastal engineering.

Artificial neural networks are designed to imitate biological neural networks. They consist of a massive number of computational units (replacement of biological neurons) (Figure 4.1a) connected together with interlayer weights (replacement of synapses) emitting signals to other units. Computational units called neurons are designed in artificial neural networks to acquire knowledge while interlayer connections or network weights between units are employed for storing the knowledge (Haykin, 1999). Through a learning process, network parameters such as neuron parameters and interlayer weights or bias values are modified and adapted. The learning process employed can be classified as supervised or unsupervised. Basically, supervised networks employ a mathematical method in the learning process. If a network does not include a learning method, or a more self-studying scheme, then it is called unsupervised. Generally, supervised learning is suitable for function estimation when input-output paired data is available, while unsupervised learning may be best in the case of pattern recognition using data without target or desired data available (Haykin, 1999). Since in this book we are dealing with a function estimation problem, supervised learning is considered appropriate and will be discussed in detail in this chapter. For more information about other learning methods and unsupervised methods in particular, please see Hagan et al. (1996) and Haykin (1999).

Take a training data set D including N input-output pairs (Haykin, 1999):

$$D = \{(p_i, t_i)\}_{i=1}^{N} \tag{4.1}$$

where p_i and t_i are input and target response vectors of the ith example.

A supervised form of learning requires layer weights and biases of the network to be computed based on the given training examples of set D, so that actual output $\widehat{y}_i$ of corresponding input x_i computed by the neural network is close enough to the corresponding desired output y_i for all i to reach the required error.

A proper statistical form of error calculation such as the mean-square error is usually used to achieve this goal (Haykin, 1999):

$$\text{MSE} = \frac{\sum_{i=1}^{\text{N}} (y_i - \hat{y}_i)^2}{N} \tag{4.2}$$

During the learning process, the computed error should be minimized to the predefined desired error.

ANNs are also classified into feed-forward and feed-backward networks. A feed-forward supervised network consists of three or more ordered layers: one input layer, one output layer, and one or more hidden layers (Figure 4.1b). Each layer consists of a number of units, linked to other units in adjacent layers. Through a feed-forward network signals travel only from input layer to output layer. Thus, there is no any kind of feedback connection or loop inside the network. Data enter the network from the input layer, passing through each hidden layer in order and finally leave the network from the output layer. Since values from the output layer are not returned back to the input layer, it is called feed-forward. Obviously, if the output values are fed back into the network, it is called a feed-backward network. To validate the model outcomes, the outputs of the network are compared with desired output values, coupled with input data, and finally, through a learning process, the network is modified to give a better prediction of values close to the desired output (Haykin, 1999). Feed-backward or unsupervised networks will not be used in this book.

Multilayer perceptrons (hereafter, MLPs) and radial basis functions (hereafter, RBFs; see Figure 4.2) are the two most commonly used types

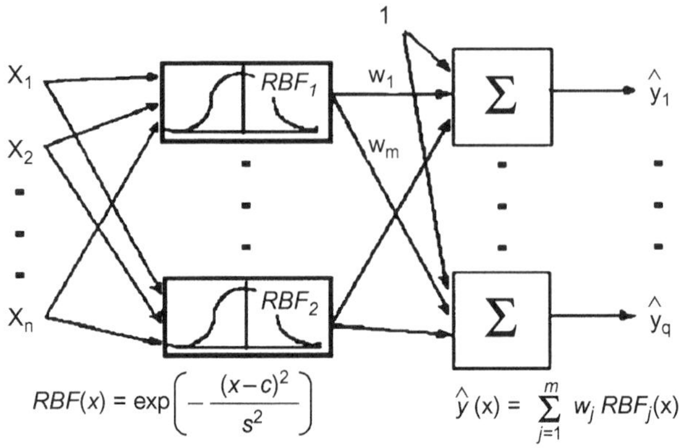

Figure 4.2 A radial-basis function (RBF) network.

of feed-forward supervised neural networks. Most methods for training MLPs can also be applied to RBF networks. In this chapter both models are described. In Chapter 11 the results of both approaches used here to simulate the wave field around submerged breakwaters are presented, validated and compared with each other and other available models along with the experimental data collected during Laboratory measurements.

Although both MLP and RBF networks are universal nonlinear layered feed-forward approximators, they have some significant differences. RBFs typically have only one hidden layer, whereas MLPs may have more than one hidden layer. Typically, RBFs are fully connected and MLPs are partially connected. In these two networks, the activation functions employed in the hidden layer are fundamentally different. (Hagan et al., 1996; Haykin, 1999). RBF network parameters are usually determined during the training process in two stages with different strategies for hidden and output layers (unsupervised and supervised), while MLP network parameters are determined in one stage for both layers with one single training strategy (only supervised). Separation of the data in the hidden layer in RBF is done by fitting each class using kernel functions, while this goal is achieved in MLP by defining hyper-planes in the input space (Haykin, 1999). Differences between the two networks are summarized in Table 4.1.

MLP Networks

As mentioned before the MLP networks may consist of a large number of neurons ordered in different layers. Thus it is possible to have several

Table 4.1 Differences of RBF and MLP networks

Items	RBF	MLP
Architecture	Simple architecture	Complex pattern of connectivity
Number of hidden layers	Single hidden layer	One or more
Interconnections	Fully connected	Fully or not fully connected
Number of neurons	More	Less
Activation functions	A specific activation function	Variety of activation functions
Neurons in layers	Different types	Common types
Representation form	Local representation	Universal representation
Activation function	Euclidean distance	Inner product
Weights determination	Two stages/One stage	One stage
Data separation	Kernel functions	Hyper-planes

hidden layer in the MLP network. However, in many cases, one hidden layer is enough for function approximating. Basically, the neurons in the input and output layers are designed to receive and distribute the signals. However, the neurons in the hidden layers are responsible for performing the required computations. This section concerns MLP networks and training algorithms commonly employed to train them.

Back-Propagation Algorithm

The back-propagation (BP) algorithm has emerged as the most popular learning algorithm for the design of MLPs (see Fausett, 1994; Haykin, 1994; Patterson, 1996). The training algorithm basically consists of two phases (Rumelhart et al., 1986; Werbos, 1974).

In the first phase, known as the forward phase, network parameters are chosen, and the input data are propagated forward through the network, layer by layer. The forward phase is completed with the output error computation (Haykin, 1999):

$$e_i = y_i - \hat{y}_i \tag{4.3}$$

where y_i is the target value and $\hat{y}_i$ is the computed output by the network corresponding to the input x_i.

In the second phase, called the backward phase, the calculated error e_i is propagated backward through the network. Then the network parameters are adjusted so as to minimize the error e_i based on a calculated statistical index of mean-square error.

The back-propagation learning algorithm is usually implemented either in on-line mode or in batch mode. In on-line mode, network parameters are adjusted on a sample-by-sample basis. The on-line mode is typically used for pattern classification. The batch mode of BP learning changes the network parameters based on an epoch-by-epoch basis. In the case of using a batch mode for training the MLP network, one epoch consists of the entire set of training samples. Batch mode is typically used for nonlinear regression (Hagan et al., 1996; Haykin, 1999).

Advantages to implementing the back-propagation learning algorithm are its simplicity and efficiency in computations. However, a disadvantage of the algorithm is its limitation on convergence and its very slow training speed, particularly in the case of complex tasks that require a massive network (Haykin, 1999). Replacing asymmetric activation functions such as the logistic function by antisymmetric activation functions like the hyperbolic tangent function might lead to a better performance in the back-propagation learning algorithm. Reordering the training samples after the presentation of each epoch or following an easy-to-learn example with a difficult one might help too. Preprocessing the input data so as to remove the mean and decorrelate the data helps. Arranging for the neurons in the different layers to learn at essentially the same rate could also cause better performance, by assigning a learning rate parameter to neurons in the last layers that is smaller than for those at the front end (Hagan et al., 1996; Haykin, 1999).

A statistical technique known as cross-validation (Haykin, 1999) is employed to determine when the training process should be stopped and the size of the hidden layer. The method splits the training data set into two parts: the *estimation subset* and the *validation subset*. The estimation subset is used to train the network, while the validation subset is used to evaluate the model performance. The model is finally tuned by using the entire set of training patterns and then tested on data not seen before.

Levenberg-Marquardt Algorithm

In function optimization problems with neural networks, the training algorithm is designed to determine the best network parameters in order to minimize network error. Various function optimization methods can be applied to ANN model training. One of these methods is the Levenberg-Marquardt algorithm. The Levenberg-Marquardt algorithm provides a numerical solution to the problem of minimizing a nonlinear function, over a space of parameters for the function. It is a popular alternative to the Gauss–Newton method of finding the minimum of a function.

The error back-propagation algorithm (BP) (Rumelhart and McClelland, 1986) has been used frequently in neural network research; however, one of its weaknesses is its very poor convergence rate. Many studies have been conducted in the past to provide some techniques for speeding up the BP algorithm. The most common methods are: *momentum*, *variable learning rate*, and *stochastic learning* (Jacobs, 1988; Miniani and Williams, 1990; Salvetti and Wilamowski, 1994; Wilamowski et al., 2001; Yu and Wilamowski, 2011). However, all these attempts could lead only to a slight improvement in the BP convergence rate. Extended studies by various researchers show that a significantly better result can be obtained if second-order approaches are applied. Several methods have been proposed, such as Newton's method, conjugate gradients, or the Levenberg-Marquardt (LM) optimization technique (Andersen and Wilamowski, 1995; Battiti, 1992).

The LM algorithm has been adopted to a great degree by engineers, researchers and scientists as an efficient method since it can provide both advantages of the Newton algorithm (speed) and the steepest descent method (stability). Consequently, it makes a good transition between these methods (Hagan and Menhaj, 1994; Wilamowski et al., 2001). The Levenberg-Marquardt algorithm (Levenberg, 1944; Marquardt, 1963) was developed by Kenneth Levenberg and Donald Marquardt. LM solves the problem of minimizing a nonlinear function numerically, very fast and with a stable convergence. The LM algorithm is well-suited for training small and medium-sized problems (Yu and Wilamowski, 2011). As the number of adjustable parameters in the network and consequently the dimensionality of the network increases, due to dealing with large Jacobians and the need to invert large matrices, computational complexity grows exponentially, which is a major disadvantage of the LM algorithm (Wilamowski et al., 2001).

Many other methods have previously been developed for neural networks training. The steepest descent algorithm is one of the earliest, also known as the BP algorithm, as mentioned in the previous section (Rumelhart et al., 1986; Werbos, 1988) and is still widely used, although it suffers from slow convergence. If the gradient is steep and step size is a constant value, logically, small step sizes should be taken. This causes a very slow convergence in places where the gradient is gentle. Another issue is the fact that the curvature of the error surface may not be the same in all directions, and may result in the slow convergence too (Yu and Wilamowski, 2011).

The convergence of the steepest descent method is improved using the Gauss–Newton algorithm (Osborne, 1992). The Gauss-Newton algorithm converges significantly faster using second-order derivatives of the error function.

The Levenberg-Marquardt algorithm combines the steepest descent (far from the optimal values) and the Gauss-Newton methods (close to the optimal value), taking advantage of the high speed of the Gauss-Newton algorithm and the high stability of the steepest descent method. Thus the Levenberg-Marquardt algorithm converges slower than the Gauss-Newton algorithm and faster than the steepest descent method (Gavin, 2011; Hagan et al., 1996; Haykin, 1999).

The algorithm derivation and training process of the Levenberg–Marquardt algorithm is presented in the next chapters.

RBF Networks

Another popular multilayer feed-forward scheme is the radial-basis function (RBF) network, which has significant universal approximation properties (Park and Sandberg, 1993). The idea of RBF derives from the theory of function approximation, and particularly solves curve-fitting problems in high-dimensional space (Broomhead and Lowe, 1988a; Poggio and Girosi, 1990). RBF networks are designed to find a multidimensional surface that provides a best fit to the training data and then the trained network will use this multidimensional surface to interpolate the test data.

Interpolation of functions in numerical modeling has emerged as a very popular solution in a wide range of applications (Rumelhart et al., 1986). For single-variable problems solved by polynomial interpolating and multivariable problems with partial differential equations, several techniques have been proposed such as linear regression and k-nearest neighbor (K-NN) methods and neural networks (Solla et al., 1988). Radial basis function networks were first formulated by Broomhead and Lowe (1988a).

RBF networks are similar to K-Means clustering, probabilistic neural networks (PNNs) and general regression neural networks (GRNNs). In contrast to the RBF networks, the PNN/GRNN networks set one neuron for each sample in the training data. However, they are capable of fast learning and have reduced sensitivity to the order of presentation of training data (Subramanian and Saraswathi, 2012). Basically, a predicted target value of a sample is almost the same as other samples that have similar values of the predictor variables (Subramanian and Saraswathi, 2012). Any function can be approximated with an arbitrary small error by an RBF network, given there are enough RBF neurons (Park and Sandberg, 1991). Although, in the case of small to medium-size training data sets, PNN/GRNN models are more practical than RBFs, these models are unrealistic for large training data sets (Subramanian and Saraswathi, 2012).

As stated already, the number of neurons in the input layer corresponds to the number of input variables. The input neurons normalize the input data and feed the values to the hidden layer. Each of the input units of the input vector is connected to the computational neurons in the hidden layer. RBFs have only one hidden layer with a variable number of neurons with Gaussian activation functions inversely related to the distance from the neuron's centers (Hagan et al., 1996; Haykin, 1999). During the training process, the optimal number of neurons in the hidden layer, the center coordinates and the spread in each dimension of the RBF, and the network weight parameters associated with the output layer are determined (Haykin, 1999). The centers may be selected randomly (Broomhead and Lowe, 1988b) or by specific methods such as the supervised method (Poggio and Girosi, 1990), and self-organized techniques (Moody and Darken, 1989). RBFs compute the Euclidean distance of data points to the center of each neuron positioned in the same dimensional space as input variables. Using the calculated distance, a radial basis function computes the network's weight parameter for each neuron. The weights have a direct relationship with distances; the further a neuron is from the point, the less influence it has (Haykin, 1999). Predictions for new data are made by summing the RBF function's outputs multiplied by weights for each neuron. A bias value of 1.0 is also multiplied by a weight factor and added to the above sum and fed to the output of the network (Hagan et al., 1996; Haykin, 1999).

A brief introduction will be provided for these numerical schemes in the next chapters.

REFERENCES

Andersen, T.J., Wilamowski, B., 1995. A Modified Regression Algorithm for Fast One Layer Neural Network Training, vol. 1. World Congress of Neural Networks, Washington, DC, USA.

Battiti, R., 1992. First- and second-order methods for learning: Between steepest descent and Newton's method. Neural Comput. 4 (2), 141–166.
Berkhoff, J.C.W., 1972. Computation of combined refraction-diffraction, Proceedings, 13th International Conf. on Coastal Engrg., vol. 1. ASCE.
Brocchini, M., Cherubim, P., Iovenitti, L., 1991. An extension of Boussinesq type model to the surf zone. Computer Modelling in Ocean Engineering 91, Rotterdam, The Netherlands, pp. 349–359.
Broomhead, D., Lowe, D., 1988a. Radial Basis Functions, Multi-Variable Functional Interpolation and Adaptive Networks. Royal Signals and Radar Establishment. Memorandum 4148.
Broomhead, D., Lowe, D., 1988b. Multivariable functional interpolation and adaptive networks. Complex Syst. 2, 321–355.
Chamberlain, P.G., Porter, D., 1995. The modified mild-slope equation. J. Fluid Mech. 291, 393–407.
Chandrasekera, C.N., Cheung, K.F., 1997. Extended linear refraction diffraction model. J. Waterway, Port Coast. Ocean Eng 123 (5), 280–286.
Eckart, C., 1952. The propagation of gravity waves from deep to shallow water. Natl. Bur. Standards, Circular 521, (National Bureau of Standards), Washington, D.C., pp. 165–173.
Fausett, L., 1994. Fundamentals of Neural Networks Architectures, Algorithms and Applications. Prentice Hall, USA.
Gavin, H., 2011. The Levenberg-Marquardt method for nonlinear least squares curve-fitting problems.
Gingold, R.A., Monaghan, J.J., 1977. Smoothed particle hydrodynamics: theory and application to non-spherical stars. Mon. Not. R. Astron. Soc. 181, 375–389.
Hagan, M.T., Menhaj, M., 1994. Training feedforward networks with the Marquardt algorithm. IEEE Trans. Neural Netw. 5 (6), 989–993.
Hagan, M.T., Demuth, H.B., Beale, M.H., 1996. Neural Network Design. PWS Publishing, Boston, MA.
Haykin, S., 1994. Neural Networks: A Comprehensive Foundation. . Macmillan.
Haykin, S., 1999. Neural Networks: A Comprehensive Foundation, second ed. Prentice-Hall, Englewood Cliffs, NJ.
Jacobs, R., 1988. Increased rate of convergence through learning rate adaptation. Neural Netw. 1, 295–307.
Karambas, Th.V., Koutitas, C., 1992. A breaking wave propagation model based on the Boussinesq equations. Coast. Eng. 18, 1–19.
Kennedy, A.B., Chen, Q., Kirby, J.T., Dalrymple, R.A., 2000. Boussinesq modeling of wave transformation, breaking, and runup. I: 1D. J. Waterw. Port Coast. Ocean Eng. 126, 39–47.
Lee, C., Park, W.S., Cho, Y.-S., Suh, K.D., 1998. Hyperbolic mild-slope equations extended to account for rapidly varying topography. Coast. Eng. 34, 243–257.
Lee, C., Kim, G., Suh, K., 2002. Extended mild-slope equation for random waves. Ocean Wave Meas. Anal. (2001), 724–733.
Lee, C.H., Kirby, J.T., 1994. Analytical comparison of time-dependent mild-slope equations. J. Korean Soc. Coast. Ocean Eng. 6, 389–396.
Levenberg, K., 1944. A method for the solution of certain problems in least squares. Q. Appl. Math. 5, 164–168.
Lucy, L.B., 1977. A numerical approach to the testing of the fission hypothesis. Astron. J. 82, 1013–1024.
Marquardt, D., 1963. An algorithm for least-squares estimation of nonlinear parameters. SIAM J. Appl. Math. 11 (2), 431–441.

Massel, S.R., 1993. Extended refraction–diffraction equation for surface waves. Coast. Eng. 19, 97–126.
Massel, S.R., 1995. Ocean Surface Waves: Their Physics and Prediction. World Scientific Publ., Singapore.
Massel, S.R., Gourlay, M.R., 2000. On the modelling of wave breaking and set-up on coral reefs. Coast. Eng. 39, 1–27.
Miniani, A.A., Williams, R.D., 1990. Acceleration of back-propagation through learning rate and momentum adaptation. Proceedings of International Joint Conference on Neural Networks, San Diego, CA 1, 676–679.
Moody, J., Darken, C.J., 1989. Fast learning in networks of locally-tunes processing units. Neural Comput. 1, 284–294.
Osborne, M.R., 1992. Fisher's method of scoring. Int. Stat. Rev. 86, 271–286.
Park, J., Sandberg, I., 1991. Universal approximation using radial-basis-function networks. Neural Comput. 3, 246–257.
Park, J., Sandberg, I., 1993. Approximation and radial-basis-function networks. Neural Comput. 5, 305–316.
Patterson, D., 1996. Artificial Neural Networks. Prentice Hall, Singapore.
Poggio, T., Girosi, F., 1990. Networks for approximation and learning. Proc. IEEE 78, 1481–1497.
Radder, A.C., Dingemans, M.W., 1985. Canonical equations for almost periodic, weakly nonlinear gravity waves. Wave Motion 7, 473–485.
Rumelhart, D., McClelland, J., 1986. Parallel Distributed Processing. MIT Press, Cambridge, Mass.
Rumelhart, D.E., Hinton, G.E., Williams, R.J., 1986. Learning representations by back-propagating errors. Nature 323, 533–536.
Salvetti, A., Wilamowski, B.M., 1994. Introducing stochastic process within the backpropagation algorithm for improved convergence. Intell. Eng. Syst. Artif. Neural Netw. 4, 205–209.
Schaffer, H.A., Madsen, P.A., Deigaard, R., 1993. A Boussinesq model for waves breaking in shallow water. Coast. Eng. 20, 185–202.
Smith, R., Sprinks, T., 1975. Scattering of waves by a conical island. J Fluid Mech. 72, 373–384.
Solla, S.A., Levin, E., Fleisher, M., 1988. Accelerated learning in layered neural networks. Complex Syst. 2, 625–639.
Subramanian, N., Saraswathi, S., 2012. Barrister groupware (instant assistance to flawless law). Bonfring Int. J. Softw. Eng. Soft Comput. 2, 34–37.
Werbos, P.J., 1974. Beyond regression: New tools for prediction and analysis in the behavioral sciences. Ph.D. thesis. Harvard University, Cambridge, MA.
Werbos, P.J., 1988. Back-propagation: past and future. In: Proceedings of International Conference on Neural Networks, San Diego, CA 1, 343–354.
Wilamowski, B., Iplikci, S., Kaynak, O., Efe, M., 2001. An algorithm for fast convergence in training neural networks. Proceedings of International Joint Conference on Neural Networks (IJCNN '01) 3, 1778–1782.
Yu, H., Wilamowski, B., 2011. second ed. Levenberg-Marquardt Training, vol. 5. Industrial Electronics Handbook.
Zelt, J.A., 1991. The run-up of nonbreaking and breaking solitary waves. Coast. Eng. 15, 205–246.

CHAPTER 5

Mathematical Modeling and Algorithm Development

In this chapter the mathematical representation of some selected models is presented.

5.1 NAVIER-STOKES EQUATIONS

The basic relation for a general description of fluid flow is a nonlinear equation of motion which is based fundamentally on the principle of balance of inertia and external forces represented by volume and surface forces, as follows (Batchelor, 1967; Foias et al., 2001):

$$\frac{\partial \rho u_i}{\partial t} + \nabla \cdot (\rho \mathbf{u} u_i) = -\frac{\partial p}{\partial x_i} + \nabla_j \cdot \mu \left[\left(\frac{\partial u_i}{\partial x_j} + \frac{\partial u_j}{\partial x_i} \right) - \frac{2}{3} \delta_{ij} \frac{\partial u_l}{\partial x_l} \right] + \rho f_i \quad (5.1)$$

where u_i are the components of the velocity vector, p is the static pressure, ρ is density, μ is the viscosity and ρf_i are the external body force components.

The conservation of mass in the flow field is also expressed by the continuity equation as follows:

$$\frac{\partial \rho}{\partial t} + \nabla \cdot (\rho \mathbf{u}) = 0 \quad (5.2)$$

For incompressible fluids ($\rho =$ constant), the basic equations are reduced to:

$$\rho \frac{\partial u_i}{\partial t} + \rho \mathbf{u} \cdot \nabla u_i = -\frac{\partial p}{\partial x_i} + \rho \nu \Delta u_i + \rho f_i \quad (5.3)$$

and

$$\nabla \cdot \mathbf{u} = 0 \quad (5.4)$$

where $\nu = \mu / \rho$ is the kinematic viscosity.

© 2016 Elsevier Ltd.
All rights reserved.

These equations are called Navier-Stokes equations.

Considering the turbulence in the flow field, the velocity vector can be decomposed to a mean value and a dynamic component as follows (Foias et al., 2001):

$$\mathbf{u} = \overline{\mathbf{u}} + \mathbf{u}' \tag{5.5}$$

where $\overline{\mathbf{u}}$ is:

$$\overline{\mathbf{u}} = \frac{1}{T}\int_0^T \mathbf{u}\,\mathrm{d}t \tag{5.6}$$

The other parameters in the equation can also be averaged (Reynolds averaging). Thus, the equation is:

$$\frac{\partial \rho \overline{u}_i}{\partial t} + \nabla \cdot (\rho \overline{\mathbf{u}} \overline{u}_i) = -\frac{\partial \overline{p}}{\partial x_i} + \nabla_j \cdot \mu \left[\left(\frac{\partial \overline{u}_i}{\partial x_j} + \frac{\partial \overline{u}_j}{\partial x_i} \right) - \frac{2}{3}\delta_{ij}\left(\frac{\partial \overline{u}_l}{\partial x_l} \right) \right] + \overline{\rho f_i} - \nabla \cdot \left(\overline{\rho \mathbf{u}' u_i'} \right) \tag{5.7}$$

For most applications in coastal and marine engineering, and particularly for submerged breakwaters, it is unnecessary to resolve the details of the turbulent fluctuations. Therefore, a turbulence model is needed. A turbulence model approximates the unknown averaged products of turbulent fluctuations. Reynolds stress is basically approximated. The two-equation turbulence models (i.e. the standard k-ε model) are the most common models used to approximate the mean flow characteristics for the numerical problem. The first variable is called turbulent kinetic energy (k) and the second variable is its turbulent dissipation (ε). The standard k-ε model is for fully turbulent flows and includes two partial differential equations (PDEs); one is for the turbulent kinetic energy k which is derived from the exact equation and one for its dissipation rate ε derived empirically as discussed before (Foias et al., 2001).

The finite volumes method can be applied to discretize the equations. Using the finite volume method, the equations can be integrated in a finite volume. The divergence theorem can be applied and then the integrals can be reduced to a finite number of points. Thus, the derivative terms are replaced using approximation differences. Other equations can also be approximated similarly.

The general form of the Navier-Stokes equations in three dimensions is presented here.

The continuity equation is:

$$\frac{\partial \rho}{\partial t} + \frac{\partial(\rho u)}{\partial x} + \frac{\partial(\rho v)}{\partial y} + \frac{\partial(\rho w)}{\partial z} = 0 \tag{5.8}$$

The momentum equation is:

$$\rho \frac{\partial u_i}{\partial t} + \rho u_j \frac{\partial u_i}{\partial x_j} + \frac{\partial p}{\partial x_i} - \rho g_i + \frac{\partial \tau_{ij}}{\partial x_j} + \rho \frac{\partial u_i' u_j'}{\partial x_j} = 0 \tag{5.9}$$

where $V = u\hat{i} + v\hat{j} + w\hat{k}$ is the velocity vector and g_i is the body forces and

$$\overline{\tau}_{ij} = -\mu\left(\frac{\partial u_i}{\partial x_j} + \frac{\partial u_j}{\partial x_i}\right) \tag{5.10}$$

where

$$u_i' u_j' = -c_\mu \frac{k^2}{\varepsilon}\left(\frac{\partial u_i}{\partial x_j} + \frac{\partial u_j}{\partial x_i}\right) + \frac{2}{3} k \delta_{ij} \tag{5.11}$$

$$-\rho \overline{u_i' u_j'} = \mu_t \left(\frac{\partial U_i}{\partial x_j} + \frac{\partial U_j}{\partial x_i}\right) - \frac{2}{3}\rho k \delta_{ij} = 2\mu_t E_{ij} - \frac{2}{3}\rho k \delta_{ij} \tag{5.12}$$

$\delta_{ij} = 1$ if $i = j$ and $\delta_{ij} = 0$ if $i \neq j$

The unknowns include velocities $(u, v, w)(x, y, z, t)$, pressure $P(x, y, z, t)$, surface $\eta(x, y, t)$ and the domain (t). When we neglect diffusive terms in the Navier-Stokes equations, we get the Euler equations for conservation of mass and momentum (Batchelor, 1967; Foias et al., 2001). The three-dimensional partial differential equations for viscous, compressible flow are:

$$\frac{\partial \rho}{\partial t} + U\frac{\partial \rho}{\partial x} + V\frac{\partial \rho}{\partial y} + W\frac{\partial \rho}{\partial z} = -\rho\left(\frac{\partial U}{\partial x} + \frac{\partial V}{\partial y} + \frac{\partial W}{\partial z}\right) \tag{5.13}$$

$$\frac{\partial U}{\partial t} + U\frac{\partial U}{\partial x} + V\frac{\partial U}{\partial y} + W\frac{\partial U}{\partial z} = -\frac{1}{\rho_0}\frac{\partial P}{\partial x} + g_x + \frac{\mu}{\rho_0}\left(\frac{\partial^2 U}{\partial x^2} + \frac{\partial^2 U}{\partial y^2} + \frac{\partial^2 U}{\partial z^2}\right) \tag{5.14}$$

$$\frac{\partial V}{\partial t} + U\frac{\partial V}{\partial x} + V\frac{\partial V}{\partial y} + W\frac{\partial V}{\partial z} = -\frac{1}{\rho_0}\frac{\partial P}{\partial y} + g_y + \frac{\mu}{\rho_0}\left(\frac{\partial^2 V}{\partial x^2} + \frac{\partial^2 V}{\partial y^2} + \frac{\partial^2 V}{\partial z^2}\right) \tag{5.15}$$

$$\frac{\partial W}{\partial t} + U\frac{\partial W}{\partial x} + V\frac{\partial W}{\partial y} + W\frac{\partial W}{\partial z} = -\frac{1}{\rho_0}\frac{\partial P}{\partial z} + g_z + \frac{\mu}{\rho_0}\left(\frac{\partial^2 W}{\partial x^2} + \frac{\partial^2 W}{\partial y^2} + \frac{\partial^2 W}{\partial z^2}\right) \tag{5.16}$$

However, if the flow is incompressible, the conservation of mass reduces to:

$$\frac{\partial U}{\partial x} + \frac{\partial V}{\partial y} + \frac{\partial W}{\partial z} = 0 \tag{5.17}$$

5.2 THE TURBULENT MODEL

Among the existing turbulent models the standard two-equation k-ε models are the most used methods to estimate the eddy viscosity. In the k-ε model the transport equations for k and ε are solved. The turbulent viscosity is computed from (Foias et al., 2001):

$$\mu_t = \rho c_\mu \frac{k^{1/2}}{l} = \rho c_\mu \frac{k^2}{\varepsilon} \tag{5.18}$$

or

$$\nu_t = \frac{\mu_t}{\rho} = c_\mu \frac{k^{1/2}}{l} = c_\mu \frac{k^2}{\varepsilon} \tag{5.19}$$

where k is the kinetic energy, ε is the dissipation rate and c_μ is a parameter that depends on the k-ε turbulence model:

$$c_\mu = 0.09 \tag{5.20}$$

The turbulent kinetic energy for three-dimensional flows is given by:

$$k = \frac{1}{2} \cdot (u^{'2} + v^{'2} + w^{'2}) \tag{5.21}$$

The governing transport equations for this model can be written as:

k-equation:

$$\frac{\partial k}{\partial t} + \frac{\partial (u_i k)}{\partial x_i} = \frac{\partial}{\partial x_i}\left(\left(\mu + \frac{\mu_t}{\sigma_{k,s}}\right)\frac{\partial k}{\partial x_i}\right) + P - \varepsilon \tag{5.22}$$

$$\begin{aligned}
&\frac{\partial k}{\partial t} + \frac{\partial (uk)}{\partial x} + \frac{\partial (vk)}{\partial y} + \frac{\partial (wk)}{\partial z} = \frac{\partial}{\partial x}\left(\left(\mu + \frac{\mu_t}{\sigma_{k,s}}\right)\frac{\partial k}{\partial x}\right) \\
&+ \frac{\partial}{\partial y}\left(\left(\mu + \frac{\mu_t}{\sigma_{k,s}}\right)\frac{\partial k}{\partial y}\right) + \frac{\partial}{\partial z}\left(\left(\mu + \frac{\mu_t}{\sigma_{k,s}}\right)\frac{\partial k}{\partial z}\right) + \\
&\left[\frac{\mu_t}{\rho}\left(2\left(\frac{\partial u}{\partial x}\right)^2 + 2\left(\frac{\partial v}{\partial y}\right)^2 + 2\left(\frac{\partial w}{\partial z}\right)^2 + \left(\frac{\partial u}{\partial y} + \frac{\partial v}{\partial x}\right)^2 + \left(\frac{\partial u}{\partial z} + \frac{\partial w}{\partial x}\right)^2\right.\right. \\
&\left.\left. + \left(\frac{\partial v}{\partial z} + \frac{\partial w}{\partial y}\right)^2\right)\right] - \varepsilon
\end{aligned} \tag{5.23}$$

ε-equation:

$$\frac{\partial(\varepsilon)}{\partial t} + \frac{\partial(u_i\varepsilon)}{\partial x_i} = \frac{\partial}{\partial x_i}\left(\left(\mu + \frac{\mu_t}{\sigma_{\varepsilon,s}}\right)\frac{\partial \varepsilon}{\partial x_i}\right) + c_{1,\varepsilon}\frac{\varepsilon}{k}P - c_{2,\varepsilon}\frac{\varepsilon^2}{k} \tag{5.24}$$

$$\begin{aligned}
&\frac{\partial \varepsilon}{\partial t} + \frac{\partial(u\varepsilon)}{\partial x} + \frac{\partial(v\varepsilon)}{\partial y} + \frac{\partial(w\varepsilon)}{\partial z} = \\
&\frac{\partial}{\partial x}\left(\left(\mu + \frac{\mu_t}{\sigma_{\varepsilon,s}}\right)\frac{\partial \varepsilon}{\partial x}\right) + \frac{\partial}{\partial y}\left(\left(\mu + \frac{\mu_t}{\sigma_{\varepsilon,s}}\right)\frac{\partial \varepsilon}{\partial y}\right) + \frac{\partial}{\partial z}\left(\left(\mu + \frac{\mu_t}{\sigma_{\varepsilon,s}}\right)\frac{\partial \varepsilon}{\partial z}\right) + \\
&c_{1,\varepsilon}\frac{\varepsilon}{k}\left[\frac{\mu_t}{\rho}\left(2\left(\frac{\partial u}{\partial x}\right)^2 + 2\left(\frac{\partial v}{\partial y}\right)^2 + 2\left(\frac{\partial w}{\partial z}\right)^2 + \left(\frac{\partial u}{\partial y} + \frac{\partial v}{\partial x}\right)^2 + \left(\frac{\partial u}{\partial z} + \frac{\partial w}{\partial x}\right)^2\right.\right. \\
&\left.\left. + \left(\frac{\partial v}{\partial z} + \frac{\partial w}{\partial y}\right)^2\right)\right] - c_{2,\varepsilon}\frac{\varepsilon^2}{k}
\end{aligned} \tag{5.25}$$

The unknown constants c_μ, $c_{1,\varepsilon}$, $c_{2,\varepsilon}$, σ_k and σ_ε are assumed to be constant for all types of flows. The c_μ constant was determined previously. The empirical constants c_k, c_ε,$c_{1,\varepsilon}$, $c_{2,\varepsilon}$ are given in Table 5.1. For further information see Batchelor, 1967; Foias et al., 2001.

$$\begin{aligned}
P &= -\overline{u_i'u_j'}\frac{\partial u_i}{\partial x_j} = \left[\frac{\mu_t}{\rho}\left(\frac{\partial u_i}{\partial x_j} + \frac{\partial u_j}{\partial x_i}\right) - \frac{2}{3}k\delta_{ij}\right]\frac{\partial u_i}{\partial x_j} \\
&= \frac{\mu_t}{\rho}\left(2\left(\frac{\partial u}{\partial x}\right)^2 + 2\left(\frac{\partial v}{\partial y}\right)^2 + \left(\frac{\partial u}{\partial y} + \frac{\partial v}{\partial x}\right)^2\right)
\end{aligned} \tag{5.26}$$

$$\begin{aligned}
P &= -\overline{u_i'u_j'}\frac{\partial u_i}{\partial x_j} = \left[\frac{\mu_t}{\rho}\left(\frac{\partial u_i}{\partial x_j} + \frac{\partial u_j}{\partial x_i}\right) - \frac{2}{3}k\delta_{ij}\right]\frac{\partial u_i}{\partial x_j} \\
&= \frac{\mu_t}{\rho}\left(2\left(\frac{\partial u}{\partial x}\right)^2 + 2\left(\frac{\partial v}{\partial y}\right)^2 + 2\left(\frac{\partial w}{\partial z}\right)^2 + \left(\frac{\partial u}{\partial y} + \frac{\partial v}{\partial x}\right)^2 + \left(\frac{\partial u}{\partial z} + \frac{\partial w}{\partial x}\right)^2\right. \\
&\quad \left. + \left(\frac{\partial v}{\partial z} + \frac{\partial w}{\partial y}\right)^2\right)
\end{aligned} \tag{5.27}$$

Table 5.1 Parameters for the standard k-ε model
Model parameters

$\sigma_k = 1$	$\sigma_\varepsilon = 1.314$	$c_{1,\varepsilon} = 1.44$	$c_{2,\varepsilon} = 1.92$

Therefore, the general form of the Navier-Stokes equations in three dimensions is:

$$\rho\frac{\partial u}{\partial t}+\rho\left(u\frac{\partial u}{\partial x}+v\frac{\partial u}{\partial y}+w\frac{\partial u}{\partial z}\right)=$$
$$\rho g_x-\frac{\partial p}{\partial x}+\mu\left(\frac{\partial^2 u}{\partial x^2}+\frac{\partial^2 u}{\partial y^2}+\frac{\partial^2 u}{\partial z^2}\right)+\rho\frac{\partial}{\partial x}\left(2v_t\frac{\partial u}{\partial x}\right) \tag{5.28}$$
$$+\rho\frac{\partial}{\partial y}\left(v_t\left(\frac{\partial u}{\partial y}+\frac{\partial v}{\partial x}\right)\right)+\rho\frac{\partial}{\partial z}\left(v_t\left(\frac{\partial u}{\partial z}+\frac{\partial w}{\partial x}\right)\right)$$

$$\rho\frac{\partial v}{\partial t}+\rho\left(u\frac{\partial v}{\partial x}+v\frac{\partial v}{\partial y}+w\frac{\partial v}{\partial z}\right)=$$
$$\rho g_y-\frac{\partial p}{\partial y}+\mu\left(\frac{\partial^2 v}{\partial x^2}+\frac{\partial^2 v}{\partial y^2}+\frac{\partial^2 v}{\partial z^2}\right)+\rho\frac{\partial}{\partial y}\left(2v_t\frac{\partial v}{\partial y}\right) \tag{5.29}$$
$$+\rho\frac{\partial}{\partial x}\left(v_t\left(\frac{\partial u}{\partial y}+\frac{\partial v}{\partial x}\right)\right)+\rho\frac{\partial}{\partial z}\left(v_t\left(\frac{\partial v}{\partial z}+\frac{\partial w}{\partial y}\right)\right)$$

$$\rho\frac{\partial w}{\partial t}+\rho\left(u\frac{\partial w}{\partial x}+v\frac{\partial w}{\partial y}+w\frac{\partial w}{\partial z}\right)=$$
$$\rho g_z-\frac{\partial p}{\partial z}+\mu\left(\frac{\partial^2 w}{\partial x^2}+\frac{\partial^2 w}{\partial y^2}+\frac{\partial^2 w}{\partial z^2}\right)+\rho\frac{\partial}{\partial z}\left(2v_t\frac{\partial w}{\partial z}\right) \tag{5.30}$$
$$+\rho\frac{\partial}{\partial x}\left(v_t\left(\frac{\partial u}{\partial z}+\frac{\partial w}{\partial x}\right)\right)+\rho\frac{\partial}{\partial y}\left(v_t\left(\frac{\partial v}{\partial z}+\frac{\partial w}{\partial y}\right)\right)$$

5.3 INITIAL AND BOUNDARY CONDITIONS

The initial and boundary conditions are also required in numerical modeling of submerged breakwaters. Depending on the problem, these conditions may vary and either one or both may be applied in the problem. For example, if we are dealing with infinite domains, we are not concerned about the boundary conditions and the problem reduces to an initial value problem. In the case of the steady conditions, the problem then reduces to a boundary value problem. The boundary conditions may be specified by variable value, gradient of the variable or relationship of the variable and its gradient. In these problems, the number of boundary equations is determined based on the order of the highest spatial derivatives in the governing equation for each coordinate space. Obviously, for an unsteady problem with finite domain, both initial and boundary conditions are needed. These problems are called initial-boundary value problems.

5.4 SHALLOW WATERS

In the model of long waves in shallow waters, the vertical velocity and acceleration of the fluid particles are assumed to be zero. Therefore, all terms containing W in the governing equations are omitted. The horizontal plane in the still water surface level is taken as the Cartesian coordinate system. Therefore, the water depth and water surface variations are determined according to this reference plane (for further information, see Mader, 2005).

Vertically averaged velocity components can be introduced by (Hansen, 1956):

$$U = \frac{1}{(h+\eta)} \int_{-h}^{\eta} u dz \tag{5.31}$$

and

$$V = \frac{1}{(h+\eta)} \int_{-h}^{\eta} v dz \tag{5.32}$$

where U is the velocity in the x direction, V is the velocity in the y direction, η is the wave height above mean water level and h is depth (Mader, 2005; Vreugdenhil, 1994).

5.5 THE EXTENDED MILD-SLOPE EQUATION

The mild-slope equation has been improved as the influence of steep slopes and bottom curvatures has been included. (Chamberlain and Porter, 1995; Massel, 1993, 1996; Porter and Staziker, 1995).

According to the expression by Massel (1989), the following restriction exists for wave propagation over mild slopes:

$$\frac{\nabla h}{kh} \cdot \frac{\nabla a}{ka} \ll 1 \tag{5.33}$$

where ∇ is the horizontal gradient operator and h, k and a are water depth, wave number and wave amplitude, respectively.

The mild-slope equation is relatively accurate for slopes to a maximum ratio of 1:3 (Booij, 1983); however, beyond this ratio the classical mild-slope equation by Berkhoff (1972) is no longer very accurate. Furthermore, for arbitrary topographies and particularly submerged breakwaters, bottom curvature ($\approx \nabla^2 h$) and the square of bottom slope ($(\nabla h)^2$) should be considered (Massel, 1993).

Massel (1993, 1996) developed an extended refraction-diffraction equation for the steep slopes based on the Galerkin-Eigenfunction method. This extended equation includes the higher-order terms.

The parameters related to the higher-order bottom effects are determined based on the expressions by Massel (1993). Some expressions given by Massel (1993) were corrected later by Suh et al. (1997).

5.6 BOUSSINESQ EQUATIONS

Boussinesq equations (Boussinesq, 1871, 1872) are a set of nonlinear partial differential equations. Boussinesq equations incorporate frequency dispersion while the shallow water equations are not frequency-dispersive. Boussinesq equations are widely applied in numerical modeling in the field of coastal engineering to model water waves in shallow seas and harbors. The Boussinesq equations are applicable for fairly long waves.

Although wave simulation in such cases is perfectly described by the Navier-Stokes equations, currently it is extremely difficult to solve the full three-dimensional equations in complicated models. Therefore, approximate models such as the Boussinesq equations can be applied. They can be used to reduce three-dimensional problems to two-dimensional states. This is typically done by assuming a linear vertical distribution of the flow field with nonhydrostatic effects which is different from the nonlinear shallow water equations (see Dingemans, 1997; Hamm et al., 1993; Kirby, 2003).

The Boussinesq equations are suitable for the simulation of wave fields in the presence of coastal structures and harbors. Finite-difference, finite-volume or finite-element techniques are applicable for the discretization of the governing equations.

5.7 SMOOTHED PARTICLES HYDRODYNAMICS

Smoothed particles hydrodynamics (SPH) (Gingold and Monaghan 1977; Lucy 1977) is a meshless Lagrangian method, where the fluid is discretized using a finite set of physical particles within the body of the fluid. The required physical quantities are computed directly and the free surface of the flow can be tracked. Each particle with given mass, density and volume can move and carry some value of u_i. SPH interpolation for field values of any scalar value $u(x, t)$ is (Dalrymple and Rogers 2006; Liu and Liu, 2003; Monaghan, 1994):

$$u(\mathbf{x}, t) = \sum_i u_i \frac{m_i}{\rho_i} W(x - x_i, h) = \sum_i u(\mathbf{x}_i) V_i W(\mathbf{x} - \mathbf{x}_i, h) \tag{5.34}$$

where m_i and ρ_i are the mass and the density of particle i, so the volume of the particle i is $V_i = m_i/\rho_i$ and we have:

$$\rho_i = \rho(\mathbf{x}_i) \tag{5.35}$$

$$u_i = u(\mathbf{x}_i) \tag{5.36}$$

W is also a kernel function on the points in the neighborhood of x. For W parametrized by its smoothing length h:

$$\begin{aligned} &\int_{R^d} W(\mathbf{x}, h) d\mathbf{x} = 1 \\ &\lim_{h \to 0} W(\cdot, h) = \delta \end{aligned} \tag{5.37}$$

where δ is the Dirac's delta distribution.

The approximation of δ with the smoothing kernel W with length h goes to zero as $h \to 0$. Depending on the distance between the particles Δp, the error may vary significantly with zero for $\Delta p \to 0$. The SPH method is consistent if $\Delta p/h \to 0$ as $h \to 0$. In practice this ratio depends on the smoothing kernel used and is between about 0.25 and 0.5 (Dalrymple and Rogers 2006; Liu and Liu, 2003; Monaghan, 1994). For further information about Smoothed particles hydrodynamics method and its application in the field of coastal engineering, see Dalrymple and Knio (2001); Dalrymple and Rogers (2006); Gingold and Monaghan (1977); Hoover (2006); Liu and Liu (2003); Lucy (1977); Monaghan, (1994).

5.8 ARTIFICIAL NEURAL NETWORKS

5.8.1 MLP Model

Algorithm Derivation

Considering the steepest descent algorithm, the Newton's algorithm and Gauss-Newton's algorithm, a brief description on the derivation of the Levenberg-Marquardt (LM) algorithm (Levenberg, 1944; Marquardt, 1963) is presented here (Yu and Wilamowski, 2011).

For the LM algorithm, the performance index to be optimized is the sum of the squares of the deviations defined to evaluate the training process. It is calculated for all training patterns and network outputs, by

$$S(w) = \frac{1}{2}\sum_{p=1}^{N} (y_p - \hat{y}_p)^2 \tag{5.38}$$

and using the equation:

$$S(w) = \frac{1}{2}\sum_{p=1}^{N} [e_p^2] \tag{5.39}$$

where w consists of all weights of the network, p is the index of patterns, from 1 to N, where N is the number of patterns, and y_p and $\hat{y}_p$ are respectively the desired and the actual output vector when applying pattern p (Gavin, 2011; Yu and Wilamowski, 2011).

The Levenberg–Marquardt algorithm is basically a combination of the steepest descent algorithm and the Gauss–Newton algorithm. The updating rule is (Yu and Wilamowski, 2011):

$$w_{k+1} = w_k - (J_k^T J_k + \lambda I)^{-1} J_k e_k \tag{5.40}$$

where λ is a non-negative damping factor and I is the identity matrix (Gavin, 2011; Marquardt, 1963; Yu and Wilamowski, 2011) and

$$J = \begin{Bmatrix} \frac{\partial e_1}{\partial w_1} & \frac{\partial e_1}{\partial w_2} & \cdots & \frac{\partial e_1}{\partial w_M} \\ \frac{\partial e_2}{\partial w_1} & \frac{\partial e_2}{\partial w_2} & \cdots & \frac{\partial e_2}{\partial w_M} \\ \vdots & \vdots & \ddots & \vdots \\ \frac{\partial e_P}{\partial w_1} & \frac{\partial e_P}{\partial w_2} & \cdots & \frac{\partial e_P}{\partial w_M} \end{Bmatrix} \tag{5.41}$$

where vector w includes weights $w_1, w_2, \ldots, w_m$ which are linear and error vector e has the form

$$e = \begin{Bmatrix} e_1 \\ e_2 \\ \vdots \\ e_P \end{Bmatrix} \tag{5.42}$$

Therefore, the Levenberg-Marquardt algorithm will be more similar to the steepest descent algorithm when the combination coefficient λ is very large or the parameters are far from their desirable value. The Levenberg-Marquardt algorithm will be more similar to the Gauss-Newton method when the combination coefficient λ is very small or parameters are close to their desirable value (see Gavin, 2011; Haykin, 1999; Marquardt, 1963; Sinha and Gupta, 1999; Yu and Wilamowski, 2011).

Transfer Function

The transfer function in an artificial neural network is an essential element of its structure. A transfer function limits the response amplitude of the unit (also known as a squashing function). It squashes the allowed amplitude of the response signal to a finite value. The most commonly used transfer functions include identity, step, binary step, unipolar sigmoid, bipolar sigmoid, hyperbolic tangent and radial basis functions (Karlik and Olgac, 2010). Usually, three conventional differentiable and monotonic transfer functions are employed for the evolution of the MLP architecture along with the LM learning algorithm. These proposed well-known and effective activation functions are bipolar sigmoid, unipolar sigmoid, and hyperbolic tangent, and are discussed next (Karlik and Olgac, 2010). The radial basis function will be discussed later, as it will be used for the RBF model.

The unipolar sigmoid transfer function is given as (Karlik and Olgac, 2010):

$$y(x) = \frac{1}{1 + e^{-x}} \tag{5.43}$$

A graphic of the sigmoid function can be seen in Figure 5.1 (Karlik and Olgac, 2010).

This function is particularly beneficial for BP algorithm as it minimizes the computational capacity for the training. The sigmoid function maps the inputs onto (0,1) (Karlik and Olgac, 2010).

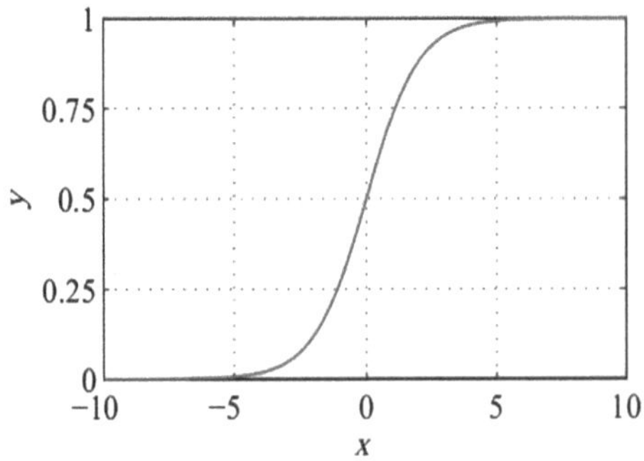

Figure 5.1 Unipolar sigmoid function.

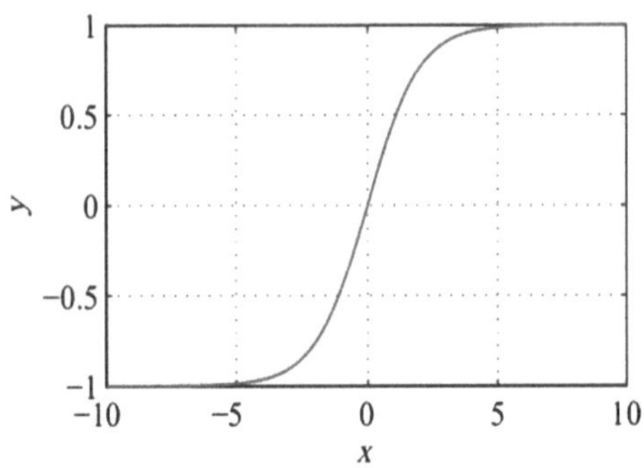

Figure 5.2 Bipolar sigmoid function.

The bipolar sigmoid transfer function is given by (Karlik and Olgac, 2010):

$$y(x) = \frac{1 - e^{-x}}{1 + e^{-x}} \tag{5.44}$$

The function is analogous to the sigmoid function. However, in the Bipolar sigmoid function output values are in the range of $[-1, 1]$ (Figure 5.2).

The function is basically the ratio between the hyperbolic sine and cosine functions. It can be also rewrite in the form of exponential functions as follows (Karlik and Olgac, 2010):

$$y(x) = \tan h(x) = \frac{\sin h(x)}{\cos h(x)} = \frac{e^{x} - e^{-x}}{e^{x} + e^{-x}} \tag{5.45}$$

The hyperbolic tangent function is similar to the sigmoid function but with an output range between -1 and 1 (as seen in Figure 5.3).

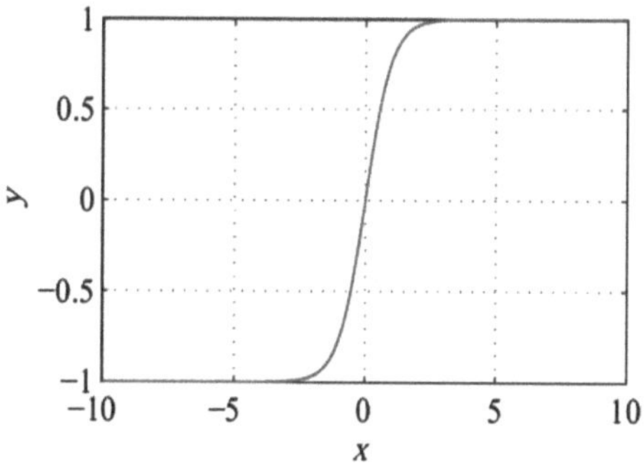

Figure 5.3 Hyperbolic tangent function.

To conduct a modeling using the neural networks, to select an appropriate transfer function for the model, all the transfer functions mentioned here should be tested using the same data and compared together. A various number of nodes in the hidden layer for all the transfer functions should be selected (for example: a number of nodes like 5, 10, 15, 20, 25, etc.). The number of iterations is typically the same for all transfer functions and the training samples are set in such a way as to be completely random for an improved learning performance. The neurons in the different models are arranged to learn at the same rate by assigning a learning rate parameter (Sinha and Gupta, 1999). Past research demonstrate that the artificial neural network performance might be improved by choosing an effective transfer function. In many cases, having compared the performances of functions, simulation results illustrate that the hyperbolic tangent (Tanh) and logistic sigmoid transfer functions offer better recognition accuracy than the other functions for various sets of hidden neurons (Karlik and Olgac, 2010). An antisymmetric hyperbolic tangent transfer function often have a slightly better performance in LM learning in comparison with an unsymmetrical logistic transfer function (Sinha and Gupta,1999). However, a serious issue to be resolved in regression problems is extrapolation effects. To resolve this problem, the transfer function in the output layer is replaced by a linear function, providing an unchanged activation level in the output layer. The linear transfer function does not saturate, so it can extrapolate further. However, a low learning rate, lower than 0.1, must be applied (Haykin, 1999; Sinha and Gupta, 1999). Depending on the cases, the model often computed better results when Tanh-Linear (combination of transfer functions for nodes in hidden and output layers, respectively) is used compared with Tanh-Tanh or Tanh-Sigmoid (Haykin, 1999; Sinha and Gupta, 1999).

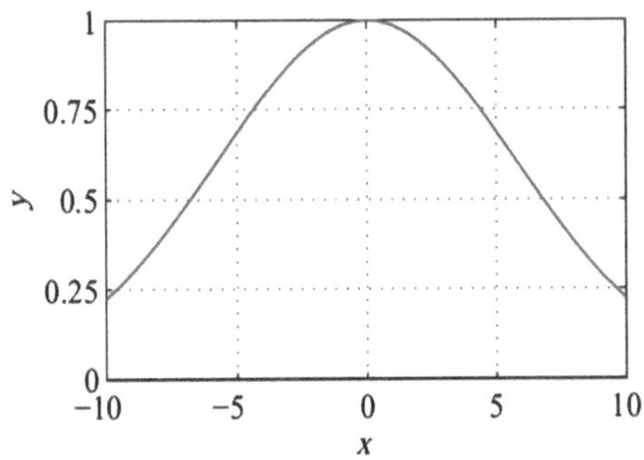

Figure 5.4 Gaussian function.

5.8.2 RBF Model

A brief summary of RBFs is presented here (for a more thorough description see Broomhead and Lowe, 1988; Haykin, 1999; Howlett and Jain, 2001a,b). A radial basis function is defined as:

$$\varphi(||x_i - c_k||) = e^{-\left(\frac{||x_i - c_k||^2}{2\sigma_k^2}\right)} \tag{5.46}$$

where φ is a Gaussian radial basis function, x_i is the input vector, c_k and σ_k represent respectively the center and the width of the kth RBF unit (Haykin, 1999).

The Gaussian function can be seen in Figure 5.4. Its output is determined by the distance between x_i and c_k. The function produces very low or high values respectively if the input is outside the range of training data and close to a neuron. The output layer of the RBF network is linear. The outputs of the hidden layer are weighted by $\{w_k\}_{k=1}^{m}$ finally providing the output function $\hat{y}(x_i)$ (Haykin, 1999):

$$\hat{y}(x_i) = \sum\nolimits_{k=1}^{m} w_k \varphi(||x_i - c_k||) + b \tag{5.47}$$

where $w_1, w_2, \ldots, w_m$ are weight factors and b is bias value.

REFERENCES

Batchelor, G.K., 1967. An Introduction to Fluid Dynamics. Cambridge University Press.

Berkhoff, J.C.W., 1972. Computation of combined refraction-diffraction, Proceedings of the 13th International Conference on Coastal Engineering, vol. 1. ASCE.

Booij, N., 1983. A note on the accuracy of the mild-slope equation. Coastal Eng. 7, 191–203.

Boussinesq, J., 1871. Théorie de l'intumescence liquide, applelée onde solitaire ou de translation, se propageant dans un canal rectangulaire. C. R. Acad. Sci. 72, 755–759.
Boussinesq, J., 1872. Théorie des ondes et des remous qui se propagent le long d'un canal rectangulaire horizontal, en communiquant au liquide contenu dans ce canal des vitesses sensiblement pareilles de la surface au fond. J. Math. Pures Appl. Deuxiéme Série 17, 55–108.
Broomhead, D.S., Lowe, D., 1988. Multivariable functional interpolation and adaptive networks. Complex Syst. 2, 321–355.
Chamberlain, P.G., Porter, D., 1995. The modified mild-slope equation. J. Fluid Mech. 291, 393–407.
Dalrymple, R., Rogers, B., 2006. Numerical modeling of water waves with the SPH method. Coast. Eng. 53 (2–3), 141–147.
Dalrymple, R.A., Knio, O., 2001. SPH modelling of water waves. In: Hanson, H., Larson, M. (Eds.), ASCE Conf. Proc, vol. 260. ASCE, Lund, Sweden, p. 80.
Dingemans, M.W., 1997. Wave propagation over uneven bottoms. Advanced Series on Ocean Engineering 13. World Scientific, Singapore, ISBN 981-02-0427-2. See Part 2, (Chapter 5).
Foias, C., Manley, O., Rosa, R., Temam, R., 2001. Navier-Stokes Equations and Turbulence. Cambridge University Press, Cambridge, UK.
Gavin, H., 2011. The Levenberg-Marquardt Method for Nonlinear Least Squares Curve-Fitting Problems. Duke University.
Gingold, R.A., Monaghan, J.J., 1977. Smoothed particle hydrodynamics: theory and application to non-spherical stars. Mon. Not. R. Astron. Soc. 181, 375–389.
Hamm, L., Madsen, P.A., Peregrine, D.H., 1993. Wave transformation in the nearshore zone: a review. Coast. Eng. 21 (1–3), 5–39.
Hansen, W., 1956. TheoriezurErrechnung des Wasserstandes und der Strömungen in TandmeerennebstAnwendungen. Tellus 8 (3).
Haykin, S., 1999. Neural Networks: A Comprehensive Foundation, second ed. Prentice-Hall, Englewood Cliffs, NJ.
Hoover, W.G., 2006. Smooth Particle Applied Mechanics: The State of the Art. World Scientific.
Howlett, R., Jain, L., 2001a. Radial basis function networks 1- Recent developments in theory and applications. Studies in Fuzzyness and Soft Computing. Physica-Verlag, New York.
Howlett, R., Jain, L., 2001b. Radial basis function networks 2 - New advances in design. Studies in Fuzzyness and Soft Computing. Physica-Verlag, New York.
Karlik, B., Olgac, A., 2010. Performance analysis of various activation functions in generalized MLP architectures of neural networks. Int. J. Artif. Intell. Expert Syst. (IJAE) 1 (4), 111–122.
Kirby, J.T., 2003. Boussinesq models and applications to nearshore wave propagation, surf-zone processes and wave-induced currents. In: Lakhan, V.C. (Ed.), Advances in Coastal Modeling. Elsevier Oceanography Series, 67. Elsevier, pp. 1–41.
Levenberg, K., 1944. A Method for the solution of certain non-linear problems in least squares. Q. Appl. Math. 2, 164–168.
Liu, G.R., Liu, M.B., 2003. Smoothed Particle Hydrodynamics: A Meshfree Particle Method. World Scientific, Singapore.
Lucy, L.B., 1977. A numerical approach to the testing of the fission hypothesis. Astron. J. 82, 1013–1024.
Mader, C.L., 2005. Numerical Modeling of Water Waves, second ed. CRC PRESS.
Marquardt, D., 1963. An algorithm for least-squares estimation of nonlinear parameters. SIAM J. Appl. Math. 11 (2), 431–441.
Massel, S.R., 1989. Hydrodynamics of Coastal Zones. Elsevier Science Publ., Amsterdam.

Massel, S.R., 1993. Extended refraction–diffraction equation for surface waves. Coastal Eng. 19, 97–126.

Massel, S.R., 1996. Ocean Surface Waves: Their Physics and Prediction. World Scientific, Singapore.

Monaghan, J.J., 1994. Simulating free surface flows with SPH. J. Comput. Phys. 110, 399–406.

Porter, D., Staziker, D.J., 1995. Extensions of the mild-slope equation. J. Fluid Mech. 300, 367–384.

Sinha, N.K., Gupta, M.M., 1999. Soft Computing and Intelligent Systems: Theory and Applications. Academic Press, New York, A volume of 25 chapters, 614 p.

Vreugdenhil, C.B., 1994. Numerical Methods for Shallow-Water Flow. Kluwer Academic Publishers.

Suh, K.D., Lee, C., Park, W.S., 1997. Time-dependent equations for wave propagation on rapidly varying topography. Coastal Eng. 32, 91–117.

Yu, H., Wilamowski, B., 2011. second ed. Levenberg-Marquardt Training, vol. 5. Industrial Electronics Handbook.

CHAPTER 6

Numerical Methods and Procedures

6.1 INTRODUCTION

The problem of numerical modeling of submerged breakwaters includes a number of major tasks which should be considered carefully. This is basically very similar to the other simulations of fluids. As a first step, a modeler needs to define the exact governing equations for the problem. Then the boundary conditions should be determined, considering the problem conditions. The turbulence equations also should be defined carefully.

The analytical form of the Navier-Stokes equations needs to be converted into a discretized form using a discretization method. There are several approaches for discretizing a given equation, such as finite difference, finite element, finite volume methods, and so forth. In the next sections a brief introduction of these methods is presented.

To conduct numerical modeling, first the computational domain or the geometry of the area of interest needs to be defined. Then the defined domain should be divided into the small sub-domains, or cells. The divided domain is also called a mesh, or grid. The cells are usually smaller in their dimensions where considerable variations are observed, while they may be larger where the changes of the flow properties are small. These generated cells, therefore, are the control volumes or elements in the model. The accuracy of the simulation strongly depends on the number of the cells in the domain. The physical processes and properties that need to be modeled are also defined. Conditions at the domain boundary are defined. Finally, the equations are solved for each node inside the cells and the flow properties of interest, such as velocity, pressure, etc., are calculated at the nodes.

As mentioned previously, several numerical techniques exist: finite difference method, finite volume method, finite element method, boundary element method, spectral method, etc. In the finite difference method, the derivatives are converted into finite difference approximations generated by the Taylor series expansions producing algebraic

© 2016 Elsevier Ltd.
All rights reserved.

equations for the unknown parameters. In contrast to the finite difference method, the finite element method uses exact derivatives rather than approximated ones, but it approximates the solution basis functions. In spectral methods, the unknown parameters of the flow are approximated and replaced in the governing equation by the Fourier series, creating the algebraic equations for the coefficients of the Fourier series. Spectral methods use global basis functions to approximate a solution across the entire domain, rather than individual elements as the finite element method does. In the finite volume method, the integral form of the governing equations is replaced by a system of algebraic equations and solved reiteratively.

A numerical method has several properties which are very important and which should be monitored carefully during the process of the simulation. These are convergent – approaching the exact solution as the grid spacing becomes zero; consistent – equality of the algebraic equations and the governing equations as the grid size becomes zero; stability – error damping over the procedure of simulation.

In contrast to the finite difference and finite element methods, in the finite volume method, if the conservation of mass, momentum and energy are satisfied in each control volume, they will also be satisfied in the whole domain. In addition, the finite volume can be applied for unstructured meshes simply. The structured grids are used for the simple domain while for the complex domain, therefore, the unstructured grids are usually preferable. Thus, in practice the finite volume may be preferred.

In comparison with other methods, the finite difference technique can be implemented in a much simpler way. Therefore, in this book, we will have more focus on the finite difference method.

6.2 FINITE DIFFERENCE METHOD

The finite difference method is one of the most commonly used numerical methods in computational fluid dynamics problems. It has also been widely used for coastal engineering problems by scientists and engineers. This method approximates the derivatives of the partial differential equations numerically using finite difference equations. In this section, we briefly review the finite difference technique for the discretization of the governing equations described in the previous chapter.

In the discretization process, the ordinary or partial derivatives are approximated using the Taylor polynomial.

A grid generation method should be applied first for the problem domain. Equally spaced nodes are used to approximate the derivatives of the governing equations. However, in some regions of the problem domain, a finer resolution may be required – for example, around rapidly varying bottoms.

Considering the Taylor polynomial as follows, the function f is estimated:

$$f(x_0+\Delta x)=f(x_0)+\frac{f'(x_0)}{1!}(\Delta x)+\frac{f''(x_0)}{2!}(\Delta x)^2+\cdots+\frac{f^n(x_0)}{n!}(\Delta x)^n+R_n(x) \tag{6.1}$$

where $R_n(x)$ is the difference between the Taylor polynomial and the function f.

For very small $R_n(x)$, the first derivative of the function f is:

$$f'(x_0)\approx\frac{f(x_0+\Delta x)-f(x_0)}{\Delta x} \tag{6.2}$$

Thus equations are expanded around a specific node using a Taylor's series, as follows:

$$f_{i+1}=f(x_i+\Delta x)=f_i+\Delta x\left(\frac{df}{dx}\right)_i+\frac{\Delta x^2}{2}\left(\frac{d^2f}{dx^2}\right)+\frac{\Delta x^3}{6}\left(\frac{d^3f}{dx^3}\right)+e \tag{6.3}$$

$$f_{i-1}=f(x_i-\Delta x)=f_i-\Delta x\left(\frac{df}{dx}\right)_i+\frac{\Delta x^2}{2}\left(\frac{d^2f}{dx^2}\right)-\frac{\Delta x^3}{6}\left(\frac{d^3f}{dx^3}\right)+e \tag{6.4}$$

where i is an internal node of the discretized domain and e is the truncation error.

The finite difference approximation to a first-order derivative of function f can be obtained by subtracting the f_{i+1} and f_{i-1} expressions and doing some rearranging as follows:

$$\left.\frac{df}{dx}\right|_i=\frac{f_{i+1}-f_{i-1}}{2\Delta x}+O(\Delta x^2) \tag{6.5}$$

$O(\Delta x^2)$ states the order of the truncation error. In practice, second-order accuracy schemes are most frequently used.

In a similar way, the finite difference approximation to a second-order derivative can also be computed by adding f_{i+1} and f_{i-1} expressions and rearranging:

$$\left.\frac{d^2 f}{dx^2}\right|_i = \frac{f_{i+1} - 2f_i + f_{i-1}}{\Delta x^2} + O(\Delta x^2) \tag{6.6}$$

Additional derivatives of function f may be similarly computed using the Taylor's series. Therefore, a finite difference approximation of the governing equations can be made. A list of some examples is presented here based on two types of approximations depending on the number of Taylor's series where used. Therefore, obviously, the second approximation in each case is more accurate as it considers more terms.

Using a forward difference scheme, we have:

For first derivative:

$$f'(x_i) = \frac{f(x_{i+1}) - f(x_i)}{\Delta x} + O(\Delta x) \tag{6.7}$$

$$f'(x_i) = \frac{-f(x_{i+2}) + 4f(x_{i+1}) - 3f(x_i)}{2\Delta x} + O(\Delta x^2) \tag{6.8}$$

For second derivative:

$$f''(x_i) = \frac{f(x_{i+2}) - 2f(x_{i+1}) + f(x_i)}{\Delta x^2} + O(\Delta x) \tag{6.9}$$

$$f''(x_i) = \frac{-f(x_{i+3}) + 4f(x_{i+2}) - 5f(x_{i+1}) + 2f(x_i)}{\Delta x^2} + O(\Delta x^2) \tag{6.10}$$

For third derivative:

$$f'''(x_i) = \frac{f(x_{i+3}) - 3f(x_{i+2}) + 3f(x_{i+1}) - f(x_i)}{\Delta x^3} + O(\Delta x) \tag{6.11}$$

$$f'''(x_i) = \frac{-3f(x_{i+4}) + 14f(x_{i+3}) - 24f(x_{i+2}) + 18f(x_{i+1}) - 5f(x_i)}{2\Delta x^3} + O(\Delta x^2) \tag{6.12}$$

For fourth derivative:

$$f''''(x_i) = \frac{f(x_{i+4}) - 4f(x_{i+3}) + 6f(x_{i+2}) - 4f(x_{i+1}) + f(x_i)}{\Delta x^4} + O(\Delta x) \tag{6.13}$$

$$f''''(x_i) = \frac{-2f(x_{i+5}) + 11f(x_{i+4}) - 24f(x_{i+3}) + 26f(x_{i+2}) - 14f(x_{i+1}) + 3f(x_i)}{\Delta x^4} + O(\Delta x^2) \tag{6.14}$$

Using a backward difference scheme, we have:
First derivative:

$$f'(x_i) = \frac{f(x_i) - f(x_{i-1})}{\Delta x} + O(\Delta x) \tag{6.15}$$

$$f'(x_i) = \frac{3f(x_i) - 4f(x_{i-1}) + f(x_{i-2})}{2\Delta x} + O(\Delta x^2) \tag{6.16}$$

Second derivative:

$$f''(x_i) = \frac{f(x_i) - 2f(x_{i-1}) + f(x_{i-2})}{\Delta x^2} + O(\Delta x) \tag{6.17}$$

$$f''(x_i) = \frac{2f(x_i) - 5f(x_{i-1}) + 4f(x_{i-2}) - f(x_{i-3})}{\Delta x^2} + O(\Delta x^2) \tag{6.18}$$

Third derivative:

$$f'''(x_i) = \frac{f(x_i) - 3f(x_{i-1}) + 3f(x_{i-2}) - f(x_{i-3})}{\Delta x^3} + O(\Delta x) \tag{6.19}$$

$$f'''(x_i) = \frac{-5f(x_i) - 18f(x_{i-1}) + 24f(x_{i-2}) - 14f(x_{i-3}) + 3f(x_{i-4})}{2\Delta x^3} + O(\Delta x^2) \tag{6.20}$$

Fourth derivative:

$$f''''(x_i) = \frac{f(x_i) - 4f(x_{i-1}) + 6f(x_{i-2}) - 4f(x_{i-3}) + f(x_{i-4})}{\Delta x^4} + O(\Delta x) \tag{6.21}$$

$$f''''(x_i) = \frac{3f(x_i) - 14f(x_{i-1}) + 26f(x_{i-2}) - 24f(x_{i-3}) + 11f(x_{i-4}) - 2f(x_{i-5})}{\Delta x^4} + O(\Delta x^2) \tag{6.22}$$

And finally, using a centered difference scheme, we have:
For first derivative:

$$f'(x_i) = \frac{f(x_{i+1}) - f(x_{i-1})}{2\Delta x} + O(\Delta x^2) \tag{6.23}$$

$$f'(x_i) = \frac{-f(x_{i+2}) + 8f(x_{i+1}) - 8f(x_{i-1}) + f(x_{i-2})}{12\Delta x} + O(\Delta x^4) \tag{6.24}$$

For second derivative:

$$f''(x_i) = \frac{f(x_{i+1}) - 2f(x_i) + f(x_{i-1})}{\Delta x^2} + O(\Delta x^2) \tag{6.25}$$

$$f''(x_i) = \frac{-f(x_{i+2}) + 16f(x_{i+1}) - 30f(x_i) + 16f(x_{i-1}) - f(x_{i-2})}{12\Delta x^2} + O(\Delta x^4) \tag{6.26}$$

For third derivative:

$$f'''(x_i) = \frac{f(x_{i+2}) - 2f(x_{i+1}) + 2f(x_{i-1}) - f(x_{i-2})}{2\Delta x^3} + O(\Delta x^2) \tag{6.27}$$

$$f'''(x_i) = \frac{-f(x_{i+3}) + 8f(x_{i+2}) - 13f(x_{i+1}) + 13f(x_{i-1}) - 8f(x_{i-2}) + f(x_{i-3})}{8\Delta x^3} + O(\Delta x^4) \tag{6.28}$$

For fourth derivative:

$$f''''(x_i) = \frac{f(x_{i+2}) - 4f(x_{i+1}) + 6f(x_i) - 4f(x_{i-1}) + f(x_{i-2})}{\Delta x^4} + O(\Delta x^2) \tag{6.29}$$

$$f''''(x_i) = \frac{-f(x_{i+3}) + 12f(x_{i+2}) + 39f(x_{i+1}) + 56f(x_i) - 39f(x_{i-1}) + 12f(x_{i-2}) + f(x_{i-3})}{6\Delta x^4} + O(\Delta x^4) \tag{6.30}$$

6.2.1 Discretization of Equations

The geometry of the interest project domain needs to be discretized in an appropriate way to provide an adequate resolution. The geometry is required to be created or imported from the available data with a proper coordinate system and units. The location of boundary conditions should be specified in the model. Then the computational domain should be divided into a very large number of cells, and the governing equations which are in partial differential forms are discretized with an appropriate method. The partial differential equations are replaced by approximating equations based on the model applied. The new equations are solved, and the main variables and their discrete distributions are estimated within the computational domain.

The size of created elements or volumes in the grid needs to be sufficiently small to obtain an adequate resolution of the geometry. On the other hand, the computing cost and time shouldn't be increased too much. This is the case particularly for 3D problems. To achieve this goal sometimes we may need to use nonuniform grids with various sizes in different areas in the computational domain. Smaller cells should be applied in critical areas. For example, in the case of submerged breakwaters, it is suggested to apply very fine grids in areas with high turbulence, gradients or large variations of water surface levels. Usually we should avoid very small and very large grid angles. Grid lines that are not aligned with the water flow direction in the model should also be avoided.

6.2.2 Grids

Different types of grids are used for discretization of a computional model, such as structures or unstructured grids, etc. Regular structured grids are one of the grid types which can be used simply for the discretization of a rectangular computational domain in a finite difference scheme. For example, for a three-dimensional problem, each cell or control volume in the computational grid can be determined using three numbers (i.e. $i = 1{:}n$, $j = 1{:}m$ and $k = 1{:}p$) where cell length (i.e. Δx, Δy and Δz) in each direction is constant. In addition to the regular grids, the structured grids can also be orthogonal, where the grid lines are normal to each other. Changing the grid lines to curved shapes will create a curvilinear grid, which is another type. However, structured grids are not very appropriate when dealing with complex computational domains.

On the other hand, unstructured grids cannot be easily used, although they are the proper kind of grids applicable for complex computational domains and for the finite elements scheme in particular. In this type of grid the number of cell sides varies from cell to cell. Therefore, we cannot use the method that is applied for the structured grids to determine the cells. Thus, information on all nodes and their coordinates is usually needed to determine the cells in an unstructured grid. However, in practical problems, unstructured grids require significantly higher storage space and computation time. Block-structured meshes which are assembled in an unstructured manner from a number of structured grid blocks are another type of grid applicable in these problems. Regardless of the type of grid, an appropriate method or tool should be applied to generate the grid for the problem and then the domain can be discretized to solve the partial differential problem over the domain using the numerical approach. In addition, boundary conditions should be specified. The initial conditions have to be specified in the case of unsteady problems.

The values calculated by numerical model should be compared and evaluated by experimental data. The physical parameters such as water surface level, wave height, velocity, pressure, etc. can be measured via different laboratory tools such as probes, velocitimeters, acoustic Doppler velocimetry (ADV), laser Doppler velocimetry (LDV), and pressure inducers. The measurements are taken at a limited number of points.

The water surface level tracking must be computed simultaneously with the flow field in the numerical model. One of the most common water surface level tracking approaches used in the coastal engineering field and, for example, in the case of submerged breakwaters, is the method of Volume of Fluid (VOF) (Hirt and Nichols, 1981). The VOF method applies mass conservation which is applicable for the case of sea water. In the VOF method, an algorithm is used to determine the water surface level from the volume fraction information for all cells.

6.2.3 The Discretizations in Time

Either explicit or implicit methods can be applied for time discretization. For the case when we apply an explicit time discretization scheme, obviously considering the nodes in the computational domain, the variables at time step $n + 1$ simply depend only on the known values at the time step n. However, stability can be a problem in the case of an explicit time discretization method. A numerical scheme is stable if errors do not grow from one time step to the next time step.

In the case of an implicit method, the variables at the time $n+1$ can depend on both values at time n and $n+1$. Therefore, an implicit method can be classified into semi-implicit or fully implicit schemes, where the variables at the time $n+1$ depend on both values at the time steps n and $n+1$, or only time step $n+1$, respectively. Thus, in both schemes of an implicit method, a system of equations must be solved, which is not the case for the explicit method. For the case of highly nonlinear equations, when all the 3D governing equations are solved simultaneously, a huge storage space might be required. However, it should be noted that a fully implicit method is unconditionally stable.

6.3 FINITE VOLUME METHOD

Assuming the integral form of the conservation equations, to set up a finite volume model, firstly we need to create an appropriate computational grid. This can be done by dividing the model domain into a large number of subdomains or discrete and separated control volumes. We need to define control volumes or cells using an appropriate grid and assign nodes to the control volumes (cells). Then the governing equations should be discretized on the cells. The equations should be satisfied for each cell and solved. The procedure is similar for both 2D and 3D problems.

Considering the following equation:

$$\frac{d}{dt}\int_V \rho f dV + \int_s \rho f(\vec{v}\cdot\vec{n})dA = -\int_s \vec{q}\cdot\vec{n}\,dA + \sum\int_V s\,dV \tag{6.31}$$

where V is discrete control volumes, assuming $F=\int_V \rho f dV$ and by averaging f over a cell we have

$$\frac{d}{dt}\int_V \rho f dV = V\frac{d\overline{F}}{dt} \tag{6.32}$$

Total flux (R) is

$$\int_s \vec{R}\cdot\vec{n}\,dA = \int_s \rho f(\vec{v}\cdot\vec{n})dA + \int_s \vec{q}\cdot\vec{n}\,dA \tag{6.33}$$

Total source term (S) is

$$S = \int_V s dV \tag{6.34}$$

Hence we have:

$$V\frac{d\overline{F}}{dt}+\int_{s}\vec{R}\cdot\vec{n}\,dA=S \tag{6.35}$$

An approximation to $f(x,y,z)$ can be made for $\overline{F}$ for each cell and boundaries and F can be evaluated. We should integrate F and compose it by integration over each cell to obtain surface and volume integrals respectively. A new value of $\overline{F}$ is then obtained. A single value for R can be computed using r. The procedure is similar for 2D problems:

$$R=\int_{S}rdA=\overline{r}S \tag{6.36}$$

To approximate the volume integrals

$$S=\int_{V}SdV \tag{6.37}$$

To approximate the volume integrals as a function of nodal values we use:

$$S_n=\int_{V}SdV=\frac{\Delta x\Delta y}{36}[16s_n+4s_b+4s_t+4s_l+4s_r+s_{br}+s_{bl}+s_{tr}+s_{tl}] \tag{6.38}$$

Values at surface locations can be interpolated. The procedure for 3D problems is similar to 2D cases.

The Finite Volume Method for the Navier-Stokes Equations

The finite volume method is a method for solving partial differential equations like the Navier-Stokes equations in the form of algebraic equations. The physical parameters are approximated at discrete nodes surrounded by finite volumes within the problem domain. Volume integrals including divergence term are replaced by surface integrals according to the divergence theorem.

$$\frac{\partial u_i}{\partial t}V+\sum_{n=1}^{N}(u_iu_jn_jA)_n+\sum_{n=1}^{N}(Pn_iA)_n-\sum_{n=1}^{N}\left(v\frac{\partial u_i}{\partial x_j}n_jA\right)_n-f_iV=0 \tag{6.39}$$

where n is the normal of the surface of the control volume and V is the volume.

$$V = \Delta x \Delta y \tag{6.40}$$

The derivatives will then be approximated using the finite difference method. The face values for u, v, and P will then be computed. A backward difference will usually be employed for the time derivative, while a central difference may be more suitable for the spatial derivatives.

6.4 ARTIFICIAL NEURAL NETWORKS MODELING

6.4.1 MLP Training Process

As stated earlier, the Levenberg-Marquardt algorithm is one of the most efficient training algorithms for neural network modeling, therefore, training a feed-forward neural network with the Levenberg-Marquardt algorithm is presented in this section. Having the updating rule (see Chapter 5), the Jacobian matrix needs to be computed and a training process should be designed.

Considering a single neuron (designated j) with p inputs in a hidden layer of a network with a total number of r neurons, $y_{j,i}$ means the ith input of neuron j, weighted by $w^h_{j,i}$, while y_j is the output of neuron j:

$$net^h_j = \sum_{i=1}^{n} (w^h_{j,i} y_{j,i} + b^h_j) \tag{6.41}$$

$$y_j = f_j(net^h_j) \tag{6.42}$$

where h is the index of the hidden layer and b^h_j, f_j and net^h_j are respectively the bias, the activation function and the sum of weighted input nodes of neuron j in the hidden layer.

The input value for the output layer equals the sum of all weighted outputs of the hidden layer plus biases. The derivative of the sum of weighted hidden nodes of neurons in the output layer then is calculated in a similar way. The slope of the activation function in the output layer is calculated similarly to the hidden layer. The network outputs are also calculated.

The process should be repeated for every output in order to calculate the Jacobian matrix. The parameter δ is calculated for each neuron j and output. Then the error is replaced by a unit value (Hagan and Menhaj, 1994; Nielsen, 1989; Yu and Wilamowski, 2011).

Considering a two-layer MLP network, the forward phase is arranged for a given pattern as follows: Firstly, net values, slopes, and outputs for all neurons in the first layer is computed. Then the outputs of the hidden layer neurons are used as the inputs of all neurons in the output layer. Then the computations as stated before are repeated. Hence, node and slope arrays (y and s respectively) are calculated for neurons within the pattern in forward phase of neural network modeling (Yu and Wilamowski, 2011).

Once the forward phase is completed, the backward phase for the given pattern is started as follows: Firstly error for output j should be computed. Then initial δ the slope of output j is determined. δ is propagated in order, first from the inputs of the output layer to the outputs of the hidden layer, then from the outputs of the hidden layer to the inputs of the hidden layer and at the end from the inputs of the hidden layer to the input layer. This process should be repeated for other outputs. Once the forward and backward phases are completed for the given pattern, the whole array and the y array are determined. $y_{j;i}$ and δ_j are computed over the forward and backward phases respectively. Hence the row elements of the Jacobian matrix are computed. The forward and backward phases as mentioned above should be repeated for other patterns. Thus, the whole elements of the Jacobian matrix are computed (Yu and Wilamowski, 2011).

As mentioned before, the Jacobian matrix includes the first order partial derivatives. For an artificial neural network, the Jacobian matrix is an $N \times M$ matrix with N row and M columns. In this matrix, N and M are the number of training patterns and network parameters, respectively. Thus, the elements of the matrix is computed and created according to the first partial derivatives of the network outputs or the activation functions in respect to the network weights (Yu and Wilamowski, 2011):

$$\frac{\partial e_p}{\partial w_{j,i}} = -\frac{\partial \hat{y}_p}{\partial y_j}\frac{\partial y_j}{\partial net_j}\frac{\partial net_j}{\partial w_{j,i}} \tag{6.43}$$

This is basically the derivative of the nonlinear function between neuron j and output.

In the Levenberg-Marquardt method, a damping factor λ is typically used in the training process which is adjusted at each iteration until the sum of squared errors decreases. Thus the algorithm solves equation (Yu and Wilamowski, 2011):

$$w_{k+1} = w_k - (J_k^T J_k + \lambda I)^{-1} J_k e_k \tag{6.44}$$

To summarize, the LM algorithm will consist of the following steps (designed for each learning iteration, or epoch) (Yu and Wilamowski, 2011):

1. The network parameters (weights; w) are initialized.
2. The sum square error (S) is calculated and evaluated.
3. The Jacobian matrix (J) is computed.
4. The error gradient (Je) is computed.
5. The Hessian is approximated using the cross-product Jacobian ($J^T J$).
6. $(J^T J + \lambda I)\Delta = Je$ is solved to find Δ.
7. The network parameters (weights; w) are adjusted using Δ.
8. The sum square error (S) is recalculated using the updated network's parameters (weights).
9. In the case of observing an increment in the value of the sum square error, the weight vector needs to be changed into the former value. λ should be stepped up and an update needs to be made by going back to the step 2.
10. In the case of observing a decrement in the value of the sum square error, the weight vector is adopted and λ should be stepped down.
11. The algorithm is repeated from step 2 according to the new weights, until the sum square error (S) becomes lower than the desired value.

6.4.2 RBF Network Training

An RBF neural network typically adapts itself into the patterns of training data with adjusting its parameters using learning rules. The training process consists of computing output from given input data, comparing the calculated values with the desired output, and adjusting the transfer function; see the following equation:

$$\hat{y}(x_i) = \sum_{k=1}^{m} w_k \varphi(\|x_i - c_k\|) + b \tag{6.45}$$

This is basically done for each neuron in order to minimize the difference between the output y_i of a given training set $T = \overline{(x_i, y_i)}_{i=1}^{p}$ and the network response $\hat{y}(x_i)$ with respect to the weights W as well as to the parameters of the basis function, c and σ (Haykin, 1999):

$$S = \sum_{i=1}^{P} ||y_i - \hat{y}(x_i)|| \rightarrow \min \tag{6.46}$$

Using unsupervised algorithms for training the RBF networks, usually a subset of the training data is used for the centers, c, and the values of the width σ are derived from the structure of the training data (Moody and Darken, 1989). Having thus predefined c and σ only the weights, W, need to be adapted during the training using a supervised algorithm. The supervised algorithms typically give better results than unsupervised algorithms, although their computational costs are often relatively high. (Bullinaria, 2004). With either approach, determining a good value for the number of hidden units remains a problem. It will generally be appropriate to compare the results for a range of different values, following the same kind of validation/cross-validation methodology used for optimizing MLPs (Bullinaria, 2004). A more detailed discussion on the selection of c and σ using unsupervised algorithms and their influence on the training process can be found in Haykin (1999). A discussion on the supervised training of the basis function parameters used in this book follows in this section.

A derivative-based algorithm, *gradient descent* (Karayiannis, 1999), is discussed here, chosen from the various algorithms proposed in the literature for training RBF networks. The gradient descent algorithm can be implemented to minimize the error after defining the error function (Bullinaria, 2004):

$$S(w, c, \sigma, b) = \sum_{j=1}^{J} (y_j - \hat{y}_j)^2 \tag{6.47}$$

where y_j indicates the desired output and $\widehat{y}_j$ indicates the RBF neural network output.

Following the training procedure, after making a decision on the number of hidden units, in order to obtain the desired output of the RBF neural network w, c, σ and b parameters are adjusted. The sum squared error S is used to evaluate the performance of the RBF network (Bullinaria, 2004):

$$\Delta w_j = -\eta_w \frac{\partial S}{\partial w_j} \tag{6.48}$$

$$\Delta c_j = -\eta_c \frac{\partial S}{\partial c_j} \tag{6.49}$$

$$\Delta \sigma_j = -\eta_\sigma \frac{\partial S}{\partial \sigma_j} \tag{6.50}$$

where η_σ is the learning rate (Liu et al., 2004).

The training procedure in the RBF network is therefore organized in the following stages (Walczak and Massart, 2000): The hidden layer initially has zero number of units. Data samples are presented to the network and signals pass through the network and at last the error is computed. The calculated output is compared to the target value of the respective sample. An RBF unit is added to the hidden layer with weights equal to the input vector with the maximum error (Bullinaria, 2004). The centres position in the hidden layer affects the network performance significantly (Simon, 2002). Then the connection weight parameters between the hidden and the output layers are adjusted to minimize the error. This error which is statistically described as the sum squared error must be lower than the desired error. At each iteration the input vector that leads to the lowest network error is used to create a radial basis unit. The error produced by the updated network is checked, and if it is lower than the desired error, training is stopped; otherwise one more neuron is added to the hidden layer (Bullinaria, 2004). The number of computational units in the hidden layer also has an important influence on the network complexity and the generalizing capability of the neural network (Liu et al., 2004). The precess needs to be repeated until at least one of the following limits is met: the desired total error or the maximum number of neurons. (Walczak and Massart, 2000).

REFERENCES

Bullinaria, J., 2004. Introduction to neural networks. University Lectures.
Hagan, M.T., Menhaj, M., 1994. Training feedforward networks with the marquardt algorithm. IEEE Trans. Neural Netw. 5 (6), 989–993.
Haykin, S., 1999. Neural Networks: A Comprehensive Foundation, second ed. Prentice-Hall, Englewood Cliffs, NJ.
Hirt, C.W., Nichols, B.D., 1981. Volume of Fluid (VOF) Method for the dynamics of free boundaries. J. Comp. Phys. 39, 201.
Karayiannis, N.B., 1999. Reformulated radial basis neural networks trained by gradient descent. IEEE Trans. Neural Netw. 10 (3), 657–671.
Liu, Y., Zheng, Q., Shi, Z., Chen, J., 2004. Training radial basis function networks with particle swarms. Lecture Notes Comput. Sci. 3173, 317–322.
Moody, J., Darken, C.J., 1989. Fast learning in networks of locally-tunes processing units. Neural Comput. 1, 284–294.
Nielsen, R.H., 1989. Theory of the back propagation neural network. Proceedings IEEE IJCNN. IEEE Press, New York.
Simon, D., 2002. Training radial basis neural networks with the extended kalmanlter. Neurocomputing 48, 455–475.
Walczak, B., Massart, D., 2000. Local modelling with radial basis function networks. Chemometrics Intell. Lab. Syst. 50 (2), 179–198.
Yu, H., Wilamowski, B., 2011. second ed. Levenberg-Marquardt Training, 5. Industrial Electronics Handbook.

CHAPTER 7

Numerical Modeling and Simulation

7.1 MODELING THE SHALLOW WATER EQUATIONS

Wave propagation around a 3D submerged breakwater is investigated numerically using the Navier-Stokes equations. A description of the model is briefly presented in the following section. The simulation process consists of three main tasks: 1) set up, 2) calibration, and 3) simulation. Each of these three tasks is described in detail in the following sections. The general descriptions of the model presented in these sections and the mathematical theories behind this model were presented in previous chapters. For further details, the reader is referred to the more general books on numerical modeling.

7.1.1 Setting Up

Basically, the first step of numerical modelling in a coastal engineering problem and particularly in our case which is specifically related to the modelling of the submerged breakwaters is setting up an appropriate model. For this purpose, a number of specific tasks needs to be completed in sequence carefully.

In this stage, firstly having a bathymetry file is essential. The bathymetry for the problem should be specified using a two dimensional matrix of water depths in the model area. The water depths can be obtained from the accurate and reliable data, maps or sources available for the area of interest. The collected data associated to the problem needs to be resolved into an appropriate spatial mesh. Thus, the governing equations can be solved numerically for the problem over its computational domain. The grid spacing and time step need to be selected carefully based on the wave conditions (i.e. wave lengths and periods) and the water depths in the domain area. The regions in the model with wave breaking should be studied with a specific attention as those areas may need smaller grid spacing. In practical applications, we should choose the time and space steps

© 2016 Elsevier Ltd.
All rights reserved.

in a way that the desired accuracy can be achieved while the computational costs are not increasing significantly.

Usually, the Courant number for the n-dimensional case defined as follows is used to control the stability in the model as in some cases instability may appear in our simulations:

$$C = \Delta t \sum_{n=1}^{N} \frac{c_n}{\Delta s_n} \leq 1 \tag{7.1}$$

where C is the Courant number, Δt is the time step, c_n and Δs_n are respectively the magnitude of the celerity and the spatial interval in nth dimension.

Basically, the grid spacing and time steps should be sufficiently small to avoid instability in the model particularly when we deal with wave breaking in the simulations. Time and space discretization are done to solve the equations numerically. A sufficient simulation period should be also selected as we need to ensure waves can reach the structure or for instance the submerged breakwater in this particular case. Appropriate time series of the water surface level can be employed.

7.1.2 Calibration

In the process of calibration, having access to various types of data may be needed; i.e. bathymetric and boundary information, submerged structure features, measured wave/current parameters or essential hydrodynamic/morphodynamics specifications at some specific points. Based on a primary selection of the calibration parameters we can run our simulations and obtain initial outcomes. Actually we may observe some differences between the outcomes and measurements. Therefore, we will need to modify the model parameters in order to decrease these differences as much as possible. Basically, the error should be lower than a desired value which is determined for that specific problem.

We need to consider generating waves along a specified generation line for our simulations typically offshore. The generated waves are propagated in the model area and can reach the submerged breakwater or other boundaries. The grid points with positive values according to the still water level (SWL) are typically considered as closed or reflective boundaries. Hence we can observe some reflected waves by the structure or closed boundaries into the model area. These waves can make the situation more complicated. Even they can re-reflected again by other boundaries and make the conditions worse. Those returning waves which reach the other model boundaries or wave generation line need to be absorbed by an absorbing system. Thus, along these boundaries and lines, we should employ a radiation boundary condition to absorb reflected waves propagating out of the computational domain.

An appropriate model or equation is considered for modelling the bottom friction. A breaker model (i.e. considering surface rollers; see Schaffer et al. 1993 and Madsen et al. 1997) also needs to be formulated carefully. The breaker model needs to handle the various type of wave breakings associated with the submerged breakwaters, particularly the plunging breakers. In addition, in some locations or boundaries and in the place of submerged structure we may need to have partial wave reflection or wave filtration through the structure body. A radiation boundary condition can be also applied at the shoreline to absorb incoming waves to a desired degree according to the available data. Hence a degree of porosity is set along the boundaries with partial wave reflection or the submerged breakwater. All these processes cause wave energy dissipation to some extent in the locations of closed boundaries or structures.

7.1.3 Simulation

Once we complete the calibration process and verifications show good agreements with measurements, the model is ready for new simulations and practical results. Results can be represented in various forms including the deterministic parameter of the surface elevation, the phase-averaged value of the significant wave height, and the wave disturbance coefficient according to the incident wave height. All these information can be presented for a specific location or along a specific line or over an arbitrary surface within the model area.

7.2 MODELING WITH NEURAL NETWORKS

A detailed description of prediction models based on artificial neural networks (ANNs) was presented in the previous chapters. The current chapter is concerned with data preparation, model set-up and calibration, performance analysis and validation.

7.2.1 Data Preparation

Data preparation is an essential stage in data-driven modeling and should be thoroughly carried out to guarantee accurate predictions by the ANN model. Some of the biggest advantages of good data preparation include speeding up the training process, reducing the network complexity and improving the generalization ability. Therefore, the degree of success in ANN modeling varies according to the quantity and quality of the data used for modelling. Data preparation consists of selecting suitable data, deciding on the number of input/output variables, splitting the data set into subsets, and data transformation (Yu et al., 2006). Techniques for dealing with each of these steps will be discussed in this section in detail.

7.2.2 Data Set and Selection

Basically, ANN models are as accurate as the selected data used for their training. For instance, in the case of submerged breakwater which is the main topic of interest in this book, to obtain accurate and reliable results from an ANN model trained by experimental laboratory data, a wide range of measured data should be collected at different scales and under various conditions in laboratory. In addition, there might be some problems in the data which need to be fixed before training, such as an unusual range of data, missing data or nonnumeric data. Therefore, selecting the appropriate data among all the collected data is an essential stage (Yu et al., 2006).

The selected data needs to properly include the whole range of input variables for a well-improved generalization within this range (interpolation). Beyond this range ANN models often produce inaccurate responses (extrapolation). Insufficient amounts of data and existence of noise in the data might lead to a reduction of the efficiency of an ANN scheme (Yu et al., 2006).

A sufficient number of training data samples are needed to fit a multidimensional response surface to the input data. The required number of data samples grows directly with the dimensionality (called the "Curse of Dimensionality" by Richard Ernest Bellman; Bellman and Rand Corporation 1957; Bellman, 1961), which refers to the exponential growth of hypervolume as a function of dimensionality (Bellman, 1961; Haykin, 1999). Based on a very simple guideline in literature and as an initial guess, the minimum size of the training data set is recommended to be at least ten times the network weights (Haykin, 1999); Sinha and Gupta, 1999). For further details see also: Fausett, (1994); Hagan et al., (1996); Yu et al., (2006). The following rule to select the size of the training data set N is also suggested (Haykin, 1999; Sinha and Gupta, 1999):

$$N = O\left(\frac{W}{e}\right) \tag{7.2}$$

where O denotes "the order of," W is the total number of weights including biases, and e denotes the fraction of errors permitted on the test data. For instance, for a total number of weights in the model of 50 and 100, for an error of 10%, the minimum size of the training data set should be approximately 500 and 1000, respectively. Generally, for most regression problems, the required size of the training data set is of the order of at least hundreds and at most thousands (Haykin, 1999). However, practically, the minimum size of the training data set is related to the existence of noise in the data and the variance of noise, the number of variables involved in the problem and its complexity. In the case we are dealing with a restricted training data set, one solution might be creating an ensemble of neural networks, training each network using a completely different resampling of the training data, and finally averaging the predictions (Haykin, 1999; Yu et al., 2006). Missing and nonnumeric values, if any, can be substituted using statistical techniques, for example using the mean value of that variable (Bishop, 1995).

7.2.3 Variable Selection

An ideal model is one that explains the problem in the easiest and simplest way using the fewest variables. It is important to identify the variables that save modeling time and space (Stein, 1993). However, variable selection is a complicated part of data preparation for neural network modeling due to lack of an exact solution for selecting effective variables. An extra variable obviously means more interlayer connections, leading to higher complexity in the network. Better performance by the ANN scheme can be obtained by excluding unnecessary input variables (Bishop, 1995; Fausett, 1994). Statistical tests might be used to make some selection of variables before starting to use neural networks in order to reduce the negative effects of dimensionality. Various configurations and combinations of input parameters can be tried. Sensitivity analysis can also be used. According to the results of sensitivity analysis, we can recognize the most important variables with respect to a particular neural network model (for further information, see Bishop, 1995; Fausett, 1994).

As discussed in previous chapters in detail, analysis of the dimensional and nondimensional parameters has highlighted that the most important physical parameters involved in the study of submerged breakwaters are limited to the variables listed here. The dimensional parameters consist of incident wave height H_i, offshore wave length L_o, water depth over the breakwater crest h_s, water depth at the toe of the breakwater h, breakwater crest width B and Cartesian coordinate system x and y. The nondimensional parameters used in numerical modeling are shown here for two- and three-dimensional simulations, respectively. Wave transmission coefficient K_t (the ratio of transmitted to incident wave height) for two cases is determined as:

$$K_{\{t,2D\}} = f(H_i, L_o, h_s, h, B) \tag{7.3}$$

$$K_{\{t,2D\}} = f(\xi_o, H_i/h, h_s/H_i, B/H_i) \tag{7.4}$$

$$K_{\{t,3D\}} = f(H_i, L_o, h_s, h, B, x, y) \tag{7.5}$$

$$K_{\{t,3D\}} = f(\xi_o, H_i/h, h_s/H_i, B/H_i, x/L_o, y/L_o) \tag{7.6}$$

7.2.4 Data Division

A cross-validation method (Haykin, 1999) which proceeds as follows was employed for generalization. This method is called the "Early-stopping" approach. The selected laboratory data are divided into two main data sets, including training data to train the ANN models and test data employed to evaluate the performance of the trained models. To improve the generalization, the training data set itself is split into two subsets: an "Estimation subset" for updating the network weights and biases, and a "Validation subset" where the error is monitored over the training process. At or near the beginning of training, calculated error in both estimation and validation subsets naturally decreases, while near the data overfitting, the computed error in the validation subset increases. The network parameters are recorded at the lowest validation subset error (see Bishop, 1995; Fausett, 1994; Haykin, 1999). The tuned networks are then tested on the test data. The philosophy behind this is to test the trained neural networks by data not used during the training process; in other words, the network experiences those inputs for the first time in order to evaluate capability of the model under new conditions and to ensure it works properly (for further information, please see Bishop, 1995; Fausett, 1994; Haykin, 1999).

Depending on the number of data points and the data set size, among the total data sets typically at least 70–80% are selected for the training (estimation and validation subsets), while the rest of the data set is kept for testing as well as verification of the model in different applications, including interpolation and extrapolation predictions, etc.

Each experiment consists of a specific number of measurement points behind the breakwater, so the total number of data samples in each data set should be calculated. The calculated number of data samples for each data set is prepared for 2D and 3D cases.

The training data set is used to train the networks and to investigate the model performance for dimensional and nondimensional parameters. The range of data used in neural network modeling in the form of dimensional and nondimensional parameters, including minimum, maximum, mean and standard deviation of each individual parameter, should be determined.

To evaluate performance of the proposed ANN model, three distinct applications of wave simulation are considered to investigate precision of the model outputs. These applications include interpolating, extrapolating and larger scale predictions, which are discussed in detail in this section. To demonstrate the influence of the input variables on the ANN predictions, a sensitivity analysis is conducted. First the dataset used for each case is described and then the analysis of results are presented and discussed.

Interpolation predictions are performed within the range of input parameters used for training, including both dimensional and nondimensional data.

The distribution of the input parameters within the data sets also plays an important role. A large concentration of samples in the database in a certain region in the input parameter spaces can affect the performance of the ANN model (Panizzo and Briganti, 2007). The frequency distribution of the nondimensional parameters and transmission coefficient used in the ANN model training and testing (interpolation) should be carefully investigated. Several classes can be defined for each nondimensional parameter in order to cover its entire range of variability.

The neural networks are trained within a specific range of input parameters and are naturally expected to return satisfactory results for new inputs within the same range as the training data. However, the proposed ANN model is expected to be employed to predict wave transmission under conditions outside the range of the training data set, so it is essential to have an additional investigation of the performance of the model outside that range, under extrapolation conditions. This is useful as it provides users with some guidelines regarding the appropriate application of the model. In-depth analysis of extrapolation predictions should be conducted for input data outside the range of the training data set including wave climates and polar coordinates where the wave transmission prediction is needed. The range of dimensional and nondimensional parameters, the frequency distribution of the nondimensional parameters and transmission coefficient used in the ANN model testing (extrapolation) used in this case (extrapolation evaluation) should be investigated. As mentioned previously, several classes may be defined for each nondimensional parameter in order to cover its entire range of variability.

A specific application of the model to simulate phenomena at a larger scale is considered, to investigate the performance of the model when the scales increase significantly. This might be critical when the neural network model is being trained with small-scale laboratory data and is going to be used to predict prototype conditions. Obviously, the evaluation data for these larger scale tests involves a scale factor of larger than unity.

Finally, a sensitivity analysis is conducted to study the effect of each input parameter on model performance.

A brief summary of these data sets is provided as follows:

1. Set 1 (Training data): A wide range of experiments with various submergence depths, water depths and wave conditions. Several breakwaters with different trapezoidal body sections and various geometries may be considered. This data set is used for ANN model training.
2. Set 2 (Interpolation-test data): The experiments within the same range of physical parameters including breakwater geometries, submergence depths, water depths and wave conditions. This data set is used to test the ANN model for interpolation prediction.
3. Set 3 (Extrapolation-test data): The experiments with a different range of physical parameters including breakwater geometries, submergence depths, water depths and wave conditions. This data set is used to test the ability of the ANN model to extrapolate beyond the range of the training data.
4. Set 4 (Larger scale-test data): This set of data includes the data collected in a physically larger wave tank. The data collected in this set is used to assess the performance of the ANN model under larger scale conditions and different laboratory facilities.

The verification data will be used to evaluate the performance of the ANN model.

7.2.5 Data Transformation

Data transformation consists of two main phases: data preprocessing and data postprocessing. It has been shown that, an artificial neural network may reproduce more accurate and reliable results by adding two extra phases of preprocessing and postprocessing into the main process. For neural networks and particularly multilayer perceptron (MLP) networks with some sort of transfer function (e.g. sigmoid function) in hidden layers, when the net input is greater than a certain value, these functions can become saturated so that the units are only sensitive to inputs within a fairly limited range. At the beginning of the training process, this leads to a very small gradient and consequently to very slow training. To resolve this issue, data normalization should be undertaken before applying the data to the network. Basically, by employing a normalization method for the input/output data sets, network will reproduce outcomes within a normalized range. Thus, the results need to be reverse transformed back into the form of the nonnormalized output data. Therefore, preprocessing and postprocessing steps respectively before and after the network are required (see Bishop, 1995; Fausett, 1994; Haykin, 1999).

To deal with different variables regardless of their scale and to ensure that the network output is in a sensible range, a specific normalization algorithm (also called a scaling algorithm) is applied. Numeric values need to be transformed into an amended scale with an appropriate range. In many circumstances, variables are scaled linearly (Haykin, 1999):

$$x_n = \frac{x - min(x)}{max(x) - min(x)} \tag{7.7}$$

where x is the value that needs to be normalized. Here all values are converted to a range of (0,1), so that all the outputs are made to lie within this range.

7.2.6 Model Set-Up and Calibration

This section is concerned with the implementation of ANN models to predict the wave field behind submerged breakwaters. Two different types of artificial neural network (see Chapter 4) are applicable to be designed and calibrated using the selected training data and compared together against the selected testing data, as explained in the previous section.

7.2.7 Multilayer Perceptrons

As discussed at length in Chapter 4, the Multilayer Perceptrons (MLP) networks (Rumelhart and McClelland, 1986) are widely applied in various function estimating problems in engineering applications. (see Bishop, 1995 for further details). The biased weighted sum of inputs for each computational unit (organized in a layered feedforward network) is passed through a transfer function to compute the corresponding output (see Haykin, 1999 and Bishop, 1995). An ANN is essentially designed by determining the interconnection pattern, the transfer functions and the learning process described before. Considering a fully connected feedforward neural network, the main task with designing an interconnection pattern is selecting the number of hidden layers and number of units in all layers. In many circumstances the inputs/outputs are specified according to the problem. However, a variable selecting process should be undertaken carefully in order for the input variables to all be meaningful and useful. This stage was discussed previously (see section 7.2.3, Variable selection). Regarding the outputs, in practical examples particularly regression applications, the neural networks often have only one neuron in their output layer. In the neural networks with multiple output variables, training process can be more complicated as multiple unknown functions need to be discovered simultaneously. This is particularly important for the computational units in the network. Therefore, practically, the best approach is to have an ensemble of smaller networks, each with a single output variable, and to run them individually (see Bishop, 1995; Hagan et al., 1996 and Haykin, 1999).

The selection of the optimum network is based on a trial-and-error scheme. This means a predefined rule does not exist for the selection of the number of hidden layers. Higher numbers of hidden layers, although increasing the generalization performance of MLP, increase the computation time when a back propagation algorithm is being used for training. Here, design of the MLP network is started first by considering only one hidden layer and then more layers are considered and compared with the first network. This has to be undertaken considering the different number of hidden units in this layer too, as will be discussed later. However, addressing all factors, in this book it is assumed that one hidden layer is sufficient to approximate the underlying function sufficiently well. For a more detailed and formal description of MLP networks, the reader is referred to Hagan et al. (1996) or Haykin (1999).

Once the number of hidden layers of the network is selected, a few parameters remain to be determined, such as the size of the hidden layer, the learning rate, the activation function and the minimum gradient. Similarly, they are usually determined experimentally, which means that often a large parameter space has to be searched. Making a decision on the sensible number of hidden units is quite complicated and time consuming. Depending on the network type, various methods can be taken to select the best number of hidden units. To achieve this, similar artificial neural networks but with different numbers of hidden neurons are built and trained. Then the optimum number of hidden units is determined from the network which produced the least error in the validation subset (Hagan et al., 1996).

Typically, both the estimation and validation errors decrease during the initial phase of training. However, as mentioned before, by increasing the number of hidden neurons, due to overlearning consequently the error in the validation set increases. In all cases, training is terminated once either the number of iterations reaches its maximum limit or the computed response error is reduced to a value less than the desired value (Haykin, 1999). The training is terminated once the number of iterations reaches a specific number (i.e. 3000) or when the response error is reduced to a desired target value (i.e. 0.01). In all cases, the network weights and biases are selected as the final model parameters at the point where minimum error in the validation subset is achieved. Therefore, the number of hidden nodes is chosen based on when the least error has been achieved by the model with that number of nodes (see Bishop, 1995; Fausett, 1994; Hagan et al., 1996; Haykin, 1999).

Sometimes it is worth to summarize the network design once the modeling is completed. Thus, the final 2D or 3D networks can be piloted with their main features. A schematic view of the final network architecture can be always very helpful for comparison purposes.

7.2.8 Radial Basis Function Networks

Radial basis function (RBF) networks are a commonly used type of artificial neural network for function approximation problems. Radial basis function networks are distinguished from other neural networks due to their universal approximation and faster learning speed. An RBF network is a type of feed forward neural network composed of three layers, namely the input layer, the hidden layer and the output layer. Each of these layers has different tasks (Haykin, 1999; Yu and He, 2006; for further details see Chapter 4). A brief description of the RBF network is presented as follows. The training of the RBF model is terminated once the calculated error reached the desired values (i.e. 0.01) or number of training iterations (i.e. 500) already was completed. An RBF network with a specific number of nodes (i.e. 10) in its hidden layer is chosen. A Gaussian function is used as the transfer function in computational units. Depending on the case, it is typically observed that the RBF network required less time to reach the end of training compared to MLP. These chosen MLP and RBF networks are later examined in the next chapters under new test conditions. The agreement between the model predictions and the experimental observations will be investigated and the results of the two models will be compared. The final model is then chosen based on the least computed error.

7.2.9 Performance Analysis and Comparison

In the next step of the modeling we are concerned with performance of the 2D/3D ANN models. A comparison of the different models (MLP/RBF) should be conducted in this stage. Models include dimensional input parameters and nondimensional parameters, as mentioned previously (see section 7.2.2). The outputs of the ANN models can be, for example, 2D wave transmission coefficient ($K_{t,\mathrm{2D}}$), 3D wave transmission coefficient ($K_{t,\mathrm{3D}}$), etc.

2D Model

Accuracy Analysis

The proposed 2D ANN models are compared with the measured 2D laboratory data. The comparison of the proposed model with the experimental data is investigated by comparing five statistical parameters including the Root Mean Square Error (RMSE), the squared multiple correlation coefficient R^2, bias or distortion β, Wilmott index I_{w} (Wilmott, 1981) and error function ε (Haller et al., 2002):

$$RMSE = \left[\frac{\sum_{i=1}^{N} \left(\frac{X_{c_i} - X_{m_i}}{X_{m_i}} \right)^2}{N} \right]^{1/2} \tag{7.8}$$

$$R^2 = \frac{\left(\sum_{i=1}^{N}(X_{m_i}-\overline{X}_m)(X_{c_i}-\overline{X}_c)\right)^2}{\sum_{i=1}^{N}(X_{m_i}-\overline{X}_m)^2\sum_{i=1}^{N}(X_{c_i}-\overline{X}_c)^2} \tag{7.9}$$

$$\beta = \frac{\sum_{i=1}^{N}\left(\frac{X_{c_i}}{X_{m_i}}\right)}{N} \tag{7.10}$$

$$I_w = 1 - \frac{\sum_{i=1}^{N}(X_{c_i}-\overline{X}_{m_i})^2}{\sum_{i=1}^{N}(|X_{c_i}-\overline{X}_m|+|X_{m_i}-\overline{X}_m|)^2} \tag{7.11}$$

$$\varepsilon = \left[\frac{\sum_{i=1}^{N}(X_{c_i}-\overline{X}_{m_i})^2}{\sum_{i=1}^{N}X_{m_i}^2}\right]^{1/2} \tag{7.12}$$

where N represents the number of data in each group, X_c is the calculated value, X_m is the measured value and the barred values demonstrate the average values of the parameters. Perfect agreement is achieved if R^2 and Wilmott index are 1.0 and the error function is zero. The values of distortion closer to 1.0 and the lower values of RMSE also show better agreement of results. R^2 is considered unsatisfactory as a measure of the goodness of fit for a multivariate regression relationship (Draper, 1984) and thus the validity of the model will be assessed mostly by the other parameters.

The statistical and graphical comparison is made and the predicted values by the ANN model are compared with the measured values for all tests with different input parameters.

The obtained error by model for different nondimensional input variables and the distribution of computed error for the whole range of nondimensional input parameters are also investigated.

Sensitivity Analysis

All parameters relevant to the wave transmission behind 2D submerged breakwaters are included simultaneously in the 2D ANN models. In order to detect the importance of each parameter to the performance of the model, networks are designed and trained using the same training data sets but with each parameter omitted in turn, in order that the most influential parameters could be identified. Assessments are made on the performance of the networks by comparison with the network results using all the statistical parameters. The most effective parameters then can be recognized. It should be mentioned that the data split into training data, validation data and test data for the sensitivity analysis is carried out in the same way as the data split for the construction of the main ANN model.

A detailed sensitivity analysis can be also presented comprising different sensitivity graphs of individual nondimensional input parameters. The sensitivity figures contain several lines. The dashed lines show the band of the 95% confidence interval, one line for the lower boundary (quartile 2.5%) and the other for the upper boundary (quartile 97.5%). The solid line is the ANN prediction (mean value). These graphs will be illustrated and discussed further in Chapter 11 with some examples. The reliability of the models then can be evaluated inside and outside the training range interval for different parameters.

Comparison with Other Empirical Formulae

In order to investigate the capability of existing empirical formulae and to highlight the importance of a new design tool, the following analysis is presented using methods selected from the literature. As explained in Chapter 3, several empirical formulas are available in the literature. In order to perform a meaningful comparison, the ANN model and the selected formulas should be calibrated on the same range of data. In addition, some of these models have specific limits to use and to calculate K_t and consequently are less accurate. For example, van der Meer et al. (2005) used lower and upper limit fixed values for K_t which reduce the accuracy of predictions. However, the ANN model is not affected by this restriction and can provide accurate predictions where the classical approaches fail. The most recent 2D empirical formulae presented by Buccino and Calabrese (2007) and Goda and Ahrens (2008) are applied. These formulations will hereinafter be referred to as BC2007 and GA2008, respectively. In order to perform a meaningful comparison, the selected approaches are tested using experimental data within the same range within which the models were calibrated. We can also quantify the accuracy of these tools by means of five statistical parameters as mentioned before.

3D Model

Accuracy Analysis

The predictions of transmission coefficients obtained by models are also evaluated against the corresponding observed data in three-dimensional cases.

The comparison between the models against the experimental data is conducted by comparing the two statistical parameters, root mean squared error (RMSE) and the squared multiple correlation coefficient R^2. Both graphical and statistical comparisons should be considered. The error distribution using notched box-and-whisker plots (Tukey, 1977 and McGill et al., 1978) determines the degree of dispersion and skewness as well as identifying outliers for the computed parameters by models. Basically, the box plots are being used by researchers and scientists to display differences between groups of data nonparametrically without making any kind of particular assumptions as to the underlying statistical distribution using five distinct numbers. The bottom and top of the box are respectively the lower quartiles (25th percentiles) and the upper quartiles (75th percentiles). Additionally, the distance between the top and bottom of the box is the interquartile range and the band near the middle of the box is the 50th percentile known median as well. In the case the median is not actually centered in the box, this indicates that skewness exists in the sample (Tukey, 1977). The whiskers are also those lines extended above and below the box. Whiskers are typically seen in an area in the graph from the ends of the interquartile ranges to the furthermost recorded data within the whisker length. Those recorded data which are not included between the whiskers are typically plotted as outliers with a "+" marker. Therefore, an outlier is a value which is basically more than 1.5 times the interquartile range away from the top or bottom of the box (Tukey, 1977). In addition, notches can display the variability of the median between the samples. The width of a notch can be computed. It is basically based on the box plots whose notches do not overlap and have different medians at the 5% significance level. This assumed significance level is obtained according to the normal distribution (McGill et al., 1978).

Similarly, graphical and descriptive comparisons can be made between the nondimensional and dimensional models as well as statistical analysis. Based on all comparisons made, a model is selected as the best model to simulate physical parameters around submerged breakwaters. This will be validated against experimental data for three different applications as mentioned before. The performance of selected model is therefore assessed. This will be discussed further in Chapter 11.

Sensitivity Analysis

All parameters relevant to the wave field around a 3D submerged breakwater are included simultaneously in the ANN models described before. As mentioned before, in order to detect the importance of each parameter to the performance of the model, networks are designed and trained using the same training data sets but with each parameter omitted in turn, in order that the most influential parameters could be identified. Assessments are made on the performance of the networks by comparison with the network results using all the statistical parameters.

This investigation illustrates the most significant parameter in the 3D wave field around the breakwater, with wave height and energy varying spatially around the structure.

Further analysis of sensitivity of the 3D ANN model to different nondimensional parameters will be presented in the next chapters.

Comparison with Available Approach

In order to show the capability of the proposed ANN3D model, a comparison with an existing design tool is carried out. The method of Vicinanza et al. (2009) as explained in Chapter 3 is applied for comparison. The two-dimensional wave transmission coefficient ($K_{t,2\mathrm{D}}$) is calculated by the Goda and Ahrens (2008) method and the wave diffraction coefficient (K_d) is calculated by the Penny and Price (1952) method. Then, according to the computed values $K_{t,2\mathrm{D}}$ and K_d, the three-dimensional wave transmission coefficient ($K_{t,3\mathrm{D}}$) will be calculated. In order to perform a meaningful comparison, the ANN and the selected approach should be calibrated on the same range of data.

The K_d is calculated by Penny and Price (1952) based on penetration of waves through a gap smaller than one wave length. When gap width is small compared with wave length, an approximate solution may be obtained by assuming that the streamlines for the motion of the water through the gap are the same as for a simple uniform streaming of water through the gap (Penny and Price, 1952). The mathematical problem then becomes identical with the problem of the penetration of sound waves through a narrow slit; this problem is treated by Lamb (1924). Using Lamb's solution, the following expression for $F(x,y)$ can be deduced for points not too near the gap (Penny and Price, 1952) based on the distance from the center of the gap, the gap width and Euler's constant, equal to 0.5772. This corresponds to waves diverging uniformly in all directions behind the breakwater from the gap as a point source, the heights of the diffracted wave at any point to incident wave height being given by modF, which is proportional to $r^{-1/2}$. The K_d (modF) can be computed (Penny and Price, 1952). The coefficient of $\sqrt{(b/\pi r)}$ in this expression is slightly greater than unity when b lies between $0.1L_o$ and $0.2L_o$. Thus, when $b = L_o/2\pi$ the value of the coefficient is 1.053. Hence for this range of values of b, the K_d is given approximately by $\sqrt{(b/\pi r)}$ (Penny and Price, 1952).

As it will be seen later in Chapter 11, the predictions of transmission coefficients obtained by ANN3D ($K_{t,3D\text{-}ANN}$) and the model by Vicinanza et al. (2009) are plotted against the corresponding observed data. The comparison of the two models with the experimental data is also investigated by comparing the five statistical parameters.

Although disagreement between the results of the Goda and Ahrens (2008) method and two-dimensional measurements ($K_{t,2D\text{-}M}$) leads to less accurate results in calculation of ($K_{t,2D}$) and consequently ($K_{t,3D}$) using the Vicinanza et al. (2009) method, ($K_{t,3D\text{-}V}$), the significant error in computation of K_d by Penny and Price (1952) ($K_{d\text{-}PP}$) is the main source of error in calculation of $K_{t,3D}$ by the Vicinanza et al. (2009) method. The K_d computed by the Penny and Price (1952) method is not able to provide full information about variation of K_d behind the structure. The method does not include the effects of different physical parameters such as incident wave height, water depth and breakwater geometry. The K_d is calculated only based on the offshore wave length and the radial distance from the center of the gap using a very simplified method.

Statistical evaluation of the nondimensional RBF model in a larger scale model with a scale factor will be also carried out.

REFERENCES

Bellman, R., Rand Corporation, 1957. Dynamic Programming. Princeton University Press.

Bellman, R., 1961. Adaptive Control Processes. Princeton University Press, Princeton, NJ.

Bishop, C.M., 1995. Neural Networks for Pattern Recognition. Oxford University Press.

Buccino, M., Calabrese, M., 2007. Conceptual approach for prediction of wave transmission at low-crested breakwaters. J. Waterway, Port, Coast. Ocean Eng. 133, 213–224.

Draper, N., 1984. The Box-Wetz criterion versus R^2. J. Stat. Soc. 147, 100–103.

Fausett, L., 1994. Fundamentals of Neural Networks. Prentice Hall, New York.

Goda, Y., Ahrens, J., 2008. New formulation for wave transmission over and through low crested structures. Proceedings of the International Conference on Coastal Engineering. ASCE, pp. 3530–3541.
Hagan, M.T., Demuth, H.B., Beale, M.H., 1996. Neural Network Design. PWS Publishing, Boston, MA.
Haller, M.C., Darlymple, R.A., Svendsen, I.A., 2002. Experimental study of nearshore dynamics on a barred beach with rip channels. J. Geophys. Res. 107 (C6), (14-1)–(14-21).
Haykin, S., 1999. Neural Networks: A Comprehensive Foundation, second ed. Prentice-Hall, Englewood Cliffs, NJ.
Lamb, H., 1924. Hydrodynamics. Cambridge University Press.
Madsen, P.A., Sorensen, O.R., Schaffer, H.A., 1997. Surf zone dynamics simulated by a Boussinesq type model. Part 1: model description and cross-shore motion of regular waves. Coast. Eng. 32 (4), 255–287.
McGill, R., Tukey, J.W., Larsen, W.A., 1978. Variations of boxplots. Am. Stat. 32, 12–16.
Panizzo, A., Briganti, R., 2007. Analysis of wave transmission behind low-crested breakwaters using neural networks. Coast. Eng. 54, 643–656.
Penny, W., Price, A., 1952. The diffraction theory of sea waves and the shelter afforded by breakwaters. Philos. Trans. R. Soc. Lond. ÙA 244, 236–253.
Rumelhart, D., McClelland, J., 1986. Parallel Distributed Processing. MIT Press, Cambridge, Mass.
Schaffer, H., Madsen, P., Deigaard, R., 1993. A Boussinesq model for waves breaking in shallow water. Coast. Eng. 20, 185–202.
Sinha, N.K., Gupta, M.M., 1999. Soft Computing and Intelligent Systems: Theory and Applications. Academic Press, New York, A volume of 25 chapters, 614 pages.
Stein, R., 1993. Selecting data for neural networks. AI Expert 8 (2), 42–47.
Tukey, J.W., 1977. Box-and-Whisker Plots, Section 2C in Exploratory Data Analysis. Addison-Wesley, Reading, MA, pp. 39–43.
van der Meer, J.W., Briganti, R., Zanuttigh, B., Wang, B., 2005. Wave transmission and reflection at low-crested structures: design formulae, oblique wave attack and spectral change. Coast. Eng. 52, 915–929.
Vicinanza, D., Caceres, I., Buccino, M., Gironella, X., Calabrese, M., 2009. Wave disturbance behind low-crested structures: diffraction and overtopping effects. Coast. Eng. 56, 1173–1185.
Wilmott, C., 1981. On the validation of models. Phys. Geogr. 2, 184–194.
Yu, L., Wang, S., Lai, K.K., 2006. An integrated data preparation scheme for neural network data analysis. IEEE Trans. Knowl. Data Eng. 18 (2), 217–230.
Yu, B., He, X., 2006. Training radial basis function networks with differential evolution. In: Proceedings of IEEE International Conference on Granular Computing, Atlanta, GA, USA, 369–372.

CHAPTER 8

Design Model Development and Analysis

This chapter discusses numerical model development and analysis for predicting the spatial wave transmission coefficient in the shadow region of a 3D submerged breakwater. A model was developed using machine learning algorithms based on radial basis function (RBF) methods. The RBF model was designed, trained and validated using new laboratory experimental data expressed in terms of nondimensional parameters, as discussed in previous chapters. The model results are tested against a new set of measured data (not used before during the training process) with different model scales and conditions. These analyses are essential to ensure the capability of the proposed model to predict real applications at larger scale and a broad range of input parameters. In this chapter, a discussion on model development for numerical modeling based on the shallow water equations (SWE) is also presented. Later in Chapter 11, the shallow water equations are also applied for comparison with the results of the three-dimensional artificial neural networks model, with particular focus on the 3D wave field behind the submerged breakwater in both small and large scale models.

This chapter describes model development and analysis of results, essential parts of model construction. The process is outlined in detail in the following sections.

8.1 EXPERIMENTAL DATA

A comprehensive series of 2D and 3D laboratory experiments has been carried out in three wave tanks with different scales. The broad range of laboratory tests undertaken is to examine the wave pattern behind submerged breakwaters caused by 2D/3D wave transformation due to overtopping and diffraction. Experiments have been carried out with the objective of providing a more thorough understanding of the hydrodynamics behavior in the vicinity of 2D and particularly 3D submerged breakwaters to prepare the required data for artificial neural network

© 2016 Elsevier Ltd.
All rights reserved.

(ANN) modeling. The objective of this section is to describe in detail the experimental set-up, equipment and facilities, and the method of data sampling throughout the tests.

8.1.1 Two-Dimensional Experiments

Tests in 0.45 m Wide Flume

The tests were carried out using small-scale models in a wave-current flume 14 m by 0.45 m by 0.75 m at University College London (UCL). Measurements were made at different positions along the flume including in front of the breakwater and wave maker, over the side slopes, over the crest and several points behind the breakwater. Offshore water depth was maintained at 400 mm for all experiments.

Several breakwaters with different trapezoidal body sections were constructed and tested in the wave flume. The breakwaters were constructed with smooth material and without permeability. The breakwaters were constructed with horizontal top and side slopes of 1V:1H in both directions.

Wave height and water level changes along the flume were measured by 13 wave probes. The locations of the probes and their distances from each other in the flume depend on the breakwater crest width employed in the test.

A total of 12 test series were conducted in this wave flume with the same wave conditions but different breakwater geometries. A total of 180 tests were performed under different combinations of 15 monochromatic wave climates, including 5 wave heights and 3 wave periods with 2 different submergence depths; 15 tests were also carried out in the flume with the same water depth and almost similar wave conditions without a breakwater in the flume, to examine the generated wave heights and beach reflection.

Further information, including the experiment set-up, data sampling, test conditions, data sets, range of tested variations in water depth, wave period and wave length, target wave height, submergence depth, etc., is presented in Chapter 11.

Tests in 1.2 m Wide Flume

These tests were carried out in a wave flume 20 m by 1.2 m wide by 1.3 m high. Measurements were made in different locations around the breakwater including in front of the breakwater and wave maker, over the side slopes, over the crest and several points behind the breakwater. Offshore water depth was maintained at 300 mm for all experiments. A 2D breakwater with trapezoidal section was constructed and tested in the wave flume. The breakwater was constructed with smooth marine plywood material and with horizontal top and side slopes of 1V:1H in both directions. Wave height and water level changes along the flume were measured by 14 wave gauges.

A series of experiments was conducted in the wave flume under various wave conditions and submergence depths. A total of 176 tests were performed under different combinations of 44 monochromatic wave climates including 11 wave heights, 4 wave periods and 4 water depths; 44 tests were also carried out with the same water depth and similar wave conditions without the breakwater in the flume to examine the generated wave heights and beach reflection.

Further information, including the experiment set-up, data sampling, test conditions, data sets, range of tested variations in water depth, wave period and wave length, target wave height, submergence depth, etc., is presented in Chapter 11.

8.1.2 Three-Dimensional Experiments

Small-Scale Model

These tests were carried out in the 1.2 m wide flume and described in detail in the previous section. A small-scale breakwater roundhead and an adjacent trapezoidal trunk section were constructed and tested in the wave flume. The breakwater was constructed with smooth marine plywood material and with horizontal top and side slopes of 1V:1H in all directions.

A series of experiments was conducted in the wave flume under various wave conditions and submergence depths. A total of 176 tests were performed under different combinations of 44 monochromatic wave climates including 11 wave heights and 4 wave period conditions with 4 different water depths.

Further information, including the experiment set-up, data sampling, test conditions, data sets, range of tested variations in water depth, wave period and wave length, target wave height, submergence depth, etc., is presented in Chapter 11.

Large-Scale Model

These tests were carried out using a large-scale model in a 20 m by 2.5 m wide by 1.3 m high wave tank. Measurements were made in different locations around the breakwater. Offshore water depth was maintained at 1000 mm for all experiments.

A large-scale breakwater roundhead and the adjacent trapezoidal trunk section were constructed and tested in the wave flume. The breakwater was constructed with smooth marine plywood material and with horizontal top and side slopes of 1V:1H in all directions.

A broad range of 3D tests was conducted in this wave flume with various wave conditions. A total of 83 tests were performed under different combinations of monochromatic wave climates including 9 wave heights and 9 wave periods with a constant water depth of 1 m; 18 tests were also carried out with the same water depth and similar wave conditions without the breakwater in the flume to examine the generated wave heights and beach reflection.

Further information, including the experiment set-up, data sampling, test conditions, data sets, range of tested variations in water depth, wave period and wave length, target wave height, submergence depth, etc., is presented in Chapter 11.

The form of the propagated waves usually varies along the flume depending on the wave steepness and relative water depth. Free secondary harmonic waves exist when waves are generated by a sinusoidal moving wave paddle (Tajziehchi, 2006) (see Biesel and Suquet, 1951 for further information). In addition, higher harmonic waves generated by wave propagation over a beach slope, coastal structure or particularly a submerged breakwater also affect the general form of observed waves in the flume (Goda et al., 1999). Therefore, studying the form of wave propagation along the flume is required to obtain a reliable wave data analysis.

8.1.3 Experiments without Breakwater

Water surface elevation data collected by wave probes in the flume without a breakwater were examined to give a better understanding of the characteristics of the generated waves. In the case of monochromatic long waves, a free second harmonic wave always exists when h/L is less than 0.1 (Flick and Guza, 1980). These secondary waves travel with a smaller speed and amplitude than primary waves (Biesel and Suquet, 1951). Therefore, looking at the water surface profile, the existence of the secondary waves may be identified. The existence of the secondary waves may be the cause of a number of problems in experimental laboratory tests for relatively long waves (Tajziehchi, 2006). Examining wave profiles and water levels at different stations along the wave flume without breakwater, depending on the distance from the wave maker, water surface levels may vary. For waves of small amplitude, the secondary waves are relatively small and may be negligible in most cases (Tajziehchi, 2006).

In this analysis, monitoring of water surface elevation and its variation in the wave flumes without breakwater were carried out by a large number of fixed wave gauges (wave probes). Analysis of water surface elevation and its variation along the wave flumes were also carried out. Spectral analysis of the recorded signals from the wave probes based on the fast Fourier transform (FFT) was conducted for a selected test condition in the flume without breakwater. In addition, the shape of the wave amplitude spectrum and results from wave probes located at different positions along the flume can be examined. It may reveal the existence of higher harmonic waves at some specific points; for instance, secondary harmonics are clearly most evident for probes close to the beach.

Regarding the reflected waves in the flumes, as mentioned before, the wave energy absorber systems were employed in the flumes to minimize wave reflection. As discussed, a number of experiments were carried out in the flumes without breakwater but with the same water depth and almost similar wave conditions to examine the generated wave heights and beach reflection. When a primary incident wave travels toward the beach, a primary reflected wave travels from the beach toward the wave maker. Then a secondary incident wave travels from the wave maker toward the shore. This process is repeated till the steady state is attained after a few reflections (Ursell et al., 1960). In the presence of a highly dissipative beach or wave energy absorber system, the secondary reflected wave is usually negligible (Tajziehchi, 2006). The calculated reflection coefficient for the tests without breakwater in the present experimental study revealed that the reflection coefficient of the flume varies between 0.03 and 0.11.

8.1.4 Experiments with Breakwater

Several experimental and numerical studies have been conducted to investigate higher harmonics generated over the submerged breakwaters (Goda et al., 1999; Longuet-Higgins, 1977; Losada et al., 1997; Massel, 1983; Tajziehchi, 2006). When waves approach a submerged obstacle, part of the wave energy is reflected by the barrier, while above the obstacle higher harmonics are generated and transmitted to the deeper water behind the barrier as free waves (Tajziehchi, 2006). The amplitude spectrum at different measurement points around the head and behind the breakwater for selected tests corresponding to tests with or without breaking wave is investigated and analyzed in detail. The form of wave spectra offshore of the submerged breakwater should be compared with those in the flume without the breakwater. The higher harmonics observed seaward of the submerged breakwater should be carefully analyzed if any.

The method of Mansard and Funke (1980) was implemented to decompose the incident and reflected wave heights using simultaneous measurements from three fixed wave probes. A MATLAB (Matlab, R2010b) script was written to read wave data and compute the incident and reflected wave heights using the Mansard and Funke (1980) method. The reflection coefficient K_r is not of primary concern in this analysis.

8.2 ANALYSIS APPROACHES FOR ANN MODEL RESULTS

Processed wave data is analyzed in detail and discussed in this section. Waves transmitted onshore of the submerged breakwater are a combination of wave overtopping, which includes higher harmonic free waves due to wave breaking over the breakwater crest and wave diffraction (Tajziehchi, 2006). Hence, transmitted wave heights vary spatially and depend on the number of significant harmonics generated and transmitted over the breakwater as well as wave diffraction patterns around the breakwater head. This phenomenon was clearly observed throughout the laboratory experimental tests and is discussed in detail in this section. Wave height transmission patterns behind 2D and 3D submerged breakwaters are analyzed and the effectiveness of several parameters (dimensional and dimensionless) are examined.

8.2.1 The Root-Mean-Square Transmitted Wave Height ($H_{t,rms}$)

In order to obtain a better understanding of the wave pattern behind the breakwater, a large number of measurement points was considered to give a better interpolation. $H_{t,rms}$ for each wave probe behind the structure was calculated with Equation 8.1 and was adopted as being representative of transmitted wave height at that specific point:

$$H_{t,rms} = \sqrt{\frac{\sum_{i=1}^{N} H_{t,n}^2}{N}} \tag{8.1}$$

with $H_{t,n}$ denoting the individual transmitted wave heights in a certain time series with a total number of N waves.

As the total wave spectrum was taken into account, the computed transmitted wave height ($H_{t,rms}$) is an average and includes the fundamental and higher harmonic waves.

8.2.2 Wave Transmission Coefficient (K_t)

The wave transmission coefficient K_t is calculated from the average transmitted wave height onshore of the breakwater of each measurement point and the incident wave height calculated in the previous sections using wave probes located offshore of the breakwater. As mentioned before, K_t is defined as the ratio of average transmitted wave height to the incident wave height (H_i):

$$K_t = \frac{H_{t,rms}}{H_i} \tag{8.2}$$

where $H_{t,\mathrm{rms}}$ is calculated using Equation 8.1.

In the case of two-dimensional experiments, transmitted wave height H_t is calculated based on the average of root mean square wave height ($\hat{H}_{t,rms}$) from the wave probes behind the structure, calculated with the following method:

$$\hat{H}_{t,rms} = \frac{\sum_{i=1}^{n} H_{t,rms}}{n} \tag{8.3}$$

with $\hat{H}_{t,rms}$ denoting the average of root mean square wave height. $H_{t,rms}$ is the standard deviation of the water surface level for the *i*th wave probe behind the breakwater, n is the number of wave probes behind the submerged breakwater. $\hat{H}_{t,rms}$ is obtained from a number of stations behind the breakwater and this average value was adopted as being representative of transmitted wave height.

Analysis of effective parameters (dimensional and nondimensional) from K_t are examined in the following sections. The purpose of this analysis is to define the dimensional and nondimensional parameters which can most effectively predict the wave transmission coefficient. Dimensional and nondimensional parameters will also be used in the numerical model for calculating K_t.

8.2.3 Dimensional Analysis

The effect of different dimensional parameters including wave characteristics, water depth and breakwater geometry on the 2D wave transmission coefficient has been thoroughly studied (Ahrens, 1987; Seabrook, 1997; Seelig, 1980; Tajziehchi, 2006; van der Meer, 1991).

In the present analysis graphical analysis was used to identify trends in the graphs to determine the significance of each parameter. In addition, statistical analysis of variables was performed for more investigation of effectiveness of each dimensional parameter on transmission coefficient. The variation of the correlation coefficient between K_t and other dimensional parameters was used as an indicator. Five main dimensional variables were selected as the most effective parameters in 2D analysis:

$$K_{t,2D} = f(H_i, L_o, h_s, h, B) \tag{8.4}$$

where H_i is incident wave height, L_o is offshore wave length, B is breakwater crest width, h_s and h are respectively water depth over the crest and at the toe of the breakwater.

The 2D dimensional analysis of the collected laboratory data on these parameters demonstrated that K_t decreases with incident wave height and breakwater crest width while offshore wave length and submergence depth have an inverse effect on K_t which was consistent with the previous studies and confirmed past research.

The wave energy dissipation coefficient is derived easily from K_t and can be used as an indicator to describe the phenomena of wave energy dissipation over the barriers. K_{ed} is defined as follows:

$$K_{ed} = \sqrt{1 - K_t^2 - K_r^2} \tag{8.5}$$

where K_r is the wave reflection coefficient calculated as follows:

$$K_r = \frac{H_r}{H_i} \tag{8.6}$$

where H_r and H_i respectively are reflected wave height and incident wave height, which were computed in the earlier sections using the decomposition method of Mansard and Funke (1980).

Typically, a scatter plot can be used in order to show the magnitude of the wave energy dissipation based on the variation of incident wave height for submergence depths. Basically, increasing the incident wave height leads to higher wave energy dissipation. The higher wave energy dissipation is often observed for lower submergence depths while the incident wave height is constant. The significance of the two parameters H_i and h_s on K_{ed} and consequently on K_t is considerable. Similar analyses can be conducted for other dimensional parameters and interesting trends were observed. The outcomes will be discussed further in Chapter 11.

However, in the case of three-dimensional analysis, two additional dimensional parameters x and y (corresponding to the Cartesian coordinate system) are also considered. Therefore, an improved form of dependency of the wave transmission coefficient on dimensional variables is defined and applied in this study:

$$K_{t,3D} = f\{H_i, L_o, h_s, h, \mathrm{B}, \mathrm{x}, \mathrm{y}\} \tag{8.7}$$

The set of variables presented in Equations (8.4) and (8.7) will be analyzed further in Chapter 11.

8.2.4 Nondimensional Analysis

The results of dimensional analysis are of great worth for the geometric design of submerged breakwaters. The guidance provided from these analyses is recommended for engineering design when wave climates are available. Nevertheless, nondimensional parameters can provide a more productive correlation of the dependent parameters (Tajziehchi, 2006).

In order to represent the combined effect of the wave steepness ($s_o = H_i/L_o$) and the seaward slope of the breakwater α, the surf similarity parameter or Iribarren number ξ_o is introduced as follows:

$$\xi_o = \frac{\tan(\alpha)}{\sqrt{\frac{H_i}{L_o}}} \tag{8.8}$$

Therefore, on the basis of the 2D dimensional analysis, the following 2D dimensionless parameters were considered for further analysis:

$$K_{t,2D} = f(\xi_o, \mathrm{H_i}/h, \mathrm{h_s}/H_i, B/H_i) \tag{8.9}$$

It should be noted that wave breaking is the primary cause for wave energy dissipation over 2D submerged breakwaters. Lower wave transmission coefficients are observed for higher wave steepness and lower submergence ratio and inverse water depth ratio. All these three nondimensional parameters are related to the breaking process over the breakwater crest. Wave steepness higher than a certain value and water depth lower than a specific ratio to wave height lead to wave breaking. The width of the breakwater crest also affects the breaking process by providing a longer shallow water area and consequently enough time to develop the breaking. Friction loss can also logically be a secondary factor in 2D wave energy dissipation over these sorts of breakwaters. However, as breakwaters with a smooth surface were employed in the tests, it is assumed that the friction loss factor is negligible. Based on the graphical interpretation, the effect of the dimensionless parameters on the wave transmission coefficient can be summarized.

Considering the three-dimensional analysis, it is assumed the 3D wave transmission coefficient can be defined as a function of the following nondimensional parameters:

$$K_{t,3D} = f\{\xi_o, H_i/h, h_s/H_i, B/H_i, x/L_o, y/L_o\} \tag{8.10}$$

Regarding the 3D effects of the nondimensional parameters, a particular analysis has been made on the impact of these variables on the wave transmission coefficient. Graphical analysis for nondimensional parameters can illustrate that variation of K_t with incident wave steepness follows the same general trend as the 2D case, while that with the submergence ratio has a slightly different pattern comparing with the 2D case. Although the general trend is similar, large values of K_t are observed for low submergence ratio values. The reason is the 3D effect of wave transformation and particularly diffraction on the wave pattern behind the structure, so that some specific cases result in higher transmitted wave height even in low submergence depths. With low submergence depths, the wave energy is significantly dissipated over the breakwater crest (2D transmission), while incident waves pass through the gap without considerable changes. Therefore, the difference between the wave height passing over the crest and the wave height passing through the gap results in diffraction. This process completely disturbs the wave pattern behind the structure, compared with the 2D case.

These 3D analyses confirm the significance and importance of 3D studies of submerged breakwaters. To conclude, the surf similarity parameter ξ_o, the submergence ratio h_s/H_i, the breakwater crest width ratio B/H_i and relative wave height H_i/h are in order the parameters most affecting the wave transmission coefficient in 2D, while x/L_o, y/L_o, the surf similarity parameter ξ_o, the submergence ratio h_s/H_i and the relative wave height H_i/h are in order the parameters most affecting transmission coefficient in 3D cases.

8.2.5 Accuracy Analysis

The comparison of the ANN model with the experimental data is investigated by comparing the five statistical parameters: the root mean squared error (RMSE), the squared multiple correlation coefficient R^2, bias or distortion *(β)*, Wilmott index I_w and error function ε as defined previously. The values of the five statistical parameters for the ANN3D model and a model based on the shallow water equations are provided and graphically compared. The proposed ANN model is verified against new experimental data measured in the wave tanks and not used during the training process. Accuracy analysis is performed under three distinct applications of the model, namely, interpolation, extrapolation, and larger scale predictions.

Interpolation

A statistical comparison can be made of wave transmission coefficients around the submerged breakwater for cases involving interpolation. Conditions are based on the dimensional form of training data set and nondimensional training data set. The correlation between predicted and measured values within different regions around the breakwater for a case with input parameters within the range of the training data set is very useful.

The values of the statistical indicators and the graphical comparison both indicate that the ANN model is able to calculate wave transmission coefficient with a high accuracy. The predicted values which give high R^2 and Wilmott number (Wilmott, 1981) I_w and low values of ε and RMSE, as well as a value of distortion close to 1.0, indicate very good accuracy. Essentially, one plot gives more information than many tables and, when presented in color, the message is easily understood. A good way of presenting the model results is using contour plots of, for example, the calculated wave disturbance coefficient or absolute error. These plots can indicate the error in different regions around the submerged breakwater. Models with low error, mostly less than 10% in most regions around the breakwater, might be acceptable. Higher values of error, up to 15% in some small regions, may be still reasonable.

These evaluations reveal the high degree of consistency between the results of the numerical model and the measurements in the interpolation cases. It will be evident whether the trained ANN model has excellent potential to predict the wave transmission coefficient in cases involving interpolation, or not.

Additional analysis can also be carried out to examine the ability of the proposed model to predict K_t in various regions behind the breakwater, as spatial prediction of K_t is the primary interest of study on wave fields around the submerged breakwaters. Results of this analysis for different points behind the structure can be studied. The variation of K_t for different submergence ratios at one specific point is analyzed. The general trend of K_t computed by the ANN model in each graph should follow the pattern of measured values reasonably. Basically, K_t increases with submergence ratio for points behind the breakwater (combined 2D transmitted and diffracted waves) and decreases at points right behind the breakwater trunk (2D transmitted waves only) due to wave dissipation by breaking over the crest. It is possible to deduce from the plots that the model produces reasonable responses under different conditions and locations behind the breakwater, particularly at points located in the shadow region of the structure, or not. The results with errors up to a highest value of 10% may be acceptable.

Extrapolation

Measurement data which was never used during the training process and not within the training data ranges is employed to test the numerical model in cases involving extrapolation. The parametric range of data in dimensional and nondimensional form of the extrapolation test data set should be considered. The correlation between predicted and measured values within different regions around the breakwater for a case with input parameters is investigated. It can be done, for example, for the case of wave period and water depth over and in front of the structure within the range of the training data but for incident wave heights outside the range of the training data.

The values of the statistical indicators and graphical comparison should indicate that the ANN3D model is able to predict wave transmission coefficients with high accuracy. The values calculated by the numerical model should provide acceptable values of R^2 and Wilmott number I_w and low value of ε and RMSE as well as a distortion value close to 1. The graph also should show good agreement in different regions around the structure.

An error contour map for the case calculated by the model can be also very useful. It shows the error in different regions around the submerged breakwater. The graph should show good agreement in different regions behind the structure including the shadow zone, behind the breakwater trunk and around the breakwater roundhead with low values of the absolute percentage error. Low errors mostly less than 10% and up to 20% in most regions can indicate relatively good agreement. Less accurate results (errors typically higher than 15%) if found in some small regions around or behind the submerged breakwater might be still reasonable with a maximum error about 20%. However, depending on the case and particular situations of the problem if any, higher values of error if observed, can clearly indicate low accuracy and reliability of the model under such conditions.

These evaluations can be very helpful to reveal consistency between the results of measured data and the numerical model in the extrapolation cases. It will be useful evidence to measure the potential of the trained model to predict the wave transmission coefficients around and behind submerged breakwaters in cases involving extrapolation.

Additional analysis can be carried out to examine the ability of the model in prediction in cases of extrapolation to various regions behind the breakwater. For example, graphs can illustrate the variation of K_t with wave steepness for one specific point. The general trend of K_t computed by the model should follow the experimental pattern reasonably well. For instance, K_t decreases with wave steepness for points behind the breakwater and points right behind the breakwater trunk (2D transmitted waves only). It can also be deduced from the plots whether the model produces reasonable responses under different conditions and locations behind the breakwater, or not. Therefore these graphs should include the extrapolation cases where the conditions are not within the range of the training data in the dimensional and nondimensional form of extrapolation test data set. The graphs may show disagreements between predicted and measured values for particular conditions as the error increases for the extrapolation cases. However, even in the extrapolation case the predictions may be still acceptable (for example, if $e < 15\%$).

Larger Scale

In the cases where measured data from small-scale experiments are used, a set of experimental data from the larger scale tests with different laboratory facilities is required to assess the performance of the numerical model. The statistical comparison can be used to investigate the correlation between predicted and measured values. They can indicate whether the results are in good agreement with measured values for different wave climates. The values of the statistical indicators and the graphical comparison can indicate whether the model is able to calculate physical parameters with reasonable and convincing accuracy, or not. The values predicted by the model should provide fairly good R^2 and Wilmott number I_u, a low value of ε and RMSE. A value of distortion close to 1.0 also points to reasonable accuracy of the predictions.

8.2.6 Sensitivity Analysis

As described before, several parameters relevant to the physical process of interest around submerged breakwaters may be included in the numerical model. In order to detect the importance of each parameter to the performance of the model, networks should be designed and trained using the same training data sets but with each parameter omitted in turn, in order to identify the most influential parameters. Assessments are made on the performance of the networks by comparison with the network results using some statistical parameters. Therefore, this investigation can illustrate the most significant parameter in that particular process with the lowest R^2 correlation coefficients and the highest error (RMSE). Bias values (β) may also be considered in the analysis. It should be noted that the division into training data, validation data and test data for the sensitivity analysis should be carried out in the same way as for the main model.

The detailed sensitivity analysis can be also done graphically by compiling different sensitivity graphs of individual dimensional and nondimensional input parameters. As described before, the sensitivity figures typically contain several lines. The dashed lines show the band of the 95% confidence interval, one line for the lower boundary (quartile 2.5%) and the other for the upper boundary (quartile 97.5%). The solid line is the model prediction (mean value). Therefore, plots reveal the range where the model gives a relatively high reliability. Outside this interval the model tends to low reliability.

8.3 DEVELOPMENT OF SHALLOW WATER EQUATIONS MODEL

Wave propagation around a 3D submerged breakwater is investigated numerically using shallow water equations (SWEs). A general description of the model is briefly presented in the following section. Then simulations of propagation of regular waves in two numerical wave flumes using the model are presented. The simulation process consists of setting up the model, calibrating the model, verifying the model results, and simulations. Each of these tasks is described in detail in the following sections.

The simulation results, comparisons and further details of analysis will be presented in Chapter 11. The mathematical theories behind this module were presented in Chapter 5. For further details, the reader is referred to the references related to the shallow water equations.

8.3.1 Description (Shallow Water Equations)

The shallow water equations are a set of partial differential equations which are derived from depth-integrating the Navier-Stokes equations which are themselves derived from the conservation equations. It is assumed that the horizontal space scale is much greater than the vertical space scale, which is actually a common condition in the coastal area. Therefore these equations are widely applicable for studies in this area. Calculation of wave transformation from offshore to the beach for studying surf zone and swash zone dynamics and calculation of wave disturbance in the presence of coastal structures can be done based on these models. The shallow water equations are employed here to simulate wave fields behind submerged breakwaters. We can conclude from the principle of conservation of mass that the vertical velocity is small. According to the momentum equation we can realize that vertical pressure gradients are virtually hydrostatic. The horizontal velocity field becomes constant vertically. Thus, we can integrate the Navier-Stokes equations throughout the depth and remove the vertical velocity term from the equations. Therefore, a simplified form of equations which are known as shallow water equations (SWE) can be derived from the original complicated equations for the problems where the horizontal dimension is extremely larger than the vertical scale (for further details see Chapter 5). The numerical solution of equations can be simply obtained by a finite difference technique with variables defined on a spatial rectangular grid.

8.3.2 Preparation (Set Up)

To set up a model, a number of tasks needs to be completed in sequence. First of all, a bathymetric map of the area of interest or a 2D matrix of water depths is needed. A 2D grid containing all the information of computational mesh used for the area of interest is therefore used. Some important points which are required to be specified are number of cells and grid spacing, time step, location and type of boundaries, etc. It should be noted that very small mesh spacing may result in a significant increase in CPU usage and processing time, while very large grids may result in outcomes with low accuracy. The time step in simulations also needs to be selected carefully. The choice of the grid spacing and time step depends on the wave conditions and the water depth in the area of interest. Therefore, it can be necessary to modify the grid spacing and time step to reach an acceptable compromise between time/cost of computation and accuracy of results. Location and type of boundaries in the model are also needed to be defined. Any other parameters and properties applied in the mode are also required to be specified in this stage.

8.3.3 Calibration (Verifying the Model)

The following data are required for calibrating and verifying the model: bathymetric data, boundary data, structural features and other model parameters. The model specifications are modified in order to reduce the differences between the simulation results and some specific measurements. The calibration data and validation data are actually the measured wave parameters at selected locations in the model area.

8.3.4 Simulation (Result)

Results of simulations can be represented in various forms as can be seen later in Chapter 11 for some examples. The main forms of presenting the simulation results include the deterministic parameter of the water surface level, the phase-averaged value of the significant wave height or the wave disturbance coefficient. The results can be represented as point series, line series or grid series.

8.4 COMPARISON OF MODELS

The SWE model and ANN model results, with particular focus on the 3D wave field behind the submerged breakwater in both small- and large-scale models can be compared based on the parameters mentioned previously. Further details about these comparisons are provided in Chapter 11 including information on the capability of two numerical models (ANN3D vs. SWE) to simulate the wave field around submerged breakwaters.

REFERENCES

Ahrens, J., 1987. Characteristics of reef breakwaters. Technical Report CERC-87-17, 45.

Biesel, F., Suquet, F., 1951. Les appareils generateurs de houle en laboratoire. La Houille Blanche 6 (2), 147–165.

Flick, R., Guza, R., 1980. Paddle generated waves in laboratory channels. J. Waterway, Port, Coast. Ocean Div., Amer. Soc. Civil Eng. 106 (WN1), 79–97.

Goda, Y., Okazaki, K., Kagawa, M., 1999. Generation and evolution of harmonic wave components by abrupt depth changes. Proc. Coastal Structure '99. ASCE.

Longuet-Higgins, M., 1977. The mean forces exerted by waves on floating or submerged bodies with applications to sand bars and wave power machines. Proc. The Royal Society of London 352, 463–480.

Losada, I., Patterson, M., Losada, M., 1997. Harmonic generation past a submerged porous step. Coast. Eng. 31 (1–4), 281–304.

Mansard, E., Funke, E., 1980. The measurement of incident and reflected spectra using a least squares method. Int. Conf. on Coastal Engineering (ICCE) Hamburg. ASCE.

Massel, S., 1983. Harmonic generation by waves propagating over a submerged step. J. Coast. Eng. 7, 357–380.

Matlab, R2010b. The MathWorks Inc.

Seabrook, S., 1997. Investigation of the performance of submerged rubblemound breakwaters. Msc. thesis. Queen's University, Kingston, Ontario, Canada, Ph.D. thesis.

Seelig, W., 1980. Two-dimensional tests of wave transmission and reflection characteristics of laboratory breakwaters. Tech. rep., Technical report, CERC, Fort Belvoir, Report No. 80-1.

Tajziehchi, M., 2006. Experimental and numerical modelling of wave-induced current and wave transformation in presence of submerged breakwaters. Ph.D. thesis. Civil & Environmental Engineering, Faculty of Engineering, UNSW.

Ursell, F., Dean, R., Yu, Y., 1960. Forced small-amplitude water waves: a comparison of theory and experiment. J. Fluid Mech. 7 (Part 3), 33–52.

van der Meer, J., 1991. Stability and transmission at low-crested structures. Tech. rep., Technical report, Delft Hydraulic, Report No. H 453. 33p.

Wilmott, C., 1981. On the validation of models. Phys. Geogr. 2, 184–194.

CHAPTER 9

Existing Simulation and Design Tools

9.1 NUMERICAL MODELS

In the analysis of hydrodynamic and morphodynamic processes as well as water quality changes in the presence of submerged breakwaters, numerical simulation is often employed to model the physical phenomena. "Numerical model" basically refers to the application of computer simulation either itself alone or integrated with physical models. In this chapter a general outline is briefly given with specific insight into the main characteristics, concepts and methodologies of the numerical models being used for the submerged breakwater analysis and design.

The models generally are applied to simulate the interaction between currents, tides and tidal currents, internal and surface waves, storm surges, tsunamis and others and the submerged breakwaters. They are basically employed to evaluate the hydrodynamic effects of the submerged breakwaters and their morphodynamic impacts on the sea bottom and the coastline. These effects are, for instance, sediment transport and corresponding changes in morphology, currents, waves, turbulence, etc.

Due to the rapid development of computing technology over recent years, a variety of computational models have been proposed for coastal hydrodynamic and morphodynamic problems. The number of numerical models and techniques applicable in coastal engineering problems, and therefore for submerged breakwaters as one of the typical coastal structures, is increasing rapidly. As discussed before, the numerical techniques are either phase-averaged or phase-resolving and are usually based on the finite difference method, finite element method, finite volume method or boundary element method with implicit or explicit time-stepping algorithms. Most of the nearshore models were typically developed based on the finite difference methods only for monochromatic wave trains rather than for directional spectra (Dean and Dalrymple, 2001). Models may be restricted to a one-dimensional spatial domains model, as a very simplified tool, or may be extended to two-dimensional models or even to the full three-dimensional models, which are the most appropriate models.

© 2016 Elsevier Ltd.
All rights reserved.

The accuracy of the results, however, to a great extent depends on the model parameters and specifications, limitation of the methods and the numerical schemes, and simplifications.

Some examples are the Reynolds Averaged Navier Stokes (RANS), the mild-slope equations, the Boussinesq equations, and spectral models. However, due to limitations of these approaches, the predictions obtained by the proposed numerical models in order to represent a physical problem need to be used carefully. For example, the accuracy of these models is basically related to the quality of the bathymetry data, model set-up and data used for calibration.

Waves generated in deep waters when approaching shallow waters are subjected to many complicated physical processes such as varying bathymetries, wave energy dissipation due to bathymetric wave breaking or bottom friction, nonlinear wave-wave interactions, wind wave generation and white-capping, shoaling, refraction, etc. Therefore, it can be seen that we have to deal with a broad spectrum of physical processes in the coastal area. An ideal model should be able to model these wide range of precesses as many as possible and as accurate as possible to provide reliable results. However, this could be extremely difficult to have a full 3D numerical model with a complete set of features. There have been great efforts over the years by many researchers in the field of coastal engineering to provide appropriate and efficient numerical models close to an ideal model. We can find many examples by reading the literature. For example, MIKE 21, Delft3D, and TELEMAC have been developed for modeling coastal hydrodynamics and coastal processes in general. Following are some examples of such models and a list of numerical models developed by coastal researchers and engineers for modeling coastal hydrodynamics and morphodynamics with possibility to be employed for the submerged breakwaters as well, which is the topic of interest in this book.

9.2 SPHysics

SPHysics is a smoothed particle hydrodynamics (SPH) model which has been developed jointly by a group of scientists and researchers from the Johns Hopkins University (U.S.A.), the University of Vigo (Spain), the University of Manchester (U.K.) and the University of Rome La Sapienza (Italy) (Gómez-Gesteira et al., 2012a,b). SPHysics is basically used to model free surface flows. It can be applied to model water surface variations, wave breaking, interaction of waves and structures, etc. (see Gómez-Gesteira et al., 2012a, b).

9.3 TELEMAC-MASCARET

9.3.1 TELEMAC

TELEMAC is an open source computer program owned by the Laboratoire National d'Hydraulique et Environnement (LNHE). LNHE itself is a part of the research and development group of Électricité de France (EDF). TELEMAC has been developed based on the finite element method. A brief description of the model is presented here.

The two-dimensional (2D) hydrodynamics module of this model (TELEMAC-2D) solves the shallow water equations (also known as the Saint Venant equations) (TELEMAC-2D, 2014). TELEMAC-2D solves the equations using the finite-element or finite-volume methods. The triangular elements are applied when the computational mesh is generated for the model. Cartesian or spherical coordinates for large domains can be applied. TELEMAC-2D takes into account the propagation of long waves where wavelengths are required to be at least 20 times the water depth. The influence of nonlinearity, turbulence, bed friction, wind and river flows is considered (TELEMAC-2D, 2014). The effect of horizontal temperature and salinity gradients on density is also considered. Sills, dikes, pipes and other natural or manmade structures and objects can also be modeled. Dry areas in the computational domain can also be considered in TELEMAC-2D. TELEMAC-2D is used in many fields of application in the marine and coastal engineering field to study the impact of various types of structures such as sills, groins, breakwaters and particularly submerged breakwaters (see TELEMAC-2D, 2014).

A 3D hydrodynamics module of TELEMAC (known as TELEMAC-3D) is also available. The 3D module similarly uses the horizontally unstructured mesh. The 3D module solves the Navier-Stokes equations. Three dimensional meshes are used (TELEMAC-3D, 2013).

9.3.2 ARTEMIS

ARTEMIS is used for modeling wave propagation in relatively small computational domains considering the dissipation processes, frequency dependence and directional spreading. ARTEMIS solves the extended mild slope equation using the finite element method have been integrated. ARTEMIS models reflection, refraction, diffraction, bottom friction, depth-induced breaking, radiation, etc. (ARTEMIS, 2014). ARTEMIS has been used in numerous applications in coastal areas, in the vicinity of different marine and coastal structures. It has been shown that the model presents reliable outcomes. The model computes the main wave characteristics such as significant wave height, orbital velocities, breaking, etc. over the problem domain. The model has benefited greatly from adapted pre- and postprocessing tools for mesh generation and results visualization. The model is applicable for different practical purposes in the coastal and marine engineering field, including structure or harbor design, coastal management, wave breaking, wave-induced current, sediment transport, etc. (See ARTEMIS, 2014).

9.3.3 TOMAWAC

TOMAWAC is a wave model that can be used in coastal areas and for design of marine and coastal structures, calculation of currents and sediment transport, etc. It solves a simplified equation of spectral action balance using a finite-element type method (TOMAWAC, 2011). TOMAWAC can take into account wind-induced waves, different kinds of wave refraction, and dissipation through different causes of wave breaking. TOMAWAC calculates significant wave height, mean and peak wave frequencies and direction, radiation stresses, wave-generated currents, etc. (TOMAWAC, 2011). It can be linked to the other hydrodynamic or sediment transport modules such as TELEMAC-2D, SISYPHE, TELEMAC-3D, etc. It has been successfully used in numerous studies and practical applications.

9.3.4 SISYPHE

SISYPHE can be used to model morphodynamics, sediment transport processes and bed evolution in coastal areas. SISYPHE is applicable to both noncohesive and cohesive sediments, uniform and nonuniform sands as well as sand-mud mixtures. The effect of bed slope and roughness can be considered. SISYPHE can be coupled to TELEMAC-2D or to TELEMAC-3D for currents only and to TOMAWAC for waves or combined waves and currents (Sisyphe, 2014). SISYPHE has been validated for a wide range of applications.

9.4 REF-DIF

REF-DIF is a phase-resolving parabolic model for ocean surface wave propagation developed by the Center for Applied Coastal Research of the University of Delaware. It was developed by Jim Kirby and Tony Dalrymple and was originally a monochromatic wave model known as REF-DIF 1 with some assumptions such as mild bottom slope, weak nonlinearity and limited wave direction (Kirby, 1986; REF/DIF, 1994). REF-DIF stands for the refraction and diffraction processes. Therefore, the model can provide good information about the transformation of the waves when approaching the shore, even over complex bathymetries. However, reflection has not been included in the model. This might therefore be a weakness of the model, particularly when dealing with structures with steep slopes such as submerged breakwaters (REF/DIF, 1994). REF-DIF provides wave height and direction, wave celerity, group speed, orbital velocity, wave dissipation rate, wave breaking index, wave mass flux, etc.

9.5 COBRAS

COBRAS (Cornell Breaking Waves and Structure) is also a two-dimensional numerical model. COBRAS can solve the Reynolds Averaged Navier-Stokes Equations (RANS) with a modified k-ε turbulence model based on the nonlinear eddy viscosity. In COBRAS, a volume of fluid (VOF) method is applied to monitor free surface level (Chang et al., 2001; Lin and Liu, 1998). The model has been verified for different applications and shown to be reliable in the modeling of wave breaking and wave-structure interactions (see Chang et al., 2001; Lin and Liu, 1998).

9.6 MIKE 21

MIKE 21 package is a two dimensional numerical model which has been developed by Danish Hydraulic Institute (DHI). The package consists of integrated modules. It can be applied for a wide range of practical applications in coastal and marine environments. It is particularly suitable for coastal engineers to simulate water waves, currents, sediment transport and morphological changes, etc in coastal areas and seas. The model is applicable for environmental impact assessment of coastal and marine structures, ports, renewable energy systems and marine infrastructures.

It can be also used for wave forecast or to measure degree of coastal flooding and shore protection, etc. MIKE 21 basically solves the full time-dependent nonlinear equations of continuity and conservation of momentum numerically (MIKE 21-BW, 2011).

Wave field around various coastal or marine structures and breakwaters can be simulated numerically using MIKE 21. It therefore can also be used to analyse influence of the submerged breakwaters on wave and current field as well as morphological changes and beach response. For instance, the Boussinesq Wave (BW) module (see MIKE 21-BW, 2011) can be applied in submerged breakwater analysis and design. A description of the model is briefly presented in this section.

The simulation process in model MIKE21 BW consists of a few main tasks. Firstly, setting up the model should be performed. Bathymetry maps need to be imported into the model and basic parameters should be set. Then, other required data should be used for calibrating and verifying the model. Parameters of the model are determined. Finally, simulations can be carried out and results are presented. Each of these tasks is described in detail in this section. The general descriptions of the MIKE 21 BW model presented in these sections are based closely on the MIKE 21 BW manual (for further information please see MIKE 21-BW, 2011).

MIKE 21 BW model numerically solves the 2D Boussinesq type equations (Madsen et al. 1991; Madsen and Sorensen, 1992). The equations include nonlinearity and frequency dispersion where the frequency dispersion is introduced in the momentum equations. Model has capability to simulate various processes in coastal area such as shoaling, refraction, diffraction, reflection, breaking, etc. (MIKE 21-BW, 2011; Madsen et al., 1997a, 1997b; Sorensen and Sorensen, 2001; Sorensen et al., 1998). MIKE 21 BW model can be employed using one or two horizontal space coordinates. In the case of two horizontal space coordinates, the model solves the enhanced Boussinesq equations by an implicit finite difference technique, while in one horizontal space coordinates, the equations are solved using a standard Galerkin finite element method (MIKE 21-BW, 2011). The one horizontal space coordinates is typically selected for calculation of wave transformation from offshore to the beach. Thus it is suitable for studying surf zone and swash zone dynamics. However, the two horizontal space coordinates is typically selected for calculation of short and long period wave disturbance. Hence, it is suitable for wave simulation in ports and harbors (MIKE 21-BW, 2011). According to these basic information and differences between these modules of MIKE21 BW, one can conclude that the two horizontal space coordinates may be a better choice to be employed for simulating the wave field around and behind the submerged breakwaters. The enhanced Boussinesq type equations are suitable for simulating the directional wave propagation from deep to shallow waters. The highest value of water depth to deep wave length (h/L_o) for the enhanced Boussinesq equations is nearly 0.5, while for the classical Boussinesq equations, this ratio is about 0.22 (MIKE 21-BW, 2011).

Throughout the set up process of modelling, the data are resolved into an appropriate computational mesh. Thus, to achieve this goal, having access to a proper bathymetry is necessary. Typically a two dimensional matrix of water depths is created. All the boundary conditions around the area of interest need to be specified carefully. According to the water depths in the computational domain, wave conditions are determined. In the model MIKE 21 BW, the shortest possible wave period which can be considered is basically specified by the largest water depth in the computational domain while the wave height without wave breaking is specified according to the smallest water depth in the computational domain (MIKE 21-BW, 2011).

Therefore, grid spacing and time step are selected according to the wave conditions and the water depth. MIKE 21 BW solves two forms of the Boussinesq equations. The ratio of the maximum water depth to the deep water wave length for the classical form of the Boussinesq equations for shallow waters must not become larger than 0.22 (Shallow waters). For the enhanced form of the Boussinesq equations with deep water terms to improve the dispersion relations, this ratio becomes 0.5 (MIKE 21-BW, 2011). The grid spacing and time step in the model can initially be selected based on a basic rule. The grid spacing and time step are basically specified based on the smallest wave length and the smallest wave period respectively. We should prevent from happening numerical instability in the computational area. The numerical instability may occur during the simulation period in the computational domain with the greatest water depth, for instance near the internal wave generation lines and at the location of open boundaries in the model area.

To prevent occurrence of instability inside the computational domain, we need to keep Courant number equal or less than unity (MIKE 21-BW, 2011):

$$C_r = c\frac{\Delta t}{\Delta x} \tag{9.1}$$

where c is the wave celerity, Δt and Δx are the temporal and spatial steps, respectively. Therefore, as we deal with wave breaking in the case of submerged breakwaters and considering the time scale of the waves, the grid spaces should be chosen relatively small with an appropriate time step so that the corresponding Courant number is lower than 1.

It is necessary to select the grid spacing and time step in a way to have both benefits of low computational costs/time and high accuracy of results. Therefore, the grid spacing and time step should be selected carefully staying always on a middle line between two extremes (MIKE 21-BW, 2011). A sufficient simulation period should be also considered to ensure the generated waves offshore can reach the submerged breakwater nearshore.

To solve the 2DH Boussinesq equations in MIKE 21 BW for the submerged breakwaters, the convective terms are also discretized. For this problem, a method for numerical integration without dissipation is available in MIKE 21 BW. Based on this method, the convective terms are numerically solved by central differences and a normal ADI algorithm with a semi-linearization of non-linear terms (Abbott and Minns, 1998; MIKE 21-BW, 2011). However, in the case of submerged breakwaters, as we deal with the large gradients, in practice, this method may be less applicable due to occurrence of instability in the model. Thus, another discretization method particularly for the submerged breakwaters with steep gradients is needed to prevent instability. An appropriate method to solve the convective terms with numerical dissipation can be employed in MIKE 21 BW to achieve this goal (For further details see MIKE 21-BW, 2011).

The following data are required for calibrating and verifying the model: bathymetric data, boundary data, simulation period, structural features of the submerged breakwater, and measured wave parameters at some specific and selected stations in the model area. Therefore, according the basic features of the model and the calibration data, we can modify the model parameters in order to reduce the computed error of the results against measured data.

A specific value (usually referred to the still water level; SWE), is typically used in MIKE 21 BW model to represent the limits of land area in the computational domain, thus, points with higher values respect to it are considered as land. Land points in the bathymetry file are also known as closed boundaries either in the form of being fully reflective or partially reflective. In MIKE 21-BW porosity layers are used to have either partial wave reflection from the submerged breakwater or wave transmission through the structure. Absorbing boundary consisting sponge layers can be also applied in the model in order to avoid wave reflection from the boundary (MIKE 21-BW, 2011). The sponge layers may be employed in MIKE 21 BW along the shorelines or model boundaries in order to absorb wave energy by a coastal structure or shoreline (for instance here when waves reaching the submerged breakwater) or to absorb waves radiating out of the model domain (for example when waves leaving a harbor area).

The bottom shear stress is formulated in MIKE 21 BW as follows according to the Chezy bed friction in order to calculate the friction at sea bed (MIKE 21-BW, 2011):

$$\tau_b = \frac{\rho g v |v|}{C^2} \tag{9.2}$$

where v is the depth-averaged velocity, ρ is the water density, g is gravity and C is the Chezy number (Svendsen and Jonsson, 1980; MIKE 21-BW, 2011).

The incorporation of wave breaking in MIKE 21 BW is based on the concept of surface rollers (MIKE 21-BW, 2011). For further information see: Madsen et al. (1997a), Ozanne et al., (2000) and Schaffer et al. (1993). However, a problem of using this breaking model in MIKE 21 BW is when dealing with plunging breakers (MIKE 21-BW, 2011). In the case of the submerged breakwaters where wave breaking occurred over the seaside slope or the crest of the breakwater, high-frequency instabilities can be removed using a numerical filter and the wave energy can be dissipated in the area where the model is not able to deal with the surface roller appropriately (MIKE 21-BW, 2011).

Results of simulation can be represented in various forms in model MIKE 21-BW, including deterministic parameter of the surface elevation, the phase-averaged value of the significant wave height or the wave disturbance coefficient. The results are presented as point series, line series or grid series data (MIKE 21-BW, 2011).

9.7 DELFT3D

The Delft3D package consists of various modules which allow for the simulation of short-crested wind-generated waves, current, sediment, ecology and water quality in channels, rivers, tidal inlets, lakes, barrier islands with tidal flats, coastal and estuarine areas, as well as harbors, coastal structures, etc. The Delft3D Wave module, also known as SWAN, an acronym for Simulating WAves Nearshore (see Booij et al., 1999; Holthuijsen et al., 1993; Ris et al., 1999), is a phase-averaged wave model. The SWAN developed at Delft University of Technology, the Netherlands and integrated in the Delft3D model suite by Deltares. SWAN is based on the discrete spectral action balance equation. It can be used to model the random wave propagation and transformation in relatively large areas from offshore to nearshore (SWAN, 2006, 2014a, 2014b). SWAN is applicable in shallow, intermediate and deep waters. It can simulate wind-generated waves, the interaction of waves and currents, wave blocking by currents, nonlinear wave-wave interactions, refraction over a bottom of variable depth and spatially varying ambient current, depth and current-induced shoaling, transmission through structures, blockage by or reflection against obstacles, wave diffraction, wave set-up, bottom friction, white capping and depth-induced breaking, etc. (SWAN, 2006, 2014a, 2014b).

The SWAN model (Holthuijsen et al., 1993) is a fully spectral mode which can predict significant wave height, mean wave period and direction, radiation stress and directional spectrum, and wave forces. The results of computations can be reproduced in terms of one- and two-dimensional wave spectra in SWAN. However, in practice a weakness of SWAN is dealing with wave simulation within harbors. For these areas phase-resolving models such as the MIKE 21 Boussinesq-type model is more applicable. SWAN is also less accurate in simulation of interaction between wave and coastal structures or steep sloping beaches (see SWAN, 2006, 2014a, 2014b).

The SWAN model does not include the limitations of the HISWA model, which is a stationary second-generation model. The SWAN applies fully implicit schemes, therefore is unconditionally stable. The SWAN model has successfully been validated with laboratory tests and field measurements (see Ris et al., 1999; SWAN, 2006, 2014a, 2014b). The SWAN model has been released under public domain. For further information see the SWAN home page.

The effect of flow on the waves, such as set-up, current refraction and enhanced bottom friction, as well as the effect of waves on current including forcing, enhanced turbulence and bed shear stress can also be taken into account.

9.8 IH2VOF

IH2VOF (Lara, 2005) solves the two-dimensional wave flow using the 2DV RANS equations outside the porous materials and the Volume-Averaged Reynolds Averaged Navier-Stokes (VARANS) equations (Hsu et al., 2002) inside the porous materials. A number of physical parameters, therefore, are considered by the model dealing with the porous materials, such as the porosity, the mean diameter of the armor stones, friction coefficients, etc. (Lara, 2005). The model decomposes the instantaneous velocity and pressure fields into mean and turbulent components. Turbulence is simulated in IH2VOF according to a k-ε model for the turbulent kinetic energy κ, and its dissipation rate ε (Lara, 2005). The water surface level is monitored with the volume of fluid (VOF) method. The model also includes a cutting cell method (for further information see Clarke et al., 1986). IH2VOF includes a variety of boundary conditions, including wave generation and absorption boundaries (see Lara, 2005; Lara et al., 2008; Losada et al., 2005; Schaffer and Klopman, 2000). IH2VOF is applicable in the surf zone simulations or modeling the coastal structures. It includes second-order wave generation and active wave absorption. IH2VOF has been validated for a wide range of coastal structures (e.g. rubble mound, vertical, composite breakwaters, etc.) (see Lara, 2005; Lara et al., 2008; Losada et al., 2005).

Internal wave generation can be employed to produce various type of waves such as monochromatic waves, Stokes waves, solitary waves, cnoidal waves, random waves, etc. Sponge layers can be employed to absorb re-reflected waves in the place of wave generating lines (see Lara et al., 2006; Lin & Liu, 1998). A wave absorbing system is used in IH2VOF to absorb wave reflection in the model. The system is basically a type of passive wave absorption or sponge layer based on the Israeli & Orszag (1981) formulation. Passive wave absorption consists of a dissipation model which is defined in the area of interest, typically with a minimum length about two wave lengths where requires an additional computational mesh in this area. Passive wave absorbers and sponge layers can be also employed in order to reduce the wave reflection on the lateral boundaries. To prevent an increase in mean water surface level, a system of active wave absorption is included in this model according to Schaffer and Klopman (2000) methodology (see Lara, 2005; Lara et al., 2008; Losada et al., 2005).

IH2VOF computes velocity components, pressure field, turbulence, and free surface level in any cell of the mesh. Further details might be obtained by the model including wave breaking, bottom shear stress, wave run-up, wave overtopping velocity, thickness and discharge, transmitted wave energy behind the structures (overtopping and filtration), horizontal and uplift forces and moments, etc. (see: Lara, 2005; Lara et al., 2008; Losada et al., 2005).

IH-3VOF (Lara, et al., 2010) has been also developed by IH Cantabria for the three dimensional and transitional flows in two phases in coastal engineering. IH-3VOF is a three-dimensional model that can be applied to design coastal and marine structures and to study the hydrodynamics of the surf zone. It has been designed in such a way as to be used particularly for solving the wave-porous structure interaction problems. A variety of wave types can be modeled in IH-3VOF. It has been successfully validated with laboratory experiments (del Jesus et al., 2012; Lara, et al., 2012).

IH-3VOF applies the finite volume method to discretize the computational domain and solve the three-dimensional Navier-Stokes equations for the velocity and pressure fields. The VOF technique, developed by Rider and Kothe (1998), is also used in the model for monitoring the water surface level. IH-3VOF can be employed for practical applications, particularly regarding wave–structure interaction (see Lara et al., 2010; del Jesus et al., 2012; Lara et al., 2012).

9.9 IHFOAM

IHFOAM is a three-dimensional numerical model based on OpenFOAM® applicable for two-phase flow conditions (Higuera et al., 2013). IHFOAM solves the three-dimensional RANS equations using a finite volume method. It also uses the volume of fluid (VOF) method. IHFOAM solves two-phase flow within porous media using VARANS equations (Higuera et al., 2014a,b). It can be used for a variety of problems in both shallow and deep waters, in coastal or offshore areas, for fixed or floating structures, and in laboratory or prototype scale. IHFOAM can be particularly used for simulation of the interaction between waves and breakwaters including the vertical or rubble mound breakwaters, emerged or submerged breakwaters. Various kinds of boundary conditions can be simulated by IHFOAM. Wave generation with different type of waves including regular waves, cnoidal, higher order Stokes waves, Streamfunction, Boussinesq solitary wave, fully 3D random and directional sea states can be modeled (Higuera et al., 2014a,b). Active absorption at the boundaries of the model also can be used simultaneously with a wave generation system to absorb incident wave energy. IHFOAM includes 2D and Quasi-3D enhanced theories (Higuera et al., 2014a,b; Schaffer and Klopman, 2000) and a full 3D theory used as the active wave absorption in the model. IHFOAM has been validated successfully for a variety of coastal engineering problems particularly regarding wave interaction with different kinds of coastal structures (see Higuera et al., 2013, 2014a,b).

9.10 GENESIS

GENESIS (GENEralized model for SImulating Shoreline change) is used to calculate wave-induced shoreline change (Gravens et al., 1991; Hanson, 1989). GENESIS is based on the one line theory, thus the bottom profile is assumed to remain unchanged in time. Therefore, the beach profile is always in equilibrium and the changes in the position of the shoreline are considered. Consequently, only longshore sand transport can be taken into account. Additionally, it has been assumed in the model that the sands move to a certain limiting depth over the beach profile and beyond that depth the bottom does not move (Gravens et al., 1991; Hanson, 1989).

The GENESIS takes into account the effects of wave refraction, diffraction, shoaling, sediment transport, and so forth. GENESIS can be applied for different coastal structures such as offshore and detached breakwaters, emerged breakwaters, low-crested structures, submerged breakwaters, seawalls, groins, jetties, etc. Any combination of these structures or other coastal structures also can be considered and modeled in GENESIS (Gravens et al., 1991; Hanson, 1989).

9.11 STWAVE

STWAVE (STeady State spectral WAVE) (STWAVE, 1999) is a numerical model for nearshore wind-wave growth and propagation. STWAVE can simulate wave refraction and shoaling under depth and current effects, wave breaking due to depth and wave steepness, wave diffraction, wave–wave interaction, whitecapping and some other physical processes, which are also the cases for submerged breakwaters. Some main assumptions in the model are: slowly varying bottom slope, negligible bottom friction and wave reflection, steady-state waves, currents, and winds, linear refraction and shoaling, and depth-uniform current (STWAVE, 1999).

9.12 ComFLOW

ComFLOW is basically a VOF model that can solve the incompressible Navier-Stokes equations with free surface variation and complex bathymetry for pressure and velocity fields. A local height function is also used in the VOF function that allows for a more accurate simulation of breaking waves (Wenneker et al., 2010). ComFLOW can be applied in both two- and three-dimensional modes, both one-phase and two-phase situations. ComFLOW uses simple rectangular Cartesian grids. A cut-cell technique is used in the model to provide possibility of modeling arbitrarily forms of structures with varying geometries (Wenneker et al., 2010).

9.13 NN_OVERTOPPING

The CLASH Neural Network Generic Overtopping Prediction Tool (called NN_OVERTOPPING) is an example of artificial neural network (ANN) models used in the field of coastal engineering and particularly breakwaters (EurOtop, 2007). It could also be applied for low-crested structures. This tool was developed during the European Community's framework: Crest Level Assessment of coastal Structures by full-scale monitoring, neural network prediction and Hazard analysis on permissible wave overtopping (CLASH). It may be helpful in the case of preliminary predictions for overtopping discharges for some specific structures (To see the type of structures and further details see EurOtop, 2007). The model is based on artificial neural networks. It can be used for the conceptual design of coastal structures. A set of laboratory data based on small-scale experimental tests was used to generate the model. Scale effects, measurement equipment and limited data range may cause low accuracy in the predictions of the model. Thus, it should be applied carefully and only for the initial design stage (EurOtop, 2007).

9.14 COULWAVE

COULWAVE or Cornell University Long and Intermediate Wave Modeling Package (COULWAVE, 2008) is a numerical model which is developed for modeling free surface waves and nearshore wave-driven hydrodynamic processes. COULWAVE solves a number of governing equations (i.e. nonlinear shallow water wave equations and dispersive Boussinesq-type equations) allowing for calculating the fully nonlinear

Table 9.1 Previous submerged breakwater numerical research (two-dimensional)

References	Model	Outputs	Breakwater type	Waves
Kobayashi and Wurjanto (1988)	Finite-amplitude shallow water equations	Wave transmission	Impermeable submerged	Regular
Rojanakamthorn et al. (1990)	The equation of waves on a porous layer as a two-dimensional elliptic equation analogous to the mild slope equation	Wave reflection and transmission coefficients	Permeable submerged	Random waves
Shen et al. (2004)	Volume Of Fluid (VOF), a two-equation k-ε turbulence model	Wave profile	Submerged	Cnoidal waves
Garcia et al. (2004)	COrnell BReaking waves And Structures (COBRAS), 2DV Reynolds Averaged Navier-Stokes (RANS) equation	Composition of the instantaneous velocity and pressure fields into mean and turbulent components	Permeable low-crested breakwaters	Regular breaking waves
Lara et al. (2006)	COrnell BReaking waves And Structures (COBRAS), 2DV Reynolds Averaged Navier-Stokes (RANS) equation	Composition of the instantaneous velocity and pressure fields into mean and turbulent components	Permeable low-crested breakwaters	Random waves

(Continued)

Table 9.1 (Continued)

References	Model	Outputs	Breakwater type	Waves
Kobayashi et al. (2007)	Time-averaged continuity, momentum, and energy equations	The mean and standard deviation of the free surface elevation and horizontal fluid velocities	Porous submerged breakwater	Irregular breaking wave
Panizzo and Briganti (2007)	Artificial neural network (training with the European Union research project DELOS data)	Wave transmission	Low-crested structures	Regular and Irregular
Christou et al. (2008)	Boundary Element Method (BEM)	Spatial water surface profiles and harmonic generation by the breakwater	Rectangular submerged breakwaters	Nonlinear regular waves
Kriezi and Karambas (2010)	Higher-order Boussinesq type of equations (improved linear dispersion characteristics)	Partial wave reflection and transmission	Monolithic rectangular submerged breakwaters	Irregular and regular, breaking and nonbreaking waves
Liu et al. (2013)	Incompressible Smoothed Particle Hydrodynamics (ISPH)	Free surface flow	Impermeable breakwater, rectangular obstacle	Linear and solitary waves, periodic wave train

Table 9.2 Previous submerged breakwater numerical research (three-dimensional)

References	Model	Outputs	Breakwater type	Waves
Hayakawa et al. (1981)	SOLA-SURF	Wave and current field	Permeable submerged breakwater	Regular
Van der Biezen et al. (1998)	Delft3D, developed by Delft Hydraulics, a phase-resolving module, based on the nonlinear shallow water equations	Velocities in the horizontal plane and free surface elevation	Rubble mound breakwater with an impermeable core, 2:3 slopes on both sides	Regular and normal incident to the beach
Johnson et al. (2005)	Phase-averaged method (2D depth averaged or quasi-3D) and a phase resolving method (2DH-Boussinesq-type model)	Wave and current field	Submerged breakwater	Regular and irregular wave
Lin (2006)	3D multiple-layer, coordinate mode	Water surface elevation	Submerged breakwater	Solitary wave
Johnson (2006)	MIKE 21 PMS-refraction/diffraction model based on the parabolic approximation to the mild slope equation	Wave transmission coefficient	Submerged breakwater	Regular and irregular wave
Zanuttigh and Martinelli (2008)	Conceptual model	Energy transmission (K_t and wave spectra)	Emergent permeable low crested structures	Regular and irregular wave
Zanuttigh and Martinelli (2008)	2DH model MIKE 21	Overtopping discharges	Detached breakwaters with a gap in between, emerged and zero-freeboard structures	Regular and irregular wave
Vicinanza et al. (2009)	Mathematical model	Wave transmission	A single detached low-crested breakwater of finite length	Regular and irregular wave
Lara et al. (2012), del Jesus (2011)	IH-3VOF-volume-averaged Navier-Stokes equations	Porous media flow, wave impact	Porous and impermeable structures	Solitary and regular wave

and dispersive water waves over variable bathymetries (COULWAVE, 2008). The numerical scheme uses finite difference scheme for the spatial derivatives. COULWAVE has possibility to model waves with directional random spectrum. COULWAVE can be applied to model various nearshore processes, such as shoaling, diffraction, refraction, rip currents, longshore currents, wave and current interaction, run up, wave-wave interactions, etc. (COULWAVE, 2008).

9.15 ADCIRC

ADCIRC is an advanced circulation model for solving time dependent, free surface circulation and transport problems in two and three dimensions. ADCIRC (ADCIRC, 2012) solves a number of governing equations including hydrostatic and Boussinesq approximations, two dimensional shallow water equations, three dimensional mass and momentum conservation equations, two dimensional sediment continuity equation, and two- and three-dimensional temperature and salinity transport equations. ADCIRC uses the finite element method to solve the equations for the spatial derivatives in complex grids. ADCIRC can be employed to model tides and wind-generated circulation, storm surge and flooding, flow and transport in coastal areas (ADCIRC, 2012).

9.16 PREVIOUS NUMERICAL RESEARCH

Previous numerical research on submerged breakwaters (two- and three-dimensional simulations) is presented in Tables 9.1 and 9.2.

REFERENCES

Abbott, M.B., Minns, A.W., 1998. Computational Hydraulics. Ashgate, Aldershot and Brookeld, USA.

ADCIRC, 2012. A (Parallel) Advanced Circulation Model for Oceanic, Coastal and Estuarine Waters; Adcirc User's Manual, by: R.A. Luettich, Jr. and J.J. Westerink <http://adcirc.org>.

ARTEMIS, 2014. ARTEMIS User Manual, TELEMAC-MASCARET <http://www.opentelemac.org>.

Booij, N., Ris, R.C., Holthuijsen, L.H., 1999. A third-generation wave model for coastal regions, Part I: model description and validation. J. Geophys. Res. 104 (C4), 7649–7666.

Chang, K.-A., Hsu, T.-J., Liu, P.L.-F., 2001. Vortex generation and evolution in water waves propagating over a submerged rectangular obstacle. Part I: solitary waves. Coast. Eng. 44 (1), 13–36.

Christou, M., Swan, C., Gudmestad, O., 2008. The interaction of surface water waves with submerged breakwaters. Coast. Eng. 55 (12), 945958.

Clarke, D., Hassan, H., Salas, M., 1986. Euler calculations for multielement airfoils using cartesian grids. AIAA J., 353–358. Available from: http://dx.doi.org/10.2514/3.9273.

COULWAVE, 2008. Modeling Wave Generation, Evolution, and Interaction with Depth-Integrated, Dispersive Wave Equations, COULWAVE Code Manual, Cornell University Long and Intermediate Wave Modeling Package, by: Patrick J. Lynett and Philip L.-F. Liu, <http://ceeserver.cee.cornell.edu/pll-group/r4coulwave.html>.

Dean, R.G., Dalrymple, R.A., 2001. Coastal Processes with Engineering Applications. Cambridge University Press.

del Jesus, M., 2011. Three-dimensional interaction of water waves with coastal structures. Ph.D. thesis. Universidad de Cantabria.

del Jesus, M., Lara, J.L., Losada, I.J., 2012. Three-dimensional interaction of waves and porous coastal structures. Part I: numerical model formulation. Coast. Eng. 64, 57–72.

EurOtop, 2007. Wave Overtopping of Sea Defences and Related Structures: Assessment Manual (Die Kuste version), EurOtop manual, Kuratorium für Forschung im Küsteningenieurwesen (KFKI) <http://www.overtopping-manual.com>.

Garcia, N., Lara, J., Losada, I., 2004. 2-D numerical analysis of near-field flow at low-crested permeable breakwaters. J. Coast. Eng. 51, 991–1020.

Gómez-Gesteira, M., Rogers, B.D., Crespo, A.J.C., Dalrymple, R.A., Narayanaswamy, M., Dominguez, J.M., 2012a. SPHysics – development of a free-surface fluid solver – part 1: theory and formulations. Comput. Geosci. 2012, http://dx.doi.org/10.1016/j.cageo.2012.02.029, <http://www.sphysics.org>.

Gómez-Gesteira, M., Crespo, A.J.C., Rogers, B.D., Dalrymple, R.A., Dominguez, J.M., Barreiro, A., 2012b. SPHysics – development of a free-surface fluid solver – Part 2: Efficiency and test cases. Comput. Geosci. 2012, http://dx.doi.org/10.1016/j.cageo.2012.02.028, <http://www.sphysics.org>.

Gravens, M.B., Kraus, N.C., Hanson, H., 1991. GENESIS: generalized model for simulating shoreline change. In: Report 2: Workbook and System User's Manual. Technical Report CERC89-19. Vicksburg, Miss: Coastal Engineering Research Center, U.S. Army Engineer Waterways Experiment Station. U.S. Army Corps of Engineers.

Hanson, H., 1989. GENESIS: a generalized shoreline change numerical model. J. Coast. Res. 5 (1), 1–27.

Higuera, P., Lara, J.L., Losada, I.J., 2014a. Three-dimensional Interaction of waves and porous coastal structures using OpenFOAM®. Part II Appl. Coast. Eng. 83, 259–270, http://dx.doi.org/10.1016/j.coastaleng.2013.09.002.

Higuera, P., Lara, J.L., Losada, I.J., 2014b. Three-dimensional interaction of waves and porous coastal structures using OpenFOAM®. Part I: formulation and validation. Coast. Eng. 83, 243–258, http://dx.doi.org/10.1016/j.coastaleng.2013.08.010.

Higuera, P., Lara, J.L., Losada, I.J., 2013. Simulating coastal engineering processes with OpenFOAM®. Coast. Eng. 71, 119–134, http://dx.doi.org/10.1016/j.coastaleng.2012.06.002.

Holthuijsen, L.H., Booij, N., Ris, R.C., 1993. A spectral wave model for the coastal zone, 2nd International Symposium on Ocean Wave Measurement and Analysis, New Orleans, Louisiana, July 25–28, 1993, New York, pp. 630–641.

Hsu, T.J., Sakakiyama, T., Liu, P.L.F., 2002. A numerical model for wave motions and turbulence flows in front of a composite breakwater. Coast. Eng. 46 (1), 25–50, http://dx.doi.org/10.1016/S0378-3839(02)00045-5.

Israeli, M., Orszag, S.A., 1981. Approximation of radiation boundary conditions. J. Comp. Phys. 41, 115–131.

Johnson, H.K., 2006. Wave modelling in the vicinity of submerged breakwaters. Coast. Eng. 53, 39–48.

Johnson, H.K., Karambas, T.V., Avgeris, I., Zanuttigh, B., Gonzalez-Marco, D., Caceres, I., 2005. Modelling of waves and currents around submerged breakwaters. Coast. Eng. 52, 949–969.

Kirby, J.T., 1986. Rational approximations in the parabolic equation method for water waves. Coast. Eng. 10, 355–378.

Kobayashi, N., Wurjanto, A., 1988. Wave transmission over submerged breakwaters. J. Waterway, Port, Coast. Ocean Eng. 115 (5), 662–680.

Kobayashi, N., Meigs, L., Ota, T., Melby, J., 2007. Irregular breaking wave transmission over-submerged porous breakwater. J. Waterway, Port, Coast. Ocean Eng. 133 (2), 104–116.

Kriezi, E., Karambas, T., 2010. Modelling wave deformation due to submerged breakwaters. Proceedings of the ICE - Maritime Engineering 163-1, 19–29.

Lara, J.L., 2005. A numerical wave flume to study the functionality and stability of coastal structures. J. Int. Navigation Assoc. (PIANC) 121, 5–29.

Lara, J., Garcia, N., Losada, I., 2006. RANS modelling applied to random wave interaction with submerged permeable structures. Coast. Eng. 53, 395–417.

Lara, J.L., Losada, I.J., Guanche, R., 2008. Wave interaction with low mound breakwaters using a RANS model, Ocean Engineering, Elsevier. 35(13), pp. 1388–1400.

Lara, J.L., Losada I.J., del Jesus, M., Barajas, G., Guanche, R., 2010. IH-3VOF: a three dimensional Navier-Stokes model for wave and structure interaction. 32st International Conference on Coastal Engineering. ASCE (American Society of Civil Engineers). Proceedings of the 32st International Conference on Coastal Engineering (ICCE), ASCE. Shanghai, China.

Lara, J.L., del Jesus, M., Losada, I.J., 2012. Three-dimensional interaction of waves and porous coastal structures. Part II: experimental validation. Coast. Eng. 64, 26–46.

Lin, P., 2006. A multiple-layer σ-coordinate model for simulation of wave-structure interaction. Comput. Fluids 35 (2), 147–167.

Lin, P., Liu, P.-F., 1998. A numerical study of breaking waves in the surf zone. J. Fluid Mech. 359, 239–264.

Liu, X., Xu, H., Shao, S., Lin, P., 2013. An improved incompressible SPH model for simulation of wave-structure interaction. Comput. Fluids 71, 113–123.

Losada, I.J., Lara, J.L., Christensen, E.D., Garcia, N., 2005. Modelling of velocity and turbulence fields around and within low-crested rubble-mound breakwaters. Coast. Eng. 52 (10–11), 887–913, Elsevier.

Madsen, P., Sorensen, O., 1992. A new form of the Boussinesq equations with improved linear dispersion characteristics. Part 2: a slowly varying bathymetry. Coast. Eng. 18, 183–204.

Madsen, P., Murray, R., Sorensen, O., 1991. A new form of the Boussinesq equations with improved linear dispersion characteristics (part 1). Coast. Eng. 15, 371–388.

Madsen, P., Sorensen, O., Schaer, H., 1997a. Surf zone dynamics simulated by a Boussinesq type model. Part 1: model description and cross-shore motion of regular waves. Coast. Eng. 32, 255–287.

Madsen, P., Sorensen, O., Schaer, H., 1997b. Surf zone dynamics simulated by a Boussinesq type model. Part 2: surf beat and swash zone oscillations for wave groups and irregular waves. Coast. Eng. 32, 289–320.

MIKE 21-BW, 2011. MIKE 21 Wave Modelling: MIKE 21 BW-Boussinesq Wave Module by DHI. <https://www.mikepoweredbydhi.com>.

Ozanne, F., Chadwick, A., Huntley, D., Simmonds, D., Lawrence, J., 2000. Velocity predictions for shoaling and breaking waves with a Boussinesq-type model. Coast. Eng. 44, 361–397.

Panizzo, A., Briganti, R., 2007. Analysis of wave transmission behind low-crested breakwaters using neural networks. Coast. Eng. 54, 643–656.

REF/DIF, 1994. Combined Refraction/Diffraction Model, Documentation and User's Manual, Center for Applied Coastal Research, Dept. of Civil Eng., Univ. of Delaware, Newark, DE, USA <https://www.udel.edu/kirby/programs/refdif/refdif.html>.

Rider, W.J., Kothe, D.B., 1998. Reconstructing volume tracking. J. Comput. Phys. 141, 112–152.

Ris, R.C., Holthuijsen, L.H., Booij, N., 1999. A third-generation wave model for coastal regions, Part II: verification. J. Geophys. Res. 104 (C4), 7667–7681.

Rojanakamthorn, S., Isobe, M., Watanabe, A., 1990. Modeling of wave transformation on submerged breakwater. Proc. 22th Int. Conf. on Coastal Engineering. ASCE.

Schäffer, H., Klopman, G., 2000. Review of multidirectional active wave absorption methods. J. Waterway Port Coast. Ocean Eng. 126, 88–97, http://dx.doi.org/10.1061/(ASCE)0733-950X(2000)126:2(88).

Schaffer, H., Madsen, P., Deigaard, R., 1993. A Boussinesq model for waves breaking in shallow water. Coast. Eng. 20, 185–202.

Shen, Y., Ng, C., Zheng, Y., 2004. Simulation of wave propagation over a submerged bar using the VOF method with a two-equation k-ε turbulence modeling. J. Coast. Eng. 31, 87–95.

Sisyphe, 2014. Sisyphe v6.3 User's Manual, EDF <http://www.opentelemac.org>.

Sorensen, O.R., Sorensen, L.S., 2001. Boussinesq type modelling using unstructured finite element technique. In Proc. 27th Coastal Eng. Conf. 190–202.

Sorensen, O.R., Madsen, P.A., Schaffer, H.A., 1998. Nearshore wave dynamics simulated by Boussinesq type models. In: ICCE.

STWAVE, 1999. STWAVE: Steady-State Spectral Wave Model, Report 1, User's Manual for STWAVE, Coastal Inlets Research Program, US Army Corps of Engineers, Waterways Experiment Station, by Jane McKee Smith, Donald T. Resio, WES Alan K. Zundel, Brigham Young University.

Svendsen, I.A., Jonsson, I.G., 1980. Hydrodynamics of coastal regions. Tech. rep., Technical University of Denmark, pp. 220–229.

SWAN, 2006. Programming rules. Delft University of Technology, Environmental Fluid Mechanics Section, available from <http://www.swan.tudelft.nl> (Version 1.3, August 2006).

SWAN, 2014a. Implementation manual. Delft University of Technology, Environmental Fluid Mechanics Section, available from <http://www.swan.tudelft.nl> (Version 41.01, April 2014).

SWAN, 2014b. Scientific and Technical documentation. Delft University of Technology, Environmental Fluid Mechanics Section, available from <http://www.swan.tudelft.nl> (Version 41.01, April 2014).

TELEMAC-2D, 2014. 2D hydrodynamics, TELEMAC-2D Software, USER MANUAL, EDF-R&D <http://www.opentelemac.org>.

TELEMAC-3D, 2013. 3D hydrodynamics, TELEMAC-3D Software, USER MANUAL, EDF-R&D <http://www.opentelemac.org>.

TOMAWAC, 2011. Software for sea state modelling on unstructured grids over oceans and coastal seas, TOMAWAC Software, USER MANUAL, EDF-R&D <http://www.opentelemac.org>.

van der Biezen, S., Roelvik, J., Graa van der, J., Schaap, J., Torrini, L., 1998. Wave transmission at low-crested structures. Proc. 26th Int. Conf. on Coastal Engineering.
Vicinanza, D., Caceres, I., Buccino, M., Gironella, X., Calabrese, M., 2009. Wave disturbance behind low-crested structures: diffraction and overtopping effects. Coast. Eng. 56, 1173–1185, ASCE, 2418–2427.
Wenneker, I., Wellens, P., Gervelas, R., 2010. Volume-of-Fluid Model Comflow Simulations of Wave Impacts on a Dike, International Conference on coastal Engineering (ICCE), 32.
Zanuttigh, B., Martinelli, L., 2008. Transmission of wave energy at permeable low-crested structures. Coast. Eng. 55 (12), 1135–1147.

CHAPTER 10

Design Algorithms and Guidelines

Rapid developments in computational fluid dynamics allow us to use numerical methods to solve partial differential equations (PDEs) in the field of coastal and marine engineering. These equations include the Navier–Stokes equations, Reynolds equations, shallow water equations, and others. For the solution of these equations, three main methods are used. As mentioned previously, the finite difference, finite volume and finite element methods are the most popular methods used by engineers and scientists in the field of fluid mechanics. Using these methods, the computational area is discretized into very small spatial subdomains and solved numerically over small time steps.

In this chapter we intend to study these partial differential equations and different governing equations and methods for solving them, with examples from simple problems to complex cases.

10.1 PARTIAL DIFFERENTIAL EQUATIONS

A partial differential equation (PDE) is an analytical expression including derivatives of an unknown function with respect to some independent variables of a physical process that occurs in time and in one or more spatial dimensions. For instance, in Eq. 10.1 f is an unknown function with respect to independent variables x and y:

$$a_1 \frac{\partial^2 f}{\partial x^2} + a_2 \frac{\partial^2 f}{\partial x \partial y} + a_3 \frac{\partial^2 f}{\partial y^2} + a_4 = 0 \tag{10.1}$$

where f as stated before is a function of variables x and y:

$$f = f(x, y) \tag{10.2}$$

and second order derivatives of f with respect to x and y can be seen in the equation. Here, a_1, a_2 and a_3 are some expressions with respect to the independent variables x and y:

$$\begin{cases} a_1(x, y) \\ a_2(x, y) \\ a_3(x, y) \end{cases} \tag{10.3}$$

© 2016 Elsevier Ltd.
All rights reserved.

and a_4 is an expression with respect to the independent variables x and y as well as function f and its first order derivatives with respect to x and y:

$$\left\{ a_4 \left(x, y, f(x, y), \frac{\partial f}{\partial x}, \frac{\partial f}{\partial y} \right) \right. \tag{10.4}$$

This PDE will be nonlinear when a_1, a_2 and a_3 include $f(x, y)$ or $\frac{\partial f}{\partial x}$ or $\frac{\partial f}{\partial y}$ or a_4 is nonlinear in f or its first derivatives.

According to what has been described here, we can conclude that this PDE equation can be written in different forms. Each form of the equation thus may describe different physical processes (see: Farlow 1993; Logan 2004 and Zachmanoglou and Thoe 1986). For example, the equation is called "elliptic" if discriminant (Δ) has a negative value:

$$\Delta = a_2^2 - 4a_1 a_3 < 0 \tag{10.5}$$

Elliptic equations typically characterize steady-state systems in time. Therefore, the equation does not include a time derivative term. The closed computational domain is typically used. The boundary conditions are in terms of $f(x,y)$, $\frac{\partial f}{\partial x}$ and $\frac{\partial f}{\partial y}$.

For example, the Laplace equation is an elliptic equation, as follows:

$$\frac{\partial^2 f}{\partial x^2} + \frac{\partial^2 f}{\partial y^2} = 0 \tag{10.6}$$

where $\Delta = -4 < 0$.

Another example is the Poisson equation:

$$\frac{\partial^2 f}{\partial x^2} + \frac{\partial^2 f}{\partial y^2} = -a_4 \left(x, y, f, \frac{\partial f}{\partial x}, \frac{\partial f}{\partial y} \right) \tag{10.7}$$

Regarding the Dirichlet boundary condition, the function f is provided along the boundary, and for the Neumann boundary condition, the derivative of f is provided along the boundary; while for a mixed boundary condition, f is provided for a part of the boundary and the derivative of f is provided for the rest of the boundary. For further information and details, please see: Farlow (1993); Logan (2004) and Zachmanoglou and Thoe (1986).

The equation is a parabolic equation if

$$\Delta = a_2^2 - 4a_1 a_3 = 0 \tag{10.8}$$

An example of a parabolic equation is the heat equation.

Parabolic equations typically have variation in both space and time. The variation in time is provided by the first derivative in time with initial values:

$$(x, y, t = 0)$$

And, for example, the Dirichlet boundary conditions are presented as follows:

$$f(x = x_1, y = y_1, t) \tag{10.9}$$

$$f(x = x_2, y = y_2, t) \tag{10.10}$$

An example of the parabolic equation is the diffusion equation:

$$\frac{\partial f}{\partial t} = C\frac{\partial^2 f}{\partial x^2} + g\left(x, f, \frac{\partial f}{\partial x}\right) \tag{10.11}$$

To solve parabolic equations numerically, for example, a finite difference approximation can be used. The algorithm used for this procedure includes:

1. Discretizing the computational domain.
2. Making the finite difference approximation.
3. Applying the initial and boundary conditions to solve the problem by going forward through time.

In explicit schemes, all future states of the function (f_i^{n+1}) in terms of $(t + \Delta t)$ (i.e. $f(x,t + \Delta t)$), are expressed in terms of present (t) and previous $(t - \Delta t)$ states, which are known, while in the implicit schemes, all future states of our function in terms of $(t + \Delta t)$ (i.e. $f(x, t + \Delta t)$), are expressed in terms of future $(t + \Delta t)$ and present (t) states. Thus, we express f_i^{n+1} in terms of f_j^{n+1}, f_i^n, and possibly also f_j^n where $j = i - 1$ and $i + 1$.

The subscripts are being applied in our formulations to indicate spatial points, while the superscripts are being used to indicate time:

$$f_i^{n+1} = f(x_i, t_{n+1}) \tag{10.12}$$

For central differences:

$$\frac{\partial^2 f}{\partial x^2} = \frac{f_{i-1}^{n+1} - 2f_i^{n+1} + f_{i+1}^{n+1}}{(\Delta x)^2} + O(\Delta x^2) \tag{10.13}$$

For backward differences:

$$\frac{\partial f}{\partial t} = \frac{f_i^{n+1} - f_i^n}{\Delta t} + O(\Delta t) \tag{10.14}$$

The initial conditions are required in the problems where $n = 0$, so in such conditions the values of the function for $n = 0$ (f_i^0) are given for all i.

In addition the boundary conditions are needed to determine the values of the boundary lines or nodes. For example, for a one-dimensional (1D) problem, they determine the values of the relevant function for the nodes at the beginning and the end of the 1D computational domain ($i = 0$ and $m + 1$) for all n.

For example, as an application of the explicit schemes for solving parabolic partial differential equations, for an equation as follows we have:

$$\frac{\partial f}{\partial t} - c\frac{\partial^2 f}{\partial x^2} = 0 \tag{10.15}$$

and substituting for the derivatives we obtain the following:

For central differences:

$$\frac{\partial^2 f}{\partial x^2} \approx \frac{f_{i-1}^n - 2f_i^n + f_{i+1}^n}{(\Delta x)^2} + O(\Delta x^2) \tag{10.16}$$

For forward differences:

$$\frac{\partial f}{\partial t} = \frac{f_i^{n+1} - f_i^n}{(\Delta t)} + O(\Delta t) \tag{10.17}$$

Solving for f_i^{n+1} we have:

$$f_i^{n+1} = f_i^n + c\frac{\Delta t}{(\Delta x)^2}(f_{i-1}^n - 2f_i^n + f_{i+1}^n) \tag{10.18}$$

However, solving parabolic equations by the implicit scheme is unconditionally stable. Depending on the desired accuracy, it is also possible to choose larger time steps. For two- and three-dimensional models, however, this scheme might face more complicated computations. It should be noted that solving parabolic equations by the explicit scheme is relatively easy and simple, although the accuracy of this scheme is quite low. For further information and details, please see: Farlow (1993); Logan (2004) and Zachmanoglou and Thoe (1986).

It is also possible to apply a technique for two- or three-dimensional states similar to that used for the one-dimensional state. In these cases the derivatives are discretized in x and y directions for the two-dimensional (2D) problems and x, y and z directions for the three-dimensional (3D) problems.

Finally, the equation is a hyperbolic equation if discriminant (Δ) has a positive value:

$$\Delta = a_2^2 - 4a_1a_3 > 0 \tag{10.19}$$

An example of this would be the wave equation.

A hyperbolic equation with second derivative in time and variation in both space and time requires two initial values and two boundary conditions, for example with a Dirichlet boundary condition:

$$\begin{cases} f(x, y, t = 0) \\ \dfrac{\partial f}{\partial t}(x, y, t = 0) \end{cases} \tag{10.20}$$

$$\begin{cases} f(x = x_1, y = y_1, t) \\ \dfrac{\partial f}{\partial t}(x = x_2, y = y_2, t) \end{cases} \tag{10.21}$$

A typical example is the wave equation:

$$\frac{\partial^2 f}{\partial x^2} - \frac{1}{C^2}\left[\frac{\partial^2 f}{\partial t^2}\right] + g\left(x, f, \frac{\partial f}{\partial x}, \frac{\partial f}{\partial t}\right) = 0 \tag{10.22}$$

For further information and details on the partial differential equations and their applications, please see: Farlow (1993); Logan (2004) and Zachmanoglou and Thoe (1986). In the rest of this chapter, development of numerical models for a various type of governing equations is presented. The main aim is to be familiar further and in detail with different methods, techniques and approaches to solve the different problems and develop numerical models for those governing equations. This is essential for numerical modeling in different engineering problems, and particularly in our case of coastal engineering and submerged breakwaters. Later, these methods and approaches for numerical modeling will be applied in Chapter 11 for the particular case of submerged breakwaters. Numerical models, approaches and simulations for wave-submerged breakwater interaction will be presented in Chapter 11. Incidentally, several physical processes related to the submerged breakwaters can be simulated according to the information presented here in this Chapter and their results can be predicted.

10.2 FINITE DIFFERENCE METHOD

The first model is based on the finite difference approach, as discussed previously. In this section development of a numerical model using this approach is presented.

The basis of the finite difference method is the Taylor series, as discussed in Chapter 7. Therefore, using the techniques presented in that chapter, it is possible to calculate all the derivatives of higher orders easily.

After determining the governing equations for our particular physical process, as discussed, we need to formulate the system of equations. Then to solve the equations numerically we have to employ a numerical method. For instance, a finite difference approximation of the system of equations can be written. Then based on the initial and boundary conditions of the problem and using direct or iterative methods we can solve the system of finite difference equations. The algorithm for this procedure is as follows:

1. Discretizing the computational domain.
2. Writing the finite difference approximations.
3. Taking to account the boundary conditions and initial values and writing the finite difference approximation for all the unknown nodes in the computational domain.
4. Generating a system of finite difference equations.
5. Solving this set of equations using a direct or iterative method as will be discussed in detail in the next section (i.e. Cramer's rule matrix inversion, Gaussian elimination and Thomas Algorithm (direct methods) or Jacobi, Gauss-Seidel (Liebmann) and successive over-relaxation (SOR) (iteration methods)). For further details and information on these methods or other methods, please see: Davis (2006); Greenbaum (1997); Kelley (2003); Kelley (1995); Olshanskii and Tyrtyshnikov (2014); Saad (2003). Regarding the further details and information on the finite difference method, please also see: Causon and Mingham (2010); LeVeque (2007); Smith (1985); Thomas (1995).

   (Using Gauss-Siedel iteratively yields the Liebmann method.)

Next, we intend to have a further look into some well-known and applicable partial differential equations, including the wave equation, Laplace equation, diffusion equation, nonlinear reaction diffusion equation, and others.

Regarding the wave equation, the form of the one-dimensional wave equation is:

$$\frac{\partial^2 f(x,t)}{\partial t^2} = c^2 \frac{\partial^2 f(x,t)}{\partial x^2} \tag{10.23}$$

where t and x are the temporal and spatial variables, respectively, and $f(x,t)$ is an unknown function.

The form of the two-dimensional (2D) wave equation is:

$$\frac{\partial^2 f(x,y,t)}{\partial t^2} = c^2 \left(\frac{\partial^2 f(x,y,t)}{\partial x^2} + \frac{\partial^2 f(x,y,t)}{\partial y^2} \right) \tag{10.24}$$

Laplace's equation is:

$$\text{for 2D cases} \Rightarrow \frac{\partial^2 f}{\partial x^2} + \frac{\partial^2 f}{\partial y^2} = 0 \tag{10.25}$$

$$\text{for 3D cases} \Rightarrow \frac{\partial^2 f}{\partial x^2} + \frac{\partial^2 f}{\partial y^2} + \frac{\partial^2 f}{\partial z^2} = 0 \tag{10.26}$$

where f depends on x, y and x, y, z for 2D and 3D cases, respectively.

Poisson's equation is:

$$\text{for 2D cases} \Rightarrow \frac{\partial^2 f}{\partial x^2} + \frac{\partial^2 f}{\partial y^2} = \varphi \tag{10.27}$$

$$\text{for 3D cases} \Rightarrow \frac{\partial^2 f}{\partial x^2} + \frac{\partial^2 f}{\partial y^2} + \frac{\partial^2 f}{\partial z^2} = \varphi \tag{10.28}$$

where f and φ depend on x, y and x, y, z for 2D and 3D cases, respectively. If $\varphi = 0$ we will have Laplace's equation.

The diffusion equation is:

$$\frac{\partial f(x,t)}{\partial t} = \delta \frac{\partial^2 \mathrm{f}(x,t)}{\partial x^2} \tag{10.29}$$

The diffusion equation in 2D and 3D cases can be rewritten as well assuming δ is constant in all directions; otherwise the individual δ parameters should be specified for each special coordinate:

$$\text{For 2D cases} \Rightarrow \frac{\partial f}{\partial t} = \delta \left(\frac{\partial^2 f}{\partial x^2} + \frac{\partial^2 f}{\partial y^2} \right) \tag{10.30}$$

$$\text{For 3D cases} \Rightarrow \frac{\partial f}{\partial t} = \delta \left(\frac{\partial^2 f}{\partial x^2} + \frac{\partial^2 f}{\partial y^2} + \frac{\partial^2 f}{\partial z^2} \right) \tag{10.31}$$

For the diffusion equation, considering the explicit form of the finite-difference approximation, the simplest difference equation is:

$$\frac{f_i^{n+1} - f_i^n}{\Delta t} = \delta \frac{f_{i-1}^n - 2f_i^n + f_{i+1}^n}{(\Delta x)^2} \tag{10.32}$$

$$f_i^{n+1} = \frac{\delta \Delta t}{(\Delta x)^2} f_{i-1}^n + \left(1 - \frac{2\delta \Delta t}{(\Delta x)^2} \right) f_i^n + \frac{\delta \Delta t}{(\Delta x)^2} f_{i+1}^n . \tag{10.33}$$

This explicit form of the finite-difference approximation is stable if:

$$\frac{\delta \Delta t}{(\Delta x)^2} \leq \frac{1}{2}$$

If $\frac{\delta \Delta t}{(\Delta x)^2} > \frac{1}{2}$ then the method is unstable.

Considering the implicit form of the finite-difference approximation, we have:

$$\frac{f_i^{n+1} - f_i^n}{\Delta t} = \delta \frac{f_{i-1}^{n+1} - 2f_i^{n+1} + f_{i+1}^{n+1}}{(\Delta x)^2} \tag{10.34}$$

$$-\frac{\delta \Delta t}{(\Delta x)^2} f_{i-1}^{n+1} + \left(1 + \frac{2\delta \Delta t}{(\Delta x)^2}\right) f_i^{n+1} - \frac{\delta \Delta t}{(\Delta x)^2} f_{i+1}^{n+1} = f_i^n \tag{10.35}$$

Solving this system of linear equations for all values of n, we will find the solution. An advantage of the implicit form is that there are no constraints on the values of Δt and Δx. Similarity the diffusion equation can be rewritten and solved numerically in 2D or 3D forms.

For the wave equation:

$$\frac{\partial^2 f(x,t)}{\partial t^2} = c^2 \frac{\partial^2 f(x,t)}{\partial x^2} \tag{10.36}$$

The wave equation in 2D and 3D cases can be rewritten as well

$$\text{For 2D cases} \Rightarrow \frac{\partial^2 f}{\partial t^2} = c^2 \left(\frac{\partial^2 f}{\partial x^2} + \frac{\partial^2 f}{\partial y^2}\right) \tag{10.37}$$

$$\text{For 3D cases} \Rightarrow \frac{\partial^2 f}{\partial t^2} = c^2 \left(\frac{\partial^2 f}{\partial x^2} + \frac{\partial^2 f}{\partial y^2} + \frac{\partial^2 f}{\partial z^2}\right) \tag{10.38}$$

Considering the explicit form of finite-difference approximation, the simplest difference equations are:

$$\frac{f_i^{n-1} - 2f_i^n + f_i^{n+1}}{(\Delta t)^2} = c^2 \frac{f_{i-1}^n - 2f_i^n + f_{i+1}^n}{(\Delta x)^2} \tag{10.39}$$

$$f_i^{n+1} = -f_i^{n-1} + \frac{c^2 \Delta t}{(\Delta x)^2}(f_{i-1}^n - 2f_i^n + f_{i+1}^n) + 2f_i^n \tag{10.40}$$

$$f_k^{n+1} = -f_k^{n-1} + \frac{c^2\Delta t}{(\Delta x)^2} f_{k-1}^n + 2\left(1 - \frac{c^2\Delta t}{(\Delta x)^2}\right) f_k^n + \frac{c^2\Delta t}{(\Delta x)^2} f_{k+1}^n \quad (10.41)$$

where for stability we have:

$$\frac{c^2\Delta t}{(\Delta x)^2} < \frac{1}{2}$$

Laplace's Equation comes up again as a steady state form of the diffusion equation or the Wave equation. For solving the Laplace equation:

$$\frac{\partial^2 f}{\partial x^2} + \frac{\partial^2 f}{\partial y^2} = 0 \quad (10.42)$$

if considering the Figure 10.1 for the computational grid in a two-dimensional state, we have:

$$\frac{\partial^2 f}{\partial x^2} = \frac{f_{i-1,j} - 2f_{i,j} + f_{i+1,j}}{(\Delta x)^2} + O(\Delta x^2) \quad (10.43)$$

$$\frac{\partial^2 f}{\partial y^2} = \frac{f_{i,j-1} - 2f_{i,j} + f_{i,j+1}}{(\Delta y)^2} + O(\Delta y^2) \quad (10.44)$$

$$\frac{\partial^2 f}{\partial x^2} + \frac{\partial^2 f}{\partial y^2} = \frac{f_{i-1,j} - 2f_{i,j} + f_{i+1,j}}{\Delta x^2} + \frac{f_{i,j-1} - 2f_{i,j} + f_{i,j+1}}{\Delta y^2} + O(\Delta x^2, \Delta y^2) \quad (10.45)$$

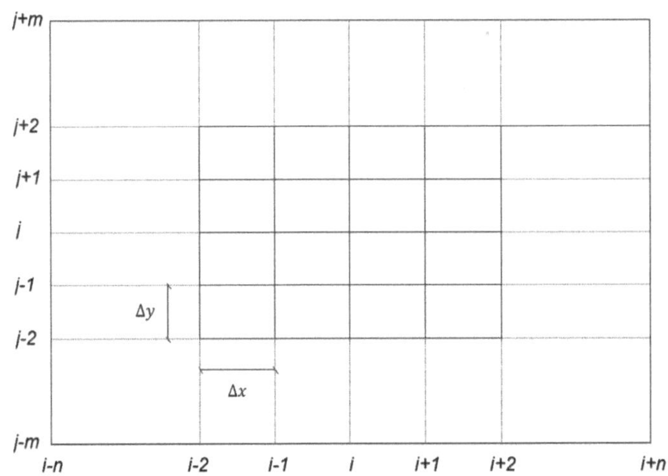

Figure 10.1 Computational grid in a two-dimensional state.

Example 10.1 Applying the finite difference approach

Apply the finite difference approach to discretize the following derivative.

$$\frac{\partial^2 f}{\partial x \partial y} \tag{10.46}$$

$$\left.\frac{\partial^2 f}{\partial x \partial y}\right|_{i,j} = \left.\frac{\partial}{\partial x}\left(\frac{\partial f}{\partial y}\right)\right|_{i,j} = \frac{\left.\frac{\partial f}{\partial y}\right|_{i+1,j} - \left.\frac{\partial f}{\partial y}\right|_{i,j}}{\Delta x} + O(\Delta x) \tag{10.47}$$

Using the central difference approximation we can also calculate the derivatives in the obtained previous expression as follows:

$$\left.\frac{\partial f}{\partial y}\right|_{i,j} = \frac{f_{i,j+1} - f_{i,j-1}}{2\Delta y} + O(\Delta y^2) \tag{10.48}$$

$$\left.\frac{\partial f}{\partial y}\right|_{i+1,j} = \frac{f_{i+1,j+1} - f_{i+1,j-1}}{2\Delta y} + O(\Delta y^2) \tag{10.49}$$

Inserting these expressions in the first equation, we have:

$$\left.\frac{\partial^2 f}{\partial x \partial y}\right|_{i,j} = \frac{1}{\Delta x}\left(\frac{f_{i,j+1} - f_{i,j-1}}{2\Delta y} - \frac{f_{i-1,j+1} - f_{i-1,j-1}}{2\Delta y}\right) + O(\Delta x, \Delta y^2) \tag{10.50}$$

where the last term describes the order of accuracy of the finite difference approximation.

Example 10.2 Further application of finite difference approach

Apply the finite difference approach to discretize the derivative in the previous example with an order of accuracy of $O(\Delta x^2, \Delta y^2)$.

$$\left.\frac{\partial^2 f}{\partial x \partial y}\right|_{i,j} = \frac{1}{\Delta x}\left(\frac{f_{i+1,j+1} - f_{i+1,j-1}}{2\Delta y} - \frac{f_{i-1,j+1} - f_{i-1,j-1}}{2\Delta y}\right) + O(\Delta x^2, \Delta y^2) \tag{10.51}$$

The unsteady 1D diffusion equation is:

$$\frac{\partial f}{\partial t} = \Gamma \frac{\partial^2 f}{\partial x^2} \quad a \le x \le b \tag{10.52}$$

with initial condition:

$$f(x, 0) = f(x) \tag{10.53}$$

and boundary conditions:

$$f(a, t) = f_1(t) \tag{10.54}$$

and

$$f(b, t) = f_2(t) \tag{10.55}$$

where in this equation $f(x,t)$ is the main variable of the problem, such as temperature or concentration. Γ is the diffusion coefficient.

Several methods are available for discretization of the different terms in this equation which will be discussed in detail in the next sections.

For further details and information on the finite difference method and its application please see: Causon and Mingham (2010); LeVeque (2007); Smith (1985); Thomas (1995). The rest of this section deals with the different methods of discretization. As mentioned in the previous example, several methods are available for discretization of the different terms in the equation of Example 10.2. Two methods are presented next: the first order explicit Euler method and the first order implicit Euler method.

10.2.1 Euler Method

In a simple approach, discretization in the spatial points ith and nth is made. The temporal term can be discretized using a first order forward approximation (explicit method) and the spatial term is discretized using a second order central approximation, as follows:

$$\left.\frac{\partial f}{\partial t}\right|_i^n = \Gamma \frac{\partial^2 f^n}{\partial x^2{}_i} \tag{10.56}$$

$$\left.\frac{\partial f}{\partial t}\right|_i^n \cong \frac{f_i^{n+1} - f_i^n}{\Delta t} \tag{10.57}$$

$$\left.\frac{\partial^2 f}{\partial x^2}\right|_i^n \cong \frac{f_{i-1}^n - 2f_i^n + f_{i+1}^n}{\Delta x^2} \tag{10.58}$$

Here the upper indexes are related to time and the lower indexes are related to space.

Finally we have:

$$f_i^{n+1} = f_i^n + r(f_{i-1}^n - 2f_i^n + f_{i+1}^n) \tag{10.59}$$

where

$$r = \Gamma \frac{\Delta t}{\Delta x^2} \tag{10.60}$$

To solve this equation, the domain needs to be divided by a number of nodes. A distance of Δx is taken between each two nodes and the time step is assumed to be Δt. Selection of Δx and Δt is very important. Increasing Δx will result in a less accurate solution. Therefore, Δx should be selected to be as small as possible so that the lower values do not affect the solution.

A first order backward approximation (implicit method) can be also used to discretize the temporal term.

$$\left.\frac{\partial f}{\partial t}\right|_i^{n+1} = \Gamma \left.\frac{\partial^2 f}{\partial x^2}\right|_i^{n+1} \tag{10.61}$$

$$\left.\frac{\partial f}{\partial t}\right|_i^{n+1} \cong \frac{f_i^{n+1} - f_i^n}{\Delta t} \tag{10.62}$$

$$\left.\frac{\partial^2 f}{\partial x^2}\right|_i^{n+1} \cong \frac{f_{i-1}^{n+1} - 2f_i^{n+1} + f_{i+1}^{n+1}}{\Delta x^2} \tag{10.63}$$

Finally we get:

$$-rf_{i-1}^{n+1} + (1+2r)f_i^{n+1} - rf_{i+1}^{n+1} = f_i^n \tag{10.64}$$

A set of equations then needs to be solved. However, this approach is always stable and does not need any additional condition for relation between Δx and Δt for stability.

10.2.2 Crank-Nicelson Method

The Crank-Nicolson scheme is the base of all the implicit finite difference methods for solving partial differential equations. This method was proposed by John Crank and Phyllis Nicolson (Crank and Nicolson, 1947). The scheme basically takes a mean of forward Euler and backward Euler finite difference approximations.

In this approach the temporal derivatives are estimated by a first order forward approximation. The spatial derivatives are estimated using the central approximation with two time steps (see: Crank and Nicolson, 1947; Crank and Nicolson, 1996):

$$\left.\frac{\partial^2 f}{\partial x^2}\right|_i^n \cong \frac{1}{2}\left[\frac{f_{i-1}^n - 2f_i^n + f_{i+1}^n}{\Delta x^2} + \frac{f_{i-1}^{n+1} - 2f_i^{n+1} + f_{i+1}^{n+1}}{\Delta x^2}\right] \tag{10.65}$$

Thus we obtain the following equation:

$$-rf_{i-1}^{n+1} + 2(1+r)f_i^{n+1} - rf_{i+1}^{n+1} = rf_{i-1}^n + 2(1-r)f_i^n + rf_{i+1}^n \tag{10.66}$$

where

$$r = \Gamma \frac{\Delta t}{\Delta x^2} \tag{10.67}$$

Example 10.3 Initial value problem (diffusion equation)

We are given a diffusing quantity (f) that varies in distance (x) and time (t): $f(x,t)$.

We are also given an initial value over the model space:

$$f(x, t = 0) \tag{10.68}$$

We want to find out how f propagates in time and space.

In one dimension, the equation is:

$$\frac{\partial f}{\partial t} = \delta \frac{\partial}{\partial x}\left(\frac{\partial f}{\partial x}\right) + \vartheta \tag{10.69}$$

where δ is diffusivity, and ϑ is a source or sink (same unit as the flux).

The continuous differential terms are approximated by the finite difference method on a discrete mesh in space with time as follows:

$$\frac{\partial f}{\partial t} \cong \frac{f_i^{n+1} - f_i^n}{\Delta t} \tag{10.70}$$

$$\frac{\partial^2 f}{\partial x^2} \cong \frac{f_{i+1}^n - 2f_i^n + f_{i-1}^n}{\Delta x^2} \tag{10.71}$$

where Δt is the time step size, f_i^{n+1} and f_i^n are the values of parameter f for node number i at time step $n+1$ and n respectively and f_{i+1}^n and f_{i-1}^n are the values of parameter f at time step n for node numbers $i+1$ *and* $i-1$ respectively.

Therefore, the finite difference approximation with the fully explicit method is:

$$\frac{f_i^{n+1} - f_i^n}{\Delta t} = \delta \frac{f_{i+1}^n - 2f_i^n + f_{i-1}^n}{\Delta x^2} + \vartheta \tag{10.72}$$

$$f_i^{n+1} = \left(\frac{\delta \Delta t}{\Delta x^2}\right) f_{i+1}^n - 2f_i^n + f_{i-1}^n + \vartheta \Delta t \tag{10.73}$$

To maintain stability we need $\lambda = \frac{2\delta \Delta t}{\Delta x^2} \leq 1$. This is the Courant condition.

Considering the finite difference approximation in terms of Taylor's series we can analyze the accuracy of the approximation:

$$f_{i-1} - f_i = -\Delta x \frac{\partial f}{\partial x} + \frac{\Delta x^2}{2!}\frac{\partial^2 f}{\partial x^2} - \frac{\Delta x^3}{3!}\frac{\partial^3 f}{\partial x^3} + O(\Delta x^4) \tag{10.74}$$

$$f_{i+1} - f_i = +\Delta x \frac{\partial f}{\partial x} + \frac{\Delta x^2}{2!}\frac{\partial^2 f}{\partial x^2} + \frac{\Delta x^3}{3!}\frac{\partial^3 f}{\partial x^3} + O(\Delta x^4) \tag{10.75}$$

where the term O indicates the magnitude of the remainder.

Then we have

$$\frac{\partial f}{\partial x} = \frac{f_{i+1} - f_i}{\Delta x} + O(\Delta x) \tag{10.76}$$

or

$$\frac{\partial f}{\partial x} = \frac{f_{i-1} - f_{i+1}}{2\Delta x} + O(\Delta x^2) \tag{10.77}$$

which are first-order and second-order accurate equations respectively.

$$\frac{\partial^2 f}{\partial x^2} = \frac{f_{i+1} - 2f_i + f_{i-1}}{\Delta x^2} + O(\Delta x)^2 \tag{10.78}$$

The second derivative is also second-order accuracy.

For the fully implicit method:

$$\frac{d^2f}{dx^2} \cong \frac{f_{i+1}^{n+1} - 2f_i^{n+1} + f_{i-1}^{n+1}}{\Delta x^2} \tag{10.79}$$

The implicit method is more stable than the explicit, but still only first-order accurate.

Based on the Crank-Nicolson method by combining the implicit and explicit methods:

$$\frac{f_i^{n+1} - f_i^n}{\Delta t} = \frac{1}{2}\delta\frac{(f_{i+1}^{n+1} - 2f_i^{n+1} + f_{i-1}^{n+1}) + (f_{i+1}^n - 2f_i^n + f_{i-1}^n)}{\Delta x^2} + \vartheta \tag{10.80}$$

or

$$f_{j-1}^{n+1} - 2\left(1 + \frac{\Delta x^2}{\delta\Delta t}\right)f_j^{n+1} + f_{j+1}^{n+1} = -2\frac{\vartheta}{\delta}\Delta x^2 - f_{j-1}^n + 2\left(1 - \frac{\Delta x^2}{\delta\Delta t}\right)f_j^n - f_{j+1}^n \tag{10.81}$$

Therefore we obtain the matrix as follows:

$$\begin{bmatrix} 1 & -2\left(1+\frac{\Delta x^2}{\delta\Delta t}\right) & 1 & 0 & 0 & 0 & 0 & 0 \\ 0 & 1 & -2\left(1+\frac{\Delta x^2}{\delta\Delta t}\right) & 1 & 0 & 0 & 0 & 0 \\ 0 & 0 & 1 & -2\left(1+\frac{\Delta x^2}{\delta\Delta t}\right) & 1 & 0 & 0 & 0 \\ \cdot & & & \cdot & & \cdot & \cdot & \cdot \\ \cdot & & & \cdot & & \cdot & \cdot & \cdot \\ \cdot & & & \cdot & & \cdot & \cdot & \cdot \\ 0 & 0 & 0 & 0 & 0 & 1 & -2\left(1+\frac{\Delta x^2}{\delta\Delta t}\right) & 1 \end{bmatrix}$$

$$\times \begin{bmatrix} f_0^{n+1} \\ f_1^{n+1} \\ f_2^{n+1} \\ \cdot \\ \cdot \\ \cdot \\ f_{N+1}^{n+1} \end{bmatrix} = \begin{bmatrix} -2\frac{\vartheta}{\delta}\Delta x^2 - f_0^n + 2\left(1-\frac{\Delta x^2}{\delta\Delta t}\right)f_1^n - f_2^n \\ -2\frac{\vartheta}{\delta}\Delta x^2 - f_1^n + 2\left(1-\frac{\Delta x^2}{\delta\Delta t}\right)f_2^n - f_3^n \\ -2\frac{\vartheta}{\delta}\Delta x^2 - f_2^n + 2\left(1-\frac{\Delta x^2}{\delta\Delta t}\right)f_3^n - f_4^n \\ \cdot \\ \cdot \\ \cdot \\ -2\frac{\vartheta}{\delta}\Delta x^2 - f_{N-1}^n + 2\left(1-\frac{\Delta x^2}{\delta\Delta t}\right)f_N^n - f_{N+1}^n \end{bmatrix} \tag{10.82}$$

In the case of a Dirichlet boundary condition with constant values for the first node we have:

$$f_0 = \text{constant (for } i = 1) \tag{10.83}$$

Thus the first equation in the matrix will be:

$$-2\left(1 + \frac{\Delta x^2}{\delta \Delta t}\right) f_1^{n+1} + f_2^{n+1} = -\frac{\vartheta \Delta x^2}{\delta} - 2f_0^n + 2\left(1 - \frac{\Delta x^2}{\delta \Delta t}\right) f_1^n - f_2^n \tag{10.84}$$

In the case of a two-dimensional diffusion equation we have:

$$\frac{\partial f}{\partial t} = \delta\left(\frac{\partial^2 f}{\partial x^2} + \frac{\partial^2 f}{\partial y^2}\right) + \vartheta \tag{10.85}$$

This equation is solved with the explicit, implicit or Crank-Nicolson methods using a similar procedure to that explained previously.

Assuming the same grid spacing in both x and y, we can apply the alternating direction implicit (ADI) technique (which will be discussed further and in detail later) with the Crank-Nicolson approach for two dimensions as follows:

$$-\frac{2(f_i^n - f_i^{n+1/2})}{\Delta t} = \frac{(f_{i,j-1}^n - 2f_{i,j}^n + f_{i,j+1}^n) + (f_{i-1,j}^{n+1/2} - 2f_{i,j}^{n+1/2} + f_{i+1,j}^{n+1/2})}{\Delta x^2}\delta + \vartheta$$
$$(\Delta x = \Delta y) \tag{10.86}$$

$$-\frac{2(f_i^{n+1/2} - f_i^{n+1})}{\Delta t} = \frac{(f_{i-1,j}^{n+1/2} - 2f_{i,j}^{n+1/2} + f_{i+1,j}^{n+1/2}) + (f_{i,j-1}^{n+1} - 2f_{i,j}^{n+1} + f_{i,j+1}^{n+1})}{\Delta x^2}\delta + \vartheta$$
$$(\Delta x = \Delta y) \tag{10.87}$$

and rearranging the equations as before we have:

$$\begin{aligned} f_{i-1,j}^{n+1/2} - 2\left(1 + \frac{\Delta x^2}{\delta \Delta t}\right) f_{i,j}^{n+1/2} + f_{i+1,j}^{n+1/2} &= -\frac{\vartheta \Delta x^2}{2\delta} - f_{i,j-1}^n \\ &+ 2\left(1 - \frac{\Delta x^2}{\delta \Delta t}\right) f_{i,j}^n - f_{i,j+1}^n \end{aligned} \tag{10.88}$$

$$
\begin{aligned}
f_{i,j-1}^{n+1} - 2\left(1+\frac{\Delta x^2}{\delta\Delta t}\right)f_{i,j}^{n+1} + f_{i,j+1}^{n+1} &= -\frac{\vartheta\Delta x^2}{2\delta} - f_{i-1,j}^{n+1/2} \\
&+ 2\left(1-\frac{\Delta x^2}{\delta\Delta t}\right)f_{i,j}^{n+1/2} - f_{i+1,j}^{n+1/2}
\end{aligned} \tag{10.89}
$$

The boundary conditions should also be considered, similar to before. For a Dirichlet condition in two-dimensional conditions for the bottom boundary we have:

$$
f_{i,0} = \text{constant} \tag{10.90}
$$

Thus we get:

$$
\begin{aligned}
f_{i-1,1}^{n+1/2} - 2\left(1+\frac{\Delta x^2}{\delta\Delta t}\right)f_{i,1}^{n+1/2} + f_{i+1,1}^{n+1/2} &= -\frac{\vartheta\Delta x^2}{2\delta} - f_{i,0}^{n} \\
&+ 2\left(1-\frac{\Delta x^2}{\delta\Delta t}\right)f_{i,1}^{n} - f_{i,2}^{n}. \quad (j=1)
\end{aligned} \tag{10.91}
$$

$$
\begin{aligned}
-2\left(1+\frac{\Delta x^2}{\delta\Delta t}\right)f_{i,1}^{n+1} + f_{i,2}^{n+1} = -\frac{\vartheta\Delta x^2}{2\delta} - f_{i,0}^{n+1/2} - f_{i-1,1}^{n+1/2} + 2\left(1-\frac{\Delta x^2}{\delta\Delta t}\right) \\
f_{i,1}^{n+1/2} - f_{i+1,1}^{n+1/2} \quad (j=1)
\end{aligned} \tag{10.92}
$$

Assuming the same grid spacing in all x, y and z directions, we can similarly extend the equations to three-dimensional ones with the implicit and explicit methods:

$$
\begin{aligned}
2f_{i-1,j,k}^{n+1/3} + 4\left(1+\frac{\Delta x^2}{\delta\Delta t}\right)f_{i,j,k}^{n+1/3} + 2f_{i+1,j,k}^{n+1/3} = 4\left(1-\frac{\Delta x^2}{\delta\Delta t}\right)f_{i,j,k}^{n} - f_{i,j-1,k}^{n} \\
- f_{i,j,k-1}^{n} - f_{i,j,k+1}^{n} - f_{i,j+1,k}^{n}(\Delta x = \Delta y = \Delta z)
\end{aligned} \tag{10.93}
$$

To solve the unsteady 2D diffusion equation, the approaches discussed before for the one-dimensional case are similarly applicable. The equation is:

$$
\frac{\partial f}{\partial t} = \Gamma_x \frac{\partial^2 f}{\partial x^2} + \Gamma_y \frac{\partial^2 f}{\partial y^2} \tag{10.94}
$$

To solve the two-dimensional diffusion equation based on a fully implicit method, a backward difference approximation is used for the temporal derivatives while the central difference approximation is used for both spatial derivatives in two directions (x and y) in the governing equation. The equation is discretized as follows:

$$\left.\frac{\partial f}{\partial t}\right|_{i,j}^{n+1} = \Gamma_x \left.\frac{\partial^2 f}{\partial x^2}\right|_{i,j}^{n+1} + \Gamma_y \left.\frac{\partial^2 f}{\partial y^2}\right|_{i,j}^{n+1} \tag{10.95}$$

$$\left.\frac{\partial f}{\partial t}\right|_{i,j}^{n+1} \cong \frac{f_{i,j}^{n+1} - f_{i,j}^{n}}{\Delta t} \tag{10.96}$$

$$\left.\frac{\partial^2 f}{\partial x^2}\right|_{i,j}^{n+1} \cong \frac{f_{i-1,j}^{n+1} - 2f_{i,j}^{n+1} + f_{i+1,j}^{n+1}}{\Delta x^2} \tag{10.97}$$

$$\left.\frac{\partial^2 f}{\partial y^2}\right|_{i,j}^{n+1} \cong \frac{f_{i,j-1}^{n+1} - 2f_{i,j}^{n+1} + f_{i,j+1}^{n+1}}{\Delta y^2} \tag{10.98}$$

where

$$r_x = \Gamma_x \frac{\Delta t}{\Delta x^2} \tag{10.99}$$

and

$$r_y = \Gamma_y \frac{\Delta t}{\Delta y^2} \tag{10.100}$$

Finally we get:

$$-r_x(f_{i-1,j}^{n+1} + f_{i+1,j}^{n+1}) + (1 - 2r_x - 2r_y)f_{i,j}^{n+1} - r_y(f_{i,j-1}^{n+1} + f_{i,j+1}^{n+1}) = f_{i,j}^{n} \tag{10.101}$$

This method is unconditionally stable.

However, according to the Crank-Nicelson method (Crank and Nicolson, 1947; Crank and Nicolson, 1996) to solve the two-dimensional diffusion equation, a first order forward approximation is used for the temporal derivative while a central approximation is applied for both spatial derivatives in two directions of x and y as follows:

$$\left.\frac{\partial f}{\partial t}\right|_{i,j}^{n} = \Gamma_x \left.\frac{\partial^2 f}{\partial x^2}\right|_{i,j}^{n} + \Gamma_y \left.\frac{\partial^2 f}{\partial y^2}\right|_{i,j}^{n} \tag{10.102}$$

$$\left.\frac{\partial f}{\partial t}\right|_{i,j}^{n} \cong \frac{f_{i,j}^{n+1} - f_{i,j}^{n}}{\Delta t} \tag{10.103}$$

$$\left.\frac{\partial^2 f}{\partial x^2}\right|_{i,j}^{n} \cong \frac{1}{2}\left[\frac{f_{i-1,j}^{n} - 2f_{i,j}^{n} + f_{i+1,j}^{n}}{\Delta x^2} + \frac{f_{i-1,j}^{n+1} - 2f_{i,j}^{n+1} + f_{i+1,j}^{n+1}}{\Delta x^2}\right] \tag{10.104}$$

$$\left.\frac{\partial^2 f}{\partial y^2}\right|_{i,j}^{n} \cong \frac{1}{2}\left[\frac{f_{i,j-1}^{n} - 2f_{i,j}^{n} + f_{i,j+1}^{n}}{\Delta y^2} + \frac{f_{i,j-1}^{n+1} - 2f_{i,j}^{n+1} + f_{i,j+1}^{n+1}}{\Delta y^2}\right] \tag{10.105}$$

where

$$r_x = \Gamma_{\mathrm{x}} \frac{\Delta \mathrm{t}}{\Delta \mathrm{x}^2} \tag{10.106}$$

and

$$r_y = \Gamma_{\mathrm{y}} \frac{\Delta \mathrm{t}}{\Delta \mathrm{y}^2} \tag{10.107}$$

Thus:

$$\begin{aligned} &-\frac{r_x}{2}\left(f_{i-1,j}^{n+1} + f_{i+1,j}^{n+1}\right) + (1 + r_x + r_y)f_{i,j}^{n+1} - \frac{r_y}{2}\left(f_{i,j-1}^{n+1} + f_{i,j+1}^{n+1}\right) \\ &= \frac{r_x}{2}\left(f_{i-1,j}^{n} + f_{i+1,j}^{n}\right) + (1 - r_x - r_y)f_{i,j}^{n} + \frac{r_y}{2}\left(f_{i,j-1}^{n} + f_{i,j+1}^{n}\right) \end{aligned} \tag{10.108}$$

10.2.3 The MacCormack Method

The governing equation for unsteady 1D convection–diffusion is also written as follows:

$$\frac{\partial f}{\partial t} + u\frac{\partial f}{\partial x} - \Gamma\frac{\partial^2 f}{\partial x^2} \tag{10.109}$$

where Γ is the diffusion coefficient and u is velocity. Similar to the diffusion equation, different discretization methods can be used. These methods are applicable for two-dimensional cases.

A well-known method which is used to solve this equation is called the MacCormack method. The MacCormack method is used for depth averaged equations in shallow waters. This approach is an explicit method which includes two steps for solving the equation. We have:

$$f_i^{n+1} = f_i^n + \frac{\overline{\partial f}}{\partial t}\Delta t \tag{10.110}$$

where

$$\frac{\overline{\partial f}}{\partial t} = \frac{1}{2}\left(\left(\frac{\partial f}{\partial t}\right)_{(1)} + \left(\frac{\partial f}{\partial t}\right)_{(2)}\right) \tag{10.111}$$

where $\left(\frac{\partial f}{\partial t}\right)_{(1)}$ and $\left(\frac{\partial f}{\partial t}\right)_{(2)}$ are related to the first and second steps of the solution respectively.

In the first step, a first order forward approach is used for the spatial derivatives of the convection part. A central approximation is also used for the diffusion part:

$$\left\langle\frac{\partial f}{\partial t}\right\rangle_{(1)} = \frac{f_{i(1)} - f_i^n}{\Delta t} = -u\frac{f_{i+1}^n - f_i^n}{\Delta x} + \Gamma\frac{f_{i-1}^n - 2f_i^n + f_{i+1}^n}{\Delta x^2} \tag{10.112}$$

Then we will have:

$$f_{i(1)} = rf_{i-1}^n + (1 + q - 2r)f_i^n + (r - q)f_{i+1}^n \tag{10.113}$$

where r is defined as before and q is:

$$q = u\frac{\Delta t}{\Delta x} \tag{10.114}$$

Then in the second step, a first order backward approach is used for spatial derivatives in the convection part. The temporal term and terms in the diffusion part are approximated as before. The known values are transferred from the first step to the second step. Thus we have:

$$\left\langle\frac{\partial f}{\partial t}\right\rangle_{(2)} = \frac{f_{i(2)} - f_{i(1)}}{\Delta t} = -u\frac{f_{i(1)} - f_{i-1(1)}}{\Delta x} + \Gamma\frac{f_{i-1(1)} - 2f_{i(1)} + f_{i+1(1)}}{\Delta x^2} \tag{10.115}$$

and then we get:

$$f_{i(2)} = (q + r)f_{i-1(1)} + (1 - q - 2r)f_{i(1)} + rf_{i+1(1)} \tag{10.116}$$

Thus we have:

$$f_i^{n+1} = \frac{f_{i_{(2)}} + f_i^n}{2} \tag{10.117}$$

So

$$f_i^{n+1} = \frac{q+r}{2} f_{i-1_{(1)}} + \frac{1-q-2r}{2} f_{i_{(1)}} + r f_{i+1_{(1)}} - \frac{1}{2} f_i^n \tag{10.118}$$

10.2.4 The Alternating Direction Implicit Method

The alternating direction implicit (ADI) method is a finite difference method and efficient numerical solution for solving parabolic and elliptic partial differential equations. For example, it is used to solve diffusion equations in two- or three-dimensional problems.

In comparison with the Crank-Nicolson method, the ADI algorithm is faster and easier to use. The ADI method also needs less memory space. (see: Causon and Mingham 2010; LeVeque 2007; Smith 1985; Thomas 1995).

In the rest of this section we implement the ADI method for 2D Laplacian Equation. Consider Laplace's equation in two dimensions as follows:

$$\nabla^2 f = 0 \tag{10.119}$$

$$\frac{\partial^2 f}{\partial x^2} + \frac{\partial^2 f}{\partial y^2} = 0 \tag{10.120}$$

Considering an irrotationally ideal fluid and continuity equation, we can write Laplace's equation for the stream and potential functions respectively as follows:

$$\frac{\partial v}{\partial x} - \frac{\partial u}{\partial y} = 0 \tag{10.121}$$

$$\frac{\partial u}{\partial x} + \frac{\partial v}{\partial y} = 0 \tag{10.122}$$

Hence, the velocity components (u and v) can be expressed in terms of both the stream function (f) and the potential function (g) as follows:

$$u = \frac{\partial f}{\partial y} = -\frac{\partial g}{\partial x} \tag{10.123}$$

$$v = -\frac{\partial f}{\partial x} = -\frac{\partial g}{\partial y} \tag{10.124}$$

Thus, obviously, by substituting these expressions respectively into the irrotationality condition and the continuity equation, we obtain Laplace's equation for the stream and potential functions.

Now replacing the values of the main equation by their finite difference approximations, we have:

$$\frac{\partial^2 f}{\partial x^2} = \frac{f_{i-1,j} - 2f_{i,j} + f_{i+1,j}}{(\Delta x)^2} \tag{10.125}$$

$$\frac{\partial^2 f}{\partial y^2} = \frac{f_{i,j-1} - 2f_{i,j} + f_{i,j+1}}{(\Delta y)^2} \tag{10.126}$$

and now we have:

$$\frac{f_{i-1,j} - 2f_{i,j} + f_{i+1,j}}{(\Delta x)^2} + \frac{f_{i,j-1} - 2f_{i,j} + f_{i,j+1}}{(\Delta y)^2} = 0 \tag{10.127}$$

and therefore we get:

$$f_{i-1,j} - 4f_{i,j} + f_{i+1,j} + f_{i,j-1} + f_{i,j+1} = 0 \tag{10.128}$$

By rearranging:

$$f_{i-1,j} - 4f_{i,j} + f_{i+1,j} = -f_{i,j-1} - f_{i,j+1} \tag{10.129}$$

or

$$f_{i,j-1} - 4f_{i,j} + f_{i,j+1} = -f_{i-1,j} - f_{i+1,j} \tag{10.130}$$

Thus the equation can be used as an iteration formula:

$$f_{i-1,j}^{(r+1)} - 4f_{i,j}^{(r+1)} + f_{i+1,j}^{(r+1)} = -f_{i,j-1}^{(r)} - f_{i,j+1}^{(r)} \tag{10.131}$$

$$f_{i,j-1}^{(r+2)} - 4f_{i,j}^{(r+2)} + f_{i,j+1}^{(r+2)} = -f_{i-1,j}^{(r+1)} - f_{i+1,j}^{(r+1)} \tag{10.132}$$

These equations can be used to compute finite difference approximations at the nodes in the computional domain.

The system of equations for the one-dimensional case always has a tridiagonal coefficient matrix. Therefore, in the case of one-dimensional problems, the Thomas algorithm can be simply used. However, in the case of higher dimensions, for instance, for the two or three-dimensional problems, the ADI method can be applied to get a tridiagonal coefficient matrix.

In this case, regarding the two-dimensional form of the Laplace equation:

$$\nabla^2 f = \frac{\partial^2 f}{\partial x^2} + \frac{\partial^2 f}{\partial y^2} = 0 \tag{10.133}$$

Using i and j indexes to identify the nodes and applying the finite difference method, for each node at iteration $n + 1$ we have

$$f_{i,j}^{n+1} = f_{i,j}^{n} + \frac{(f_{i,j-1}^{n} - 2f_{i,j}^{n} - f_{i,j+1}^{n}) + (f_{i-1,j}^{n} - 2f_{i,j}^{n} + f_{i+1,j}^{n})}{4} \tag{10.134}$$

The two-step procedure of iteration is performed for $n + \frac{1}{2}$ and n + 1 levels as follows:

$$f_{i,j}^{n+\frac{1}{2}} = f_{i,j}^{n} + \frac{\left(f_{i,j-1}^{n+\frac{1}{2}} - 2f_{i,j}^{n+\frac{1}{2}} - f_{i,j+1}^{n+\frac{1}{2}}\right) + (f_{i-1,j}^{n} - 2f_{i,j}^{n} + f_{i+1,j}^{n})}{4} \tag{10.135}$$

$$f_{i,j}^{n+1} = f_{i,j}^{n+\frac{1}{2}} + \frac{\left(f_{i,j-1}^{n+\frac{1}{2}} - 2f_{i,j}^{n+\frac{1}{2}} - f_{i,j+1}^{n+\frac{1}{2}}\right) + (f_{i-1,j}^{n+1} - 2f_{i,j}^{n+1} + f_{i+1,j}^{n+1})}{4} \tag{10.136}$$

Then a tridiagonal matrix at the $n + \frac{1}{2}$ level is generated which can be solved using the two-step iteration. Hence we can compute the new values for $f_{i,j}^{n+1}$ using the algorithm.

Example 10.4 2D Laplace's equation

Assuming a two-dimensional Laplace's equation, determine the values of the physical parameter f at nodes 6, 7, 10 and 11 in Figure 10.2.

$$\begin{aligned} i &= 1, 2, 3 \text{ and } 4 \\ i &= 1, 2, 3 \text{ and } 4 \end{aligned} \tag{10.137}$$

$$\begin{aligned} &f_{1,2}^{n} = 1000, f_{1,3}^{n} = 1000, f_{2,1}^{n} = 1000, \text{and } f_{2,4}^{n} = 2000 \\ &f_{3,1}^{n} = 1000, f_{3,4}^{n} = 2000, f_{4,2}^{n} = 3000, \text{and } f_{4,3}^{n} = 3000 \end{aligned} \tag{10.138}$$

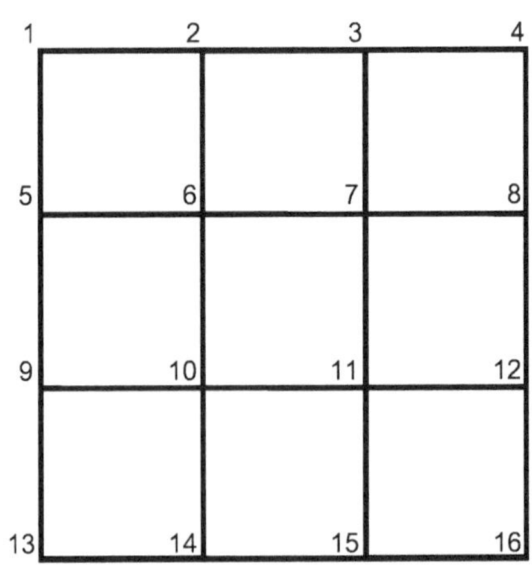

Figure 10.2 The 2D computational domain in a steady-state condition.

As mentioned before, the 2D governing equation for steady state is:

$$\frac{\partial^2 f}{\partial x^2} + \frac{\partial^2 f}{\partial y^2} = 0 \tag{10.139}$$

Applying the ADI method we have:

$$f_{i,j}^{n+1} - f_{i,j}^{n} = \frac{1}{4}\left\{\left[f_{i,j+1}^{n} - 2f_{i,j}^{n} + f_{i,j-1}^{n}\right] + \left[f_{i+1,j}^{n} - 2f_{i,j}^{n} + f_{i-1,j}^{n}\right]\right\} \tag{10.140}$$

and for the first step:

$$f_{i,j}^{n+\frac{1}{2}} - f_{i,j}^{n+\frac{1}{2}} = \frac{1}{4}\left\{\left[f_{i,j+1}^{n+\frac{1}{2}} - 2f_{i,j}^{n+\frac{1}{2}} + f_{i,j-1}^{n+\frac{1}{2}}\right] + \left[f_{i+1,j}^{n} - 2f_{i,j}^{n} + f_{i-1,j}^{n}\right]\right\} \tag{10.141}$$

Assuming $f_{2,2}^{n} = 1000$, $f_{2,3}^{n} = 1500$, $f_{3,2}^{n} = 1500$, and $f_{3,3}^{n} = 2500$, for nodes (2,2) and (2,3) we can write:

$$\text{Node }(2,2)\text{: } f_{2,2}^{n+\frac{1}{2}} - 1000 = \frac{1}{4}\left\{\left[f_{2,3}^{n+\frac{1}{2}} - 2f_{2,2}^{n+\frac{1}{2}} + 1000\right] + [1500 - 2(1000) + 1000]\right\} \tag{10.142}$$

Simplifying:

$$1.5f_{2,2}^{n+\frac{1}{2}} - 0.25f_{2,3}^{n+\frac{1}{2}} = 1375 \tag{10.143}$$

and

$$\text{Node }(2,3)\text{: } f_{2,3}^{n+\frac{1}{2}} - 1500 = \frac{1}{4}\left\{\left[2000 - 2f_{2,3}^{n+\frac{1}{2}} + f_{2,2}^{n+\frac{1}{2}}\right] + [2500 - 2(1500) + 1000]\right\} \tag{10.144}$$

Simplifying:

$$-0.25f_{2,2}^{n+\frac{1}{2}} + 1.5f_{2,3}^{n+\frac{1}{2}} = 2125 \tag{10.145}$$

These two equations in row 2 can be simply solved. Solving the system of two equations we have:

$$f_{2,2}^{n+\frac{1}{2}} = 1185.71, \text{and } f_{2,3}^{n+\frac{1}{2}} = 1614.29 \tag{10.146}$$

In addition considering the nodes (3,2) and (3.3), we can write:

$$\text{Node }(3,2)\ f_{3,2}^{n+\frac{1}{2}} - 1500 = \frac{1}{4}\left\{\left[f_{3,3}^{n+\frac{1}{2}} - 2f_{3,2}^{n+\frac{1}{2}} + 1000\right] + [3000 - 2(1500) + 1000]\right\} \tag{10.147}$$

Simplifying:

$$1.5f_{3,2}^{n+\frac{1}{2}} - 0.25f_{3,3}^{n+\frac{1}{2}} = 2000 \tag{10.148}$$

and

$$\text{Node }(3,3)\ f_{3,3}^{n+\frac{1}{2}} - 2500 = \frac{1}{4}\left\{\left[2000 - f_{3,3}^{n+\frac{1}{2}} + f_{3,2}^{n+\frac{1}{2}}\right] + [3000 - 2(2500) + 1500]\right\} \tag{10.149}$$

Simplifying:

$$-0.25f_{3,2}^{n+\frac{1}{2}} + 1.25f_{3,3}^{n+\frac{1}{2}} = 2875 \quad (10.150)$$

These two equations can be solved similarly. Thus we calculate:

$$f_{3,2}^{n+\frac{1}{2}} = 1775.86, \text{and } f_{3,3}^{n+\frac{1}{2}} = 2655.87 \quad (10.151)$$

In the second step we have:

$$f_{i,j}^{n+1} - f_{i,j}^{n+\frac{1}{2}} = \frac{1}{4}\left\{\left[f_{i,j+1}^{n+\frac{1}{2}} - 2f_{i,j}^{n+\frac{1}{2}} + f_{i,j-1}^{n+\frac{1}{2}}\right] + \left[f_{i+1,j}^{n+1} - 2f_{i,j}^{n+1} + f_{i-1,j}^{n+1}\right]\right\} \quad (10.152)$$

Then using a similar approach to that in the first step, we can rewrite the equations for nodes (2,2) and (3,2) as follows:

$$\text{Node } (2,2)\ f_{2,2}^{n+1} - 1185.71 = \frac{1}{4}\left\{[1614.29 - 2(1185.71) + 1000] + \left[f_{3,2}^{n+1} - 2f_{2,2}^{n+1} + 1000\right]\right\} \quad (10.153)$$

which can be easily simplified to:

$$1.5f_{2,2}^{n+1} - 0.25f_{3,2}^{n+1} = 1496.43 \quad (10.154)$$

and

$$\text{Node}(3,2)\ f_{3,2}^{n+1} - 1775.86 = \frac{1}{4}\left\{[2657.17 - 2(1775.86) + 1000] + \left[3000 - 2f_{3,2}^{n+1} + f_{2,2}^{n+1}\right]\right\} \quad (10.155)$$

Simplifying:

$$-0.25f_{2,2}^{n+1} + 1.5f_{3,2}^{n+1} = 2552.22 \quad (10.156)$$

Therefore, these two equations can also be solved easily:

$$f_{2,2}^{n+1} = 1317.81, \text{and } f_{3,2}^{n+1} = 1921.12 \quad (10.157)$$

For nodes (2,3) and (3,3) we can write:

$$\text{Node } (2,3)\ f_{2,3}^{n+1} - 1614.29 = \frac{1}{4}\left\{[2000 - 2(2655.17) + 1185.71] + \left[f_{3,3}^{n+1} - 2f_{2,3}^{n+1} + 1000\right]\right\} \quad (10.158)$$

By simplifying:

$$1.5f_{2,3}^{n+1} - 0.25f_{3,3}^{n+1} = 1333.13 \quad (10.159)$$

and

$$\text{Node } (3,3)\ f_{2,2}^{n+1} - 1185.71 = \frac{1}{4}\left\{[1614.29 - 2(1185.71) + 1000] + \left[f_{3,2}^{n+1} - 2f_{2,2}^{n+1} + 1000\right]\right\} \quad (10.160)$$

which is simplified to:

$$-0.25f_{2,3}^{n+1} + 1.5f_{3,3}^{n+1} = 1496.43 \tag{10.161}$$

Solving this set of two equations, we calculate the answers as follows:

$$f_{2,3}^{n+1} = 1258.89, \text{and } f_{3,3}^{n+1} = 2220.85 \tag{10.162}$$

For further details and information on the finite difference method and its application please see: Causon and Mingham (2010); LeVeque (2007); Smith (1985); Thomas (1995).

Example 10.5 Analyzing streamlines using polar coordinates

Here is another example of application of the finite difference method to solve the PDEs (in this case by considering the Laplace's equation) numerically, however using the polar coordinates. Considering the streamlines for incompressible, inviscid, irrotational, two-dimensional flow, we can analyze the flow around a structure. The streamlines can clearly illustrate variation of pressure and velocity field around the structure. The polar coordinates are more appropriate in many practical problems with complex geometries and can typically be used to avoid complexity, for example near the boundary around the head of breakwaters and coastal structures. This is, for example, the case regarding the study of wave diffraction around the breakwater roundhead. The stream function is specified according to an elliptic linear PDE. This equation is known as Laplace's equation expressed in polar coordinates as:

$$\frac{\partial^2 f}{\partial r^2} + \frac{1}{r}\frac{\partial f}{\partial r} + \frac{1}{r^2}\frac{\partial^2 f}{\partial \theta^2} = 0 \tag{10.163}$$

Firstly, we need to define the nodes for our computational mesh in that specific problem which needs to be solved. The nodes are selected at the intersection of equally spaced semicircles around the head of the breakwater with straight lines originating from the center of the breakwater roundhead. The polar coordinates are applied. A finite difference method is also applied to solve the governing equation. A matrix for the stream function is obtained. A matrix is obtained. The Gauss-Seidel iterative technique is applied to solve the matrix. Then Laplace's equation can be solved numerically. We consider uniform flow velocity (equals unity). One boundary is located at the boundary of the breakwater roundhead ($f = 0$) which is given by

$$\sin\theta\left(\frac{r^2 - c^2}{r}\right) = 0 \tag{10.164}$$

where c is radius. This flow streamline is satisfied by $\theta = 0$, π, or $r = c$.

The other boundary is located at a specific distance away from the breakwater roundhead and is a function of the angular position around the breakwater roundhead.

We use an r-θ plane with $\theta = 0°$ to $180°$ and $r = r_1$ to r_2. Δr and $\Delta\theta$ are selected so that we have:

$$r = r_1 + i\Delta r \quad \text{for} \quad i = 0, 1, 2, 3, \ldots, n \tag{10.165}$$

and

$$\theta = j\Delta\theta \quad \text{for} \quad j = 0, 1, 2, 3, \ldots, m \tag{10.166}$$

This results in a total of $(n + 1)(m + 1)$ nodes. We can write:

$$\frac{\partial^2 f}{\partial r^2} = \frac{f_{i-1,j} - 2f_{i,j} + f_{i+1,j}}{\Delta r^2} \tag{10.167}$$

$$\frac{\partial f}{\partial r} = -\frac{f_{i-1,j} - f_{i+1,j}}{2\Delta r} \tag{10.168}$$

$$\frac{\partial^2 f}{\partial \theta^2} = \frac{f_{i,j-1} - 2f_{i,j} + f_{i,j+1}}{\Delta \theta^2} \tag{10.169}$$

Substituting into Laplace's equation we have:

$$-\frac{1}{c + i\Delta r}\frac{f_{i-1,j} - f_{i+1,j}}{2\Delta r} + \frac{f_{i-1,j} - 2f_{i,j} + f_{i+1,j}}{\Delta r^2} + \frac{1}{(c+i\Delta r)^2}\frac{f_{i,j-1} - 2f_{i,j} + f_{i,j+1}}{\Delta \theta^2} = 0 \tag{10.170}$$

And then

$$f_{i-1,j}\left[\frac{1}{\Delta r^2} - \frac{1}{(c + i\Delta r)(2\Delta r)}\right] + f_{i,j}\left[\frac{-2}{\Delta r^2} - \frac{2}{\Delta\theta^2(c+i\Delta r)^2}\right] + f_{i,j-1}\left[\frac{1}{\Delta\theta^2(c+i\Delta r)^2}\right]$$
$$+ f_{i+1,j}\left[\frac{1}{\Delta r^2} + \frac{1}{(c + i\Delta r)(2\Delta r)}\right] + f_{i,j+1}\left[\frac{1}{\Delta\theta^2(c+i\Delta r)^2}\right] = 0 \tag{10.171}$$

Therefore, the finite difference approximations are written for the unknown nodes. The function is known at the boundaries. Thus, we will obtain the same number of finite difference equations as the unknowns. A set of linear equations is generated which can be solved by the direct or iterative methods (for example methods such as Thomas Algorithm, Jacobi, Gauss-Seidel, etc.) as will be discussed later. A stopping criteria for the iterative procedure is considered according to the computed required convergence factor. Then the values of the stream functions at each node are calculated. The methods for solving systems of linear algebraic equations including the direct or iterative methods will be discussed in detail in the next section.

10.3 SOLUTION OF DISCRETIZED EQUATIONS

There are two types of methods for solving a system of linear algebraic equations: direct or iterative. Direct methods include, for example, Cramer's rule matrix inversion, Gaussian elimination and Thomas Algorithm methods. Iterative methods include, for instance, Jacobi, Gauss-Seidel (Liebmann method) and successive over-relaxation (SOR) iteration methods. There are many other methods available to solve a system of linear equations. In addition, many methods have been proposed for a system of nonlinear equations. For further details and information about other methods for both systems of linear and nonlinear equations please see: Davis (2006); Greenbaum (1997); Kelley (2003); Kelley (1995); Olshanskii and Tyrtyshnikov (2014); Saad (2003).

Since the systems of equations in many practical problems are often very large, i.e. several thousand and more (sometimes even an order of 10^5 and 10^6), direct methods need a huge storage space and are therefore less applicable for practical reasons. In these cases, iterative methods are more applicable. The iterative methods only save the nonzero numbers, so they can save significant memory space.

10.3.1 Direct Methods

The tridiagonal matrix algorithm, or Thomas algorithm, is an example of a direct method (Thomas, 1949). The Thomas algorithm is actually a simplified form of the Gaussian elimination method. It can be employed to solve tridiagonal systems of equations. The method can be applied iteratively for solving multidimensional problems. It is widely applied in different fluid dynamics problems.

Using the Gaussian elimination method for a set of linear equations with a tridiagonal coefficient matrix obviously reduces the computational time and costs considerably. Therefore, it is preferred to work with a tridiagonal coefficient matrix rather than a general set of equations. Consider the following tridiagonal system of equations:

$$\begin{bmatrix} a_{11} & a_{12} & 0 & 0 & . & . & . & 0 \\ a_{21} & a_{22} & a_{23} & 0 & . & . & . & 0 \\ 0 & a_{32} & a_{33} & a_{34} & . & . & . & 0 \\ . & . & . & . & . & . & . & . \\ . & . & . & . & . & . & . & . \\ . & . & . & . & . & . & . & . \\ 0 & . & . & . & 0 & 0 & a_{nn-1} & a_{nn} \end{bmatrix} \begin{bmatrix} x_1 \\ x_2 \\ x_3 \\ . \\ . \\ x_n \end{bmatrix} = \begin{bmatrix} b_1 \\ b_2 \\ b_3 \\ . \\ . \\ b_n \end{bmatrix} \tag{10.172}$$

or

$$AX = B \tag{10.173}$$

where a_{ij} is the diagonal element in the ith column and jth row of the matrix A. Based on the Gaussian elimination, x_1 can be eliminated from the system of equations by using the first equation to eliminate x_1 from the second equation. In a similar manner, the second equation is applied to eliminate x_2 from the third equation. The process is repeated until the nth equation, where there are only two unknowns x_{n-1} and x_n in this equation. Thus, the elimination of x_{n-1} provides the answer for x_n. Back substitution to the $(n-1)$th equation provides us the answer for x_{n-1}. Similarly, the solution for x_{n-2} can be obtained from knowing x_{i-1}. The process is repeated until the first equation where the unknown x_1 is determined finally. This procedure is called the Thomas algorithm. For further details, please see: Davis (2006).

Gaussian Elimination

Gaussian elimination is the name of a method to solve a system of linear equations. For example:

$$\mathrm{a}_1.x_1 + \mathrm{b}_1.x_2 = y_1 \tag{10.174}$$

$$\mathrm{a}_2.x_1 + \mathrm{b}_2.x_2 = y_2 \tag{10.175}$$

The solutions for x_1 and x_2 could be written in determinant form as:

$$x_1 = \frac{\begin{vmatrix} y_1 & b_1 \\ y_2 & b_2 \end{vmatrix}}{\begin{vmatrix} a_1 & b_1 \\ a_2 & b_2 \end{vmatrix}} \tag{10.176}$$

and

$$x_2 = \frac{\begin{vmatrix} a_1 & y_1 \\ a_2 & y_2 \end{vmatrix}}{\begin{vmatrix} a_1 & b_1 \\ a_2 & b_2 \end{vmatrix}} \tag{10.177}$$

We could extend this approach to a system of three equations as follows:

$$\begin{cases} a_1.x_1 + b_1.x_2 + c_1.x_3 = y_1 \\ a_2.x_1 + b_2.x_2 + c_3.x_3 = y_2 \\ a_3.x_1 + b_3.x_2 + c_3.x_3 = y_3 \end{cases} \tag{10.178}$$

Using the format for a 2×2 determinant that we introduced previously, this could be written as:

$$x_1 = \frac{\begin{vmatrix} y_1 & b_1 & c_1 \\ y_2 & b_2 & c_2 \\ y_3 & b_3 & c_3 \end{vmatrix}}{\begin{vmatrix} a_1 & b_1 & c_1 \\ a_2 & b_2 & c_2 \\ a_3 & b_3 & c_3 \end{vmatrix}} = \frac{y_1 . \begin{vmatrix} b_2 & c_2 \\ b_3 & c_3 \end{vmatrix} - b_1 . \begin{vmatrix} y_2 & c_2 \\ y_3 & c_3 \end{vmatrix} + c_1 . \begin{vmatrix} y_2 & b_2 \\ y_3 & b_3 \end{vmatrix}}{a_1 . \begin{vmatrix} b_2 & c_2 \\ b_3 & c_3 \end{vmatrix} - b_1 . \begin{vmatrix} a_2 & c_2 \\ a_3 & c_3 \end{vmatrix} + c_1 . \begin{vmatrix} a_2 & b_2 \\ a_3 & b_3 \end{vmatrix}} \tag{10.179}$$

and

$$x_2 = \frac{\begin{vmatrix} a_1 & y_1 & c_1 \\ a_2 & y_2 & c_2 \\ a_3 & y_3 & c_3 \end{vmatrix}}{\begin{vmatrix} a_1 & b_1 & c_1 \\ a_2 & b_2 & c_2 \\ a_3 & b_3 & c_3 \end{vmatrix}} = \frac{a_1 . \begin{vmatrix} y_2 & c_2 \\ y_3 & c_3 \end{vmatrix} - y_1 \begin{vmatrix} a_2 & c_2 \\ a_3 & c_3 \end{vmatrix} + c_1 . \begin{vmatrix} a_2 & y_2 \\ a_3 & y_3 \end{vmatrix}}{a_1 . \begin{vmatrix} b_2 & c_2 \\ b_3 & c_3 \end{vmatrix} - b_1 . \begin{vmatrix} a_2 & c_2 \\ a_3 & c_3 \end{vmatrix} + c_1 . \begin{vmatrix} a_2 & b_2 \\ a_3 & b_3 \end{vmatrix}} \tag{10.180}$$

and

$$x_1 = \frac{\begin{vmatrix} a_1 & b_1 & y_1 \\ a_2 & b_2 & y_2 \\ a_3 & b_3 & y_3 \end{vmatrix}}{\begin{vmatrix} a_1 & b_1 & c_1 \\ a_2 & b_2 & c_2 \\ a_3 & b_3 & c_3 \end{vmatrix}} = \frac{a_1 . \begin{vmatrix} b_2 & y_2 \\ b_3 & y_3 \end{vmatrix} - b_1 . \begin{vmatrix} a_2 & y_2 \\ a_3 & y_3 \end{vmatrix} + y_1 . \begin{vmatrix} a_2 & b_2 \\ a_3 & b_3 \end{vmatrix}}{a_1 . \begin{vmatrix} b_2 & c_2 \\ b_3 & c_3 \end{vmatrix} - b_1 . \begin{vmatrix} a_2 & c_2 \\ a_3 & c_3 \end{vmatrix} + c_1 . \begin{vmatrix} a_2 & b_2 \\ a_3 & b_3 \end{vmatrix}} \tag{10.181}$$

This is basically Cramer's rule for solving linear equations, which is only applicable for a square matrix.

The determinant for a matrix of size $N \times N$ is evaluated from determinants of N smaller matrices of size $(N-1) \times (N-1)$ by excluding the row and column of the corresponding cell. The evaluation is carried out by taking any element in the matrix, and multiplying it by the corresponding determinant and with an appropriate sign as follows: If the row number is an odd number, a + sign for the first element, a − sign for the second element, a + sign for the third element, a − sign for the fourth element, and so on should be used, and if the row number is an even number, a − sign for the first element, a + sign for the second element, a − sign for the third element, a + sign for the fourth element, etc. should be selected.

Each element of the matrix A is subscripted and identified by the row and column indexes as follows:

$$\begin{cases} a_1.x_1 + b_1.x_2 + c_1.x_3 + \ldots = y_1 \\ a_2.x_1 + b_2.x_2 + c_2.x_3 + \ldots = y_2 \\ a_3.x_1 + b_3.x_2 + c_3.x_3 + \ldots = y_3 \\ \qquad\vdots \\ a_n.x_1 + b_n.x_2 + c_n.x_3 + \ldots = y_n \end{cases} \tag{10.182}$$

This could be written $Mx = y$ where x and y are defined as:

$$A = \begin{bmatrix} a_{11} & a_{21} & a_{31} & . & . & . & a_{N1} \\ a_{12} & a_{22} & a_{32} & . & . & . & a_{N2} \\ a_{13} & a_{23} & a_{33} & . & . & . & a_{N3} \\ . & . & . & . & . & . & . \\ . & . & . & . & . & . & . \\ . & . & . & . & . & . & . \\ a_{1N} & a_{2N} & a_{3N} & . & . & . & a_{NN} \end{bmatrix}, \quad \mathrm{x} = \begin{bmatrix} x_1 \\ x_2 \\ x_3 \\ . \\ . \\ . \\ x_N \end{bmatrix}, \quad \mathrm{y} = \begin{bmatrix} y_1 \\ y_2 \\ y_3 \\ . \\ . \\ . \\ y_n \end{bmatrix} \tag{10.183}$$

With this notation, the Cramer's rule solution is simply used. Considering a linear equation expressed in matrix form as $Ax = y$, to calculate the vector of solutions, x, we require the inverse matrix, A^{-1}. The inverse matrix has the property that:

$$\mathrm{A}^{-1}\mathrm{A} = \mathrm{I} \tag{10.184}$$

or

$$\mathrm{IA} = \mathrm{A} \tag{10.185}$$

where I is the identity matrix.

The content of I is simply a matrix as follows:

$$I = \begin{bmatrix} 1 & 0 & 0 & . & . & . & 0 \\ 0 & 1 & 0 & . & . & . & 0 \\ 0 & 0 & 1 & . & . & . & 0 \\ . & . & . & . & . & . & . \\ . & . & . & . & . & . & . \\ . & . & . & . & . & . & . \\ 0 & 0 & 0 & . & . & . & 1 \end{bmatrix} \tag{10.186}$$

Calculating A^{-1}, then the solution is:

$$x = A^{-1}y \tag{10.187}$$

Thus,

$$x_1 = a_{11}^{-1}.y_1 + a_{12}^{-1}.y_2 + a_{13}^{-1}.y_3 + \cdots + a_{1n}^{-1}.y_n \tag{10.188}$$

$$x_1 = (y_1.A_{11} + y_2.A_{21} + y_3.A_{31} + y_4.A_{41} + \cdots y_n.A_{n1})/|A| \tag{10.189}$$

$$x_2 = (y_1.A_{12} + y_2.A_{22} + y_3.A_{32} + y_4.A_{42} + \cdots y_n.A_{n2})/|A| \tag{10.190}$$

etc.

Separating the **y** vector, we get:

$$\begin{bmatrix} x_1 \\ x_2 \\ x_3 \\ \cdot \\ \cdot \\ \cdot \\ x_N \end{bmatrix} = \frac{\begin{bmatrix} A_{11} & A_{21} & A_{31} & \cdot & \cdot & \cdot & A_{n1} \\ A_{12} & A_{22} & A_{32} & \cdot & \cdot & \cdot & A_{n2} \\ A_{13} & A_{23} & A_{33} & \cdot & \cdot & \cdot & A_{n3} \\ \cdot & \cdot & \cdot & \cdot & \cdot & \cdot & \cdot \\ \cdot & \cdot & \cdot & \cdot & \cdot & \cdot & \cdot \\ \cdot & \cdot & \cdot & \cdot & \cdot & \cdot & \cdot \\ A_{1N} & A_{2N} & A_{3N} & \cdot & \cdot & \cdot & A_{NN} \end{bmatrix} \begin{bmatrix} y_1 \\ y_2 \\ y_3 \\ \cdot \\ \cdot \\ \cdot \\ y_N \end{bmatrix}}{|A|} \tag{10.191}$$

Then the determinant is found by

$$|A| = \sum_j a_{ij} A_{ij} \tag{10.192}$$

where A_{ij} is the cofactor of element (a_{ij}) at row i and column j of the matrix A.

The determinant of the numerator above could be written as:

$y_1.A_{11} + y_2.A_{21} + y_3.A_{31} + y_4.A_{41} \ldots y_n.A_{n1}$. (where A_{11}, A_{21}, etc. are cofactors of a_{11}, a_{21}, etc.)

This gives us $x_1 = (y_1.A_{11} + y_2.A_{21} + y_3.A_{31} + y_4.A_{41} \ldots y_n.A_{n1})/|A|$ as the solution for x_1 using Cramer's Rule. We will use this result later in the formula for inversion. For further details, please see: Davis (2006).

The Thomas Algorithm

Consider a tridiagonal system of equations as follows:

$$\begin{cases} f_0 & & & & & & = \delta_0 \\ -\alpha_1 f_0 & \beta_1 f_1 & -\gamma_1 f_2 & & & & = \delta_0 \\ & -\alpha_2 f_1 & \beta_2 f_2 & -\gamma_2 f_3 & & & = \delta_0 \\ & & -\alpha_3 f_2 & \beta_3 f_3 & -\gamma_3 f_4 & & = \delta_0 \\ & & \cdot & \cdot & & & \cdot \quad \cdot \\ & & & \cdot & \cdot & & \cdot \quad \cdot \\ & & & & \cdot \quad \cdot & & \cdot \quad \cdot \\ & & & & -\alpha_{n-1} f_{n-2} & \beta_{n-1} f_{n-1} \quad -\gamma_{n-1} f_n & = \delta_{n-1} \\ & & & & & f_n & = \delta_n \end{cases} \tag{10.193}$$

where f_0 and f_n are known boundary values.

The general form is

$$-\alpha_{i-1} f_{i-2} + \beta_{i-1} f_{i-1} - \gamma_{i-1} f_i = \delta_{i-1} \tag{10.194}$$

Or we can write it as:

$$f_1 = \frac{\alpha_1}{\beta_1} f_0 + \frac{\gamma_1}{\beta_1} f_2 + \frac{\delta_1}{\beta_1} \tag{10.195}$$

$$f_2 = \frac{\alpha_2}{\beta_2} f_1 + \frac{\gamma_2}{\beta_2} f_3 + \frac{\delta_2}{\beta_2} \tag{10.196}$$

$$f_3 = \frac{\alpha_3}{\beta_3} f_2 + \frac{\gamma_3}{\beta_3} f_4 + \frac{\delta_3}{\beta_3} \tag{10.197}$$

$$f_{n-1} = \frac{\alpha_{n-1}}{\beta_{n-1}} f_{n-2} + \frac{\gamma_{n-1}}{\beta_{n-1}} f_n + \frac{\delta_{n-1}}{\beta_{n-1}} \tag{10.198}$$

We may rewrite the equation

$$f_{i-1} = a_{i-1} + b_{i-1} f_i \tag{10.199}$$

where

$$a_{i-1} = \frac{\alpha_{i-1} a_{i-2} + \delta_{i-1}}{-\alpha_{i-1} b_{i-2} + \beta_{i-1}} \tag{10.200}$$

and

$$b_{i-1} = \frac{\gamma_{i-1}}{-\alpha_{i-1} b_{i-2} + \beta_{i-1}} \tag{10.201}$$

For the boundary condition at point $i = 0$ we have

$$a_0 = f_0 \tag{10.202}$$

and

$$b_0 = 0 \tag{10.203}$$

For the boundary condition at point $i = n$ we have

$$a_n = f_n \tag{10.204}$$

and

$$b_n = 0 \tag{10.205}$$

Considering the boundary condition at point $i = n$, f_n is known, therefore we can estimate f_i in a reverse order; firstly for $i = n - 1$ (f_{n-1}), second for $i = n - 2$ (f_{n-2}), and continuing to reach $i = 1(f_1)$. For further details, please see: Davis (2006).

The method is also applicable to two-dimensional and three-dimensional problems. The method can be similarly used based on a line-by-line scheme and a line-by-line plus plane-by-plane scheme in computations for 2D and 3D cases, respectively. The method can be applied using an iterative process, to 2D and 3D problems. For further details, please see the next section or see: Greenbaum (1997); Kelley (2003); Kelley (1995); Olshanskii and Tyrtyshnikov (2014); Saad (2003).

Example 10.6 Discretization of equations

$$\frac{\partial f}{\partial t} = C\frac{\partial^2 f}{\partial x^2} \tag{10.206}$$

In this example, for the left term, which is the temporal derivative, we use the forward difference method, and for the right term, which is the special derivative, the central difference is used because it is more accurate. So:

$$\frac{f_i^n - f_i^{n+1}}{\Delta t} = -C\frac{f_{i-1}^m - 2f_i^m + f_{i+1}^m}{(\Delta x)^2} \tag{10.207}$$

But for the special derivative three modes can be considered:

1. Explicit method: $m = n$:

$$\frac{f_i^n - f_i^{n+1}}{\Delta t} = -C\frac{f_{i-1}^n - 2f_i^n + f_{i+1}^n}{(\Delta x)^2} \tag{10.208}$$

or

$$f_i^{n+1} = \frac{c\Delta t}{(\Delta x)^2} f_{i-1}^n + \left(1 - 2\frac{c\Delta t}{(\Delta x)^2}\right) f_i^n + \frac{c\Delta t}{(\Delta x)^2} f_{i+1}^n \tag{10.209}$$

Although this method is simple as it contains only one equation and one unknown parameter, the stability should be checked. For the numerical solution to be stable, all the equations should be positive. So the stability can be achieved in this way:

$$0 < 1 - 2\frac{c\Delta t}{(\Delta x)^2} \tag{10.210}$$

$$\frac{c\Delta t}{(\Delta x)^2} < \frac{1}{2} \tag{10.211}$$

2. Implicit method: $m = n + 1$

$$\frac{f_i^n - f_i^{n+1}}{\Delta t} = -c\frac{f_{i-1}^{n+1} - 2f_i^{n+1} + f_{i+1}^{n+1}}{(\Delta x)^2} \tag{10.212}$$

or

$$-\frac{c\Delta t}{(\Delta x)^2} f_{i-1}^{n+1} + \left(1 + 2\frac{c\Delta t}{(\Delta x)^2}\right) f_i^{n+1} - \frac{c\Delta t}{(\Delta x)^2} f_{i+1}^{n+1} = f_i^n \tag{10.213}$$

The implicit method is always stable so, as mentioned before, a system of equations should be solved.

Since this set of equations is a tridiagonal matrix, the Thomas method can be used:

$$f_{i-1}^{n+1} = a_{n-1} - b_{n-1} f_i^{n+1} \tag{10.214}$$

where

$$a_i = \frac{-\propto_i a_{i-1} + \delta_i}{-\propto_i b_{i-1} + \beta_i} \tag{10.215}$$

$$b_i = \frac{\gamma_i}{-\propto_i b_{i-1} + \beta_i} \tag{10.216}$$

where

$$\propto_i = -\frac{c\Delta t}{(\Delta x)^2} \tag{10.217}$$

$$\beta_i = 1 + 2\frac{c\Delta t}{(\Delta x)^2} \tag{10.218}$$

$$\gamma_i = -\frac{c\Delta t}{(\Delta x)^2} \tag{10.219}$$

$$\delta_i = f_i^n \tag{10.220}$$

Then the equations can be solved, for example:

$$f_2 = \frac{\alpha_2\left(\frac{\alpha_1}{\beta_1}f_0 + \frac{\delta_1}{\beta_1}\right) + \delta_2}{\beta_2 - \alpha_2\frac{\gamma_1}{\beta_1}} + \frac{\gamma_2}{\beta_2 - \alpha_2\frac{\gamma_1}{\beta_1}}f_3 \tag{10.221}$$

Rewriting:

$$f_2 = \frac{\alpha_2 a_1 + \delta_2}{-\alpha_2 b_1 + \beta_2} + \frac{\gamma_2}{-\alpha_2 b_1 + \beta_2}f_3 \tag{10.222}$$

where

$$a_1 = \frac{\alpha_1}{\beta_1}f_0 + \frac{\delta_1}{\beta_1} \tag{10.223}$$

and

$$b_1 = \frac{\gamma_1}{\beta_1} \tag{10.224}$$

or

$$f_2 = b_2 f_3 + a_2 \tag{10.225}$$

where

$$a_2 = \left(\frac{\alpha_2 a_1 + \delta_2}{-\alpha_2 b_1 + \beta_2}\right) \tag{10.226}$$

and

$$b_2 = \frac{\gamma_2}{-\alpha_2 b_1 + \beta_2} \tag{10.227}$$

3. Semi-implicit method (Crank-Nicolson method):

$$\frac{f_i^{n+1} - f_i^n}{\Delta t} = \frac{1}{2}C\frac{(f_{i-1}^n + f_{i-1}^{n+1}) - 2(f_i^n + f_i^{n+1}) + (f_{i+1}^n + f_{i+1}^{n+1})}{(\Delta x)^2} \tag{10.228}$$

$$\begin{aligned} -f_{i-1}^n\left(\frac{1}{2}\frac{c}{(\Delta x)^2}\right) - f_i^n\left(\frac{1}{\Delta t} - \frac{c}{(\Delta x)^2}\right) - f_{i+1}^n\left(\frac{1}{2}\frac{c}{(\Delta x)^2}\right) - f_{i-1}^{n+1}\left(\frac{1}{2}\frac{c}{(\Delta x)^2}\right) \\ + f_i^{n+1}\left(\frac{1}{\Delta t} + \frac{c}{(\Delta x)^2}\right) - f_{i+1}^{n+1}\left(\frac{1}{2}\frac{c}{(\Delta x)^2}\right) = 0 \end{aligned} \tag{10.229}$$

So according to the Thomas method:

$$\alpha_i = -\frac{c\Delta t}{2(\Delta x)^2} \tag{10.230}$$

$$\beta_i = 1 + \frac{c\Delta t}{(\Delta x)^2} \tag{10.231}$$

$$\gamma_i = -\frac{c\Delta t}{2(\Delta x)^2} \tag{10.232}$$

$$\delta_i = \left(1 - \frac{c\Delta t}{(\Delta x)^2}\right) f_i^n + \frac{1}{2}\frac{c\Delta t}{(\Delta x)^2} f_{i-1}^n + \frac{1}{2}\frac{c\Delta t}{(\Delta x)^2} f_{i+1}^n \tag{10.233}$$

The stability of this method is also conditional, which is:

$$\frac{c\Delta t}{(\Delta x)^2} < \frac{1}{2} \tag{10.234}$$

10.3.2 Iterative Methods

Gauss-Seidel (Liebmann) Method

If we have a large set of equations, direct methods might be inappropriate and less efficient. In preference, iterative methods could be more advantageous. One advantage of the iterative methods comparing with the direct methods is related to the degree of total round-off error (see: Greenbaum 1997; Kelley 2003; Kelley 1995; Olshanskii and Tyrtyshnikov 2014; Saad 2003). For instance, in such cases, Gaussian elimination method basically faces to a large round-off error, while in the Gauss-Seidel method, the round-off error could be monitored and managed properly. Additionally, for iterative methods, much less storage space might be needed over the computations.

In this iterative solution, we begin with arbitrary values and follow a sequence of iterates to calculate new values. Considering a system of N equations as follows and assuming that the diagonal elements have non-zero values

$$a_{11}x_1 + a_{12}x_2 + a_{13}x_3 + a_{14}x_4 + \cdots + a_{1N}x_N = b_1 \tag{10.235}$$

$$a_{21}x_1 + a_{22}x_2 + a_{23}x_3 + a_{2N}x_N + \cdots + a_{2N}x_N = b_2 \tag{10.236}$$

$$\begin{matrix} \cdot & \cdot \\ \cdot & \cdot \\ \cdot & \cdot \end{matrix}$$

$$a_{N1}x_1 + a_{N2}x_2 + a_{N3}x_3 + a_{N4}x_4 + \cdots + a_{NN}x_N = b_N \tag{10.237}$$

or we can simply write this system of equations as follows

$$\mathrm{AX} = \mathrm{B} \tag{10.238}$$

or

$$\begin{bmatrix} a_{11} & a_{12} & a_{13} & a_{14} & \cdot & \cdot & \cdot & a_{1N} \\ a_{21} & a_{22} & a_{23} & a_{24} & \cdot & \cdot & \cdot & a_{2N} \\ a_{31} & a_{32} & a_{33} & a_{34} & \cdot & \cdot & \cdot & a_{3N} \\ \cdot & \cdot & \cdot & \cdot & \cdot & \cdot & \cdot & \cdot \\ \cdot & \cdot & \cdot & \cdot & \cdot & \cdot & \cdot & \cdot \\ \cdot & \cdot & \cdot & \cdot & \cdot & \cdot & \cdot & \cdot \\ a_{N1} & a_{N2} & a_{N3} & a_{N4} & \cdot & \cdot & \cdot & a_{NN} \end{bmatrix} \begin{bmatrix} x_1 \\ x_2 \\ x_3 \\ \cdot \\ \cdot \\ \cdot \\ x_N \end{bmatrix} = \begin{bmatrix} b_1 \\ b_2 \\ b_3 \\ \cdot \\ \cdot \\ \cdot \\ b_N \end{bmatrix} \tag{10.239}$$

where

$$A = \begin{bmatrix} a_{11} & a_{21} & a_{31} & \cdot & \cdot & \cdot & a_{N1} \\ a_{12} & a_{22} & a_{32} & \cdot & \cdot & \cdot & a_{N2} \\ a_{13} & a_{23} & a_{33} & \cdot & \cdot & \cdot & a_{N3} \\ \cdot & \cdot & \cdot & \cdot & \cdot & \cdot & \cdot \\ \cdot & \cdot & \cdot & \cdot & \cdot & \cdot & \cdot \\ \cdot & \cdot & \cdot & \cdot & \cdot & \cdot & \cdot \\ a_{1N} & a_{2N} & a_{3N} & \cdot & \cdot & \cdot & a_{NN} \end{bmatrix}, \mathrm{x} = \begin{bmatrix} x_1 \\ x_2 \\ x_3 \\ \cdot \\ \cdot \\ \cdot \\ x_N \end{bmatrix}, B = \begin{bmatrix} b_1 \\ b_2 \\ b_3 \\ \cdot \\ \cdot \\ \cdot \\ b_n \end{bmatrix} \tag{10.240}$$

The system of equations can be rewritten as follows

$$x_1 = \frac{b_1 - a_{12}x_2 - a_{13}x_3 - \cdots - a_{1N}x_N}{a_{11}} \tag{10.241}$$

$$x_2 = \frac{b_2 - a_{21}x_1 - a_{23}x_3 - \cdots - a_{2N}x_N}{a_{22}} \tag{10.242}$$

$$x_3 = \frac{b_3 - a_{31}x_1 - a_{32}x_3 - \cdots - a_{3N}x_N}{a_{33}} \tag{10.243}$$

$$\cdot$$
$$\cdot$$
$$\cdot$$

$$x_i = \frac{b_i - a_{i1}x_1 - a_{i2}x_2 - \cdots - a_{ii-1}x_{i-1} - a_{ii+1}x_{i+1} - \cdots - a_{iN}x_N}{a_{ii}} \tag{10.244}$$

$$\cdot$$
$$\cdot$$
$$\cdot$$

$$x_N = \frac{b_N - a_{N1}x_1 - a_{N2}x_2 - \cdots - a_{NN-1}x_{N-1}}{a_{NN}} \tag{10.245}$$

or

$$x_1 = \frac{b_1 - \sum_{\substack{j=1 \\ j \neq 1}}^{N} a_{1j} x_j}{a_{11}} \tag{10.246}$$

$$x_2 = \frac{b_2 - \sum_{\substack{j=1 \\ j \neq 2}}^{N} a_{2j} x_j}{a_{22}} \tag{10.247}$$

$$\vdots$$

$$x_i = \frac{b_i - \sum_{\substack{j=1 \\ j \neq i}}^{N} a_{ij} x_j}{a_{ii}} \tag{10.248}$$

$$\vdots$$

$$x_N = \frac{b_N - \sum_{\substack{j=1 \\ j \neq N}}^{N} a_{Nj} x_j}{a_{NN}} \tag{10.249}$$

Initial guesses can be made for the x_i to calculate the new estimates and then the new estimates are used to calculate the next estimates for x_i:

$$x_i^{(k+1)} = \frac{1}{a_{ii}} \left[b_i - \sum_{j=1}^{i-1} a_{ij} x_j^{(k+1)} - \sum_{j=i+1}^{N} a_{ij} x_j^{(k)} \right] \quad k = 0, 1, \ldots \tag{10.250}$$

or

$$x_i^{k+1} = \frac{b_i - \sum_{j<i} a_{ij} x_j^{k+1} - \sum_{j>i} a_{ij} x_j^{k}}{a_{ii}} \quad k = 0, 1, \ldots \tag{10.251}$$

At the end of each iteration, the approximation error for each x_i can be computed as follows. The convergence criterion should be satisfied. To test for convergence:

$$|e| = \frac{\left| x_i^{k+1} - x_i^{k} \right|}{\left| x_i^{k} \right|} \tag{10.252}$$

All the x_i values are required to be evaluated according to this convergence test. The iterative process can be terminated once the calculated error for each x_i is less than a desired value (ε):

$$|e| \leq \varepsilon \tag{10.253}$$

Jacobi's Method

Jacobi method is nearly similar to Gauss-Seidel method, except that each x-value is improved using the most recent approximations to the values of the other variables.

Considering similar set of equations as Gauss-Seidel method, we can similarly define matrix A as before by assuming that the diagonal terms of matrix A have non-zero values, then we can rewrite

$$x_i^{k+1} = \frac{b_i - \sum_{j \neq i} a_{ij} x_j^k}{a_{ii}}, k = 0, 1, \ldots \tag{10.254}$$

The iterative process is terminated when a convergence criterion is satisfied.

Unlike the Gauss-Seidel method, the previous estimations are not instantly replaced by the new values in Jacobi method, thus the storage space required is twice the Gauss-Seidel method and the convergence rapidness is lower.

Successive Over-Relaxation (SOR) Method

The successive over-relaxation (SOR) method is a different form of the Gauss–Seidel method. Considering the Gauss-Seidel method, in SOR, we apply a weight factor called relaxation parameter. Basically, $(1-\omega)$ portion of the previous estimation for x_i and ω portion of the new estimation are adopted to write the iterative process equation:

$$x_i^{(k+1)} = (1-\omega)x_i^{(k)} + \frac{\omega}{a_{ii}}\left[-\sum_{j=1}^{i-1} a_{ij}x_j^{(k+1)} - \sum_{j=i+1}^{N} a_{ij}x_j^{(k)} + b_i\right] \tag{10.255}$$

The relaxation parameter ω is typically smaller than 2.0. For $0 < \omega < 1$, we have under-relaxation method, while for $1 < \omega < 2$, we have the over-relaxation method. If $\omega = 1$, the SOR method is changed to the Gauss-Seidel method. SOR method is able to converge more quickly than the Gauss–Seidel method. In SOR method, components are replaced immediately in the calculations, thus, the storage space required is half the Jacobi method. For further details, please see: Greenbaum (1997); Kelley (2003); Kelley (1995); Olshanskii and Tyrtyshnikov (2014); Saad (2003).

The finite difference method is also used to solve ordinary differential equations called boundary value problems. In this section, we solve second order ordinary differential equations as boundary value problems.

Example 10.7

Solve the equation

$$\frac{d^2y}{dx^2} - y - x(1 - x) = 0 \tag{10.256}$$

Approximating the derivative $\frac{d^2y}{dx^2}$ at node i by the central divided difference approximation:

$$\frac{d^2y}{dx^2} \approx \frac{y_{i+1} - 2y_i + y_{i-1}}{(\Delta x)^2} \tag{10.257}$$

Therefore, a set of linear equations can be obtained at different nodes that can be solved by using methods such as Gaussian elimination, the Gauss–Seidel method, etc.

Therefore the finite difference approximation is:

$$\frac{y_{i+1} - 2y_i + y_{i-1}}{(\Delta x)^2} - y_i - x_i(1 - x_i) = 0 \tag{10.258}$$

Considering a spatial step for x equals 1 and boundary condition at $x = 0$, we obtain $y_1 - 0$.

For the next nodes:

$$\frac{y_3 - 2y_2 + y_1}{(1)^2} - y_2 - x_2(1 - x_2) = 0 \tag{10.259}$$

$$y_1 - 3y_2 + y_3 - (1)(1 - 1) = 0$$

$$y_1 - 3y_2 + y_3 = 0 \tag{10.260}$$

$$\frac{y_4 - 2y_3 + y_2}{(1)^2} - y_3 - x_3(1 - x_3) = 0 \tag{10.261}$$

$$y_2 - 3y_3 + y_4 - (2)(1 - 2) = 0$$

$$y_2 - 3y_3 + y_4 = -2 \tag{10.262}$$

And at boundary condition at $x = 3$, we obtain $y_4 = 0$.

Thus, equations can be written in matrix form as:

$$\begin{bmatrix} 1 & 0 & 0 & 0 \\ 1 & -3 & 1 & 0 \\ 0 & 1 & -3 & 1 \\ 0 & 0 & 0 & 1 \end{bmatrix} \begin{bmatrix} y_1 \\ y_2 \\ y_3 \\ y_4 \end{bmatrix} = \begin{bmatrix} 0 \\ 0 \\ -2 \\ 0 \end{bmatrix} \tag{10.263}$$

These equations have a coefficient matrix that is tridiagonal. Therefore, we can use Thomas's algorithm to solve the equations.

Solving the equations we get:

$$\begin{bmatrix} y_1 \\ y_2 \\ y_3 \\ y_4 \end{bmatrix} = \begin{bmatrix} 0.00 \\ -0.25 \\ +0.75 \\ 0.00 \end{bmatrix} \tag{10.264}$$

Example 10.8

Solve the differential equation given here:

$$\frac{d^2f}{dr^2} - \frac{2}{r}\frac{df}{dr} + \frac{2}{r^2}f = 0 \tag{10.265}$$

with the boundary conditions of:

$$f(r_0) = 100, f(r_n) = 400 \tag{10.266}$$

where $r_0 = 10$ m and $r_n = 20$ m.

To solve this equation numerically and to find variable f, we need first to divide the problem area into a limited number of parts. Therefore, we will be able to calculate the variable f at each node in the computational domain. If we divide the radial domain into n equally spaced parts, then we should have $n + 1$ nodes. As discussed before, we can write at an arbitrary node i:

$$\frac{d^2f}{dr^2} \approx \frac{f_{i+1} - 2f_i + f_{i-1}}{(\Delta r)^2} \tag{10.267}$$

$$\frac{df}{dr} \approx \frac{f_{i+1} - f_i}{\Delta r} \tag{10.268}$$

Substituting these finite difference approximations in the first equation we have:

$$\frac{f_{i+1} - 2f_i + f_{i-1}}{(\Delta r)^2} - \frac{2}{r_i}\frac{f_{i+1} - f_i}{\Delta r} + \frac{2}{r_i^2}f_i = 0 \tag{10.269}$$

$$\left(\frac{1}{(\Delta r)^2} - \frac{2}{r_i \Delta r}\right)f_{i+1} - \left(\frac{2}{(\Delta r)^2} - \frac{2}{r_i \Delta r} - \frac{2}{r_i^2}\right)f_i + \frac{1}{(\Delta r)^2}f_{i-1} = 0 \tag{10.270}$$

Breaking the computational domain into $n + 1$ nodes, we have $n + 1$ unknowns. where $r = r_0$ is node $i = 0$ and $r = r_n$ is node $i = n$.

Writing this equation for nodes $1, \ldots, n-1$ will give us $n-1$ equations. At the nodes, $i=0$ and $i=n$, we use the boundary conditions of $f(r_1)$ and $f(r_2)$ respectively. Thus, a total of $n+1$ linear equations with $n+1$ unknowns is obtained, which can be solved simultaneously by the numerical method described before.

For a total number of 11 nodes and consequently 10 intervals:

$$\Delta r = \frac{r_{10} - r_0}{n} = \frac{20 - 10}{10} = 1 \tag{10.271}$$

At node $i = 0; r = r_0 = 10, f_0 = 100$

At node $i = 1; r_1 = r_0 + \Delta r = 10 + 1 = 11$

$$\left(\frac{1}{(1)^2} - \frac{2}{(11)(1)}\right)f_2 - \left(\frac{2}{(1)^2} - \frac{2}{(11)(1)} - \frac{2}{11^2}\right)f_1 + \frac{1}{(1)^2}f_0 = 0$$

$$0.8182f_2 - 1.8016f_1 + f_0 = 0 \tag{10.272}$$

At node $i = 2; r_2 = r_1 + \Delta r = 11 + 1 = 12$

$$\left(\frac{1}{(1)^2} - \frac{2}{(12)(1)}\right)f_3 - \left(\frac{2}{(1)^2} - \frac{2}{(12)(1)} - \frac{2}{12^2}\right)f_2 + \frac{1}{(1)^2}f_1 = 0$$

$$0.8333f_3 - 1.8194f_2 + f_1 = 0 \tag{10.273}$$

At node $i = 3; r_3 = r_2 + \Delta r = 12 + 1 = 13$

$$\left(\frac{1}{(1)^2} - \frac{2}{(13)(1)}\right)f_4 - \left(\frac{2}{(1)^2} - \frac{2}{(13)(1)} - \frac{2}{13^2}\right)f_3 + \frac{1}{(1)^2}f_2 = 0$$

$$0.8461f_4 - 1.8343f_3 + f_2 = 0 \tag{10.274}$$

At node $i = 4; r_4 = r_3 + \Delta r = 13 + 1 = 14$

$$\left(\frac{1}{(1)^2} - \frac{2}{(14)(1)}\right)f_5 - \left(\frac{2}{(1)^2} - \frac{2}{(14)(1)} - \frac{2}{14^2}\right)f_4 + \frac{1}{(1)^2}f_3 = 0$$

$$0.8571f_5 - 1.8469f_4 + f_3 = 0 \tag{10.275}$$

At node $i = 5; r_5 = r_4 + \Delta r = 14 + 1 = 15$

$$\left(\frac{1}{(1)^2} - \frac{2}{(15)(1)}\right)f_6 - \left(\frac{2}{(1)^2} - \frac{2}{(15)(1)} - \frac{2}{15^2}\right)f_5 + \frac{1}{(1)^2}f_4 = 0$$

$$0.8666f_6 - 1.8578f_5 + f_4 = 0 \tag{10.276}$$

At node $i = 6; r_6 = r_5 + \Delta r = 15 + 1 = 16$

$$\left(\frac{1}{(1)^2} - \frac{2}{(16)(1)}\right)f_7 - \left(\frac{2}{(1)^2} - \frac{2}{(16)(1)} - \frac{2}{16^2}\right)f_6 + \frac{1}{(1)^2}f_5 = 0$$

$$0.8750f_7 - 1.8672f_6 + f_5 = 0 \tag{10.277}$$

At node $i = 7; r_7 = r_6 + \Delta r = 16 + 1 = 17$

$$\left(\frac{1}{(1)^2} - \frac{2}{(17)(1)}\right)f_8 - \left(\frac{2}{(1)^2} - \frac{2}{(17)(1)} - \frac{2}{17^2}\right)f_7 + \frac{1}{(1)^2}f_6 = 0$$

$$0.8823f_8 - 1.8754f_7 + f_6 = 0 \quad (10.278)$$

At node $i = 8; r_8 = r_7 + \Delta r = 17 + 1 = 18$

$$\left(\frac{1}{(1)^2} - \frac{2}{(18)(1)}\right)f_9 - \left(\frac{2}{(1)^2} - \frac{2}{(18)(1)} - \frac{2}{18^2}\right)f_8 + \frac{1}{(1)^2}f_7 = 0$$

$$0.8889f_9 - 1.8827f_8 + f_7 = 0 \quad (10.279)$$

At node $i = 9; r_9 = r_8 + \Delta r = 18 + 1 = 19$

$$\left(\frac{1}{(1)^2} - \frac{2}{(19)(1)}\right)f_{10} - \left(\frac{2}{(1)^2} - \frac{2}{(19)(1)} - \frac{2}{19^2}\right)f_9 + \frac{1}{(1)^2}f_8 = 0$$

$$0.8947f_{10} - 1.8892f_9 + f_8 = 0 \quad (10.280)$$

At node $i = 10; r_{10} = r_9 + \Delta r = 19 + 1 = 20; f_{10} = 400$
So we have

$$0.8182f_2 - 1.8016f_1 + f_0 = 0$$

$$0.8333f_3 - 1.8194f_2 + f_1 = 0$$

$$0.8461f_4 - 1.8343f_3 + f_2 = 0$$

$$0.8571f_5 - 1.8469f_4 + f_3 = 0$$

$$0.8666f_6 - 1.8578f_5 + f_4 = 0$$

$$0.8750f_7 - 1.8672f_6 + f_5 = 0$$

$$0.8823f_8 - 1.8754f_7 + f_6 = 0$$

$$0.8889f_9 - 1.8827f_8 + f_7 = 0$$

$$0.8947f_{10} - 1.8892f_9 + f_8 = 0$$

Writing equations in matrix form gives:

$$
\begin{bmatrix}
1.0000 & 0.0000 & 0.0000 & 0.0000 & 0.0000 & 0.0000 & 0.0000 & 0.0000 & 0.0000 & 0.0000 & 0.0000 \\
1.0000 & -1.8016 & 0.8182 & 0.0000 & 0.0000 & 0.0000 & 0.0000 & 0.0000 & 0.0000 & 0.0000 & 0.0000 \\
0.0000 & 1.0000 & -1.8194 & 0.8333 & 0.0000 & 0.0000 & 0.0000 & 0.0000 & 0.0000 & 0.0000 & 0.0000 \\
0.0000 & 0.0000 & 1.0000 & -1.8343 & 0.8461 & 0.0000 & 0.0000 & 0.0000 & 0.0000 & 0.0000 & 0.0000 \\
0.0000 & 0.0000 & 0.0000 & 1.0000 & -1.8469 & 0.8571 & 0.0000 & 0.0000 & 0.0000 & 0.0000 & 0.0000 \\
0.0000 & 0.0000 & 0.0000 & 0.0000 & 1.0000 & -1.8578 & 0.8666 & 0.0000 & 0.0000 & 0.0000 & 0.0000 \\
0.0000 & 0.0000 & 0.0000 & 0.0000 & 0.0000 & 1.0000 & -1.8672 & 0.8750 & 0.0000 & 0.0000 & 0.0000 \\
0.0000 & 0.0000 & 0.0000 & 0.0000 & 0.0000 & 0.0000 & 1.0000 & -1.8754 & 0.8823 & 0.0000 & 0.0000 \\
0.0000 & 0.0000 & 0.0000 & 0.0000 & 0.0000 & 0.0000 & 0.0000 & 1.0000 & -1.8827 & 0.8889 & 0.0000 \\
0.0000 & 0.0000 & 0.0000 & 0.0000 & 0.0000 & 0.0000 & 0.0000 & 0.0000 & 1.0000 & -1.8892 & 0.8947 \\
0.0000 & 0.0000 & 0.0000 & 0.0000 & 0.0000 & 0.0000 & 0.0000 & 0.0000 & 0.0000 & 0.0000 & 1.0000
\end{bmatrix}
$$

$$
\times \begin{bmatrix} f_0 \\ f_1 \\ f_2 \\ f_3 \\ f_4 \\ f_5 \\ f_6 \\ f_7 \\ f_8 \\ f_9 \\ f_{10} \end{bmatrix} = \begin{bmatrix} 0 \\ 0 \\ 0 \\ 0 \\ 0 \\ 0 \\ 0 \\ 0 \\ 0 \\ 0 \\ 0 \end{bmatrix} \tag{10.281}
$$

We can use the Thomas algorithm to solve this tridiagonal system of equations:

$$
\begin{aligned}
f_0 &= 100.0000 \\
f_1 &= 120.3831 \\
f_2 &= 142.8529 \\
f_3 &= 167.4348 \\
f_4 &= 194.1529 \\
f_5 &= 223.0150 \\
f_6 &= 254.0554 \\
f_7 &= 287.2655 \\
f_8 &= 322.6592 \\
f_9 &= 360.2261 \\
f_{10} &= 400.0000
\end{aligned} \tag{10.282}
$$

The approximation in this example is first order accurate, so that the true error is of $O(\Delta r)$ because it is a mixing of a first order and a second order accordant approximation $O((\Delta r)^2)$.

Using a

$$
\frac{df}{dr} \cong -\frac{f_{i-1} - f_{i+1}}{2(\Delta r)} \tag{10.283}
$$

approximation will result in a more accurate approximation, because this equation is second order accurate and mixing two second order accurate approximations results in a second order accurate approximation.

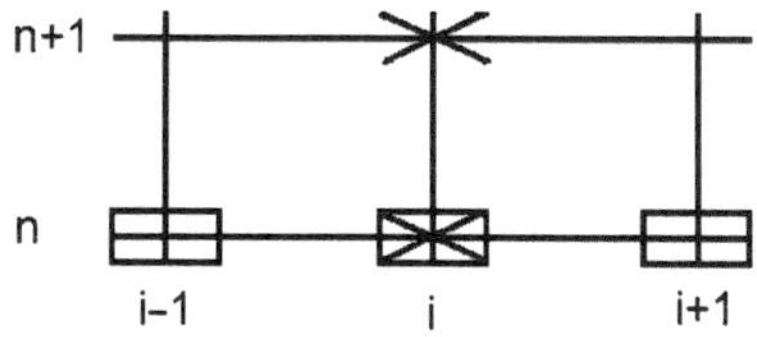

Figure 10.3 A sketch for the explicit method.

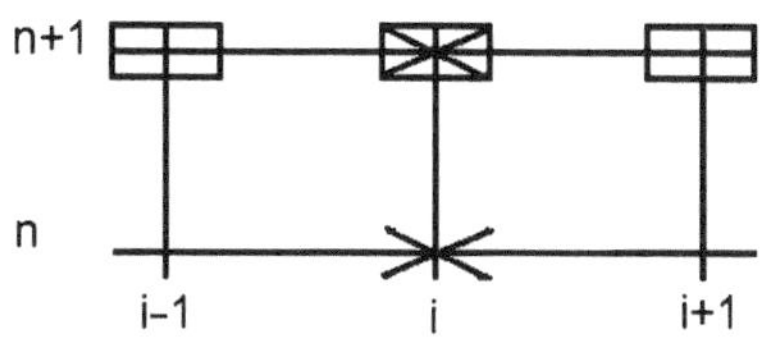

Figure 10.4 A sketch for the implicit method.

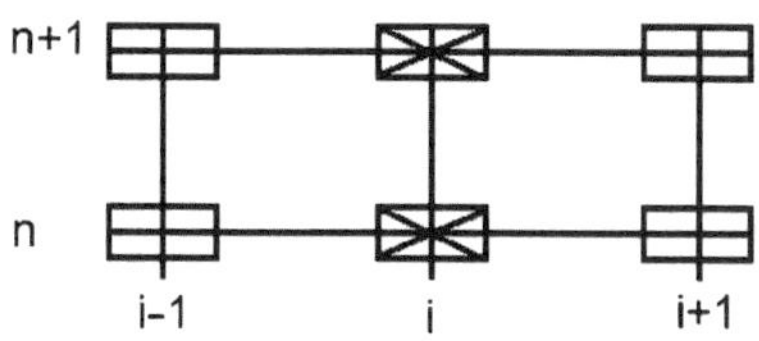

Figure 10.5 A sketch for the Crank-Nicolson method.

Figures 10.3, 10.4 and 10.5 show the explicit, implicit and Crank-Nicolson methods.

The $\times$ signs show the grid points involved in the time difference while the $\square$ signs illustrate the grid points involved in the space difference.

If $\lambda = 1$ we then apply an explicit method while an implicit method is used if $\lambda = 0$. In fact λ can be any value between 0 and 1.

For the Crank-Nicolson method $\lambda = 0.5$. For further details and information on the finite difference method and its application, please see: Causon and Mingham (2010); LeVeque (2007); Smith (1985); Thomas (1995). In addition, for further information about the discussed methods in this section or other methods for solving both systems of linear and nonlinear equations, please see: Davis (2006); Greenbaum (1997); Kelley (2003); Kelley (1995); Olshanskii and Tyrtyshnikov (2014); Saad (2003).

Example 10.9

Use the Crank-Nicolson method to solve the unsteady state one-dimensional partial differential equation:

$$\frac{\partial f}{\partial t} - 500\frac{\partial^2 f}{\partial x^2} = 0 \qquad (10.284)$$

The initial conditions are:

$$f(x,0) = x^2 - 1E+4 \quad \text{for} \quad 0 \le x \le 100 \tag{10.285}$$

and the boundary conditions:

$$\begin{cases} f(0,t) = (t-10) * 1E+3 \\ f(100,t) = 1000t \end{cases} \tag{10.286}$$

With the spatial and temporal steps as follows:

$$\Delta x = 10$$

$$\Delta t = 1.0$$

The finite difference form of the governing equation is now given with the spatial second derivative evaluated from a combination of the derivatives at time steps (n) and ($n+1$) as follows:

$$\frac{f_i^{n+1} - f_i^n}{\Delta t} = \frac{f_{i-1}^n - 2f_i^n + f_{i+1}^n}{2(\Delta x)^2} + \frac{f_{i-1}^{n+1} - 2f_i^{n+1} + f_{i+1}^{n+1}}{2(\Delta x)^2} \tag{10.287}$$

Arranging the equation and applying these equations to all the nodes:

$$-\frac{\Delta t}{2(\Delta x)^2} f_{i-1}^{n+1} + \left(1 + \frac{2\Delta t}{(\Delta x)^2}\right) f_i^{n+1} - \frac{\Delta t}{2(\Delta x)^2} f_{i+1}^{n+1} = \frac{\Delta t}{2(\Delta x)^2} f_{i-1}^n + \left(1 - \frac{2\Delta t}{(\Delta x)^2}\right) f_i^n + \frac{\Delta t}{2(\Delta x)^2} f_{i+1}^n \tag{10.288}$$

or

$$-0.05 f_{i-1}^{n+1} + 1.2 f_i^{n+1} - 0.05 f_{i+1}^{n+1} = 0.05 f_{i-1}^n + 0.8 f_i^n + 0.05 f_{i+1}^n \tag{10.289}$$

For each node we have:

$i = 2$:

$$-0.05 f_1^{n+1} + 1.2 f_2^{n+1} - 0.05 f_3^{n+1} = 0.05 f_1^n + 0.8 f_2^n + 0.05 f_3^n \tag{10.290}$$

$i = 3$

$$-0.05 f_2^{n+1} + 1.2 f_3^{n+1} - 0.05 f_4^{n+1} = 0.05 f_2^n + 0.8 f_3^n + 0.05 f_4^n \tag{10.291}$$

$i = 4$

$$-0.05 f_3^{n+1} + 1.2 f_4^{n+1} - 0.05 f_5^{n+1} = 0.05 f_3^n + 0.8 f_4^n + 0.05 f_5^n \tag{10.292}$$

$i = 5$

$$-0.05 f_4^{n+1} + 1.2 f_5^{n+1} - 0.05 f_6^{n+1} = 0.05 f_4^n + 0.8 f_5^n + 0.05 f_6^n \tag{10.293}$$

$i = 6$

$$-0.05 f_5^{n+1} + 1.2 f_6^{n+1} - 0.05 f_7^{n+1} = 0.05 f_5^n + 0.8 f_6^n + 0.05 f_7^n \tag{10.294}$$

$i = 7$

$$-0.05f_6^{n+1} + 1.2f_7^{n+1} - 0.05f_8^{n+1} = 0.05f_6^n + 0.8f_7^n + 0.05f_8^n \tag{10.295}$$

$i = 8$

$$-0.05f_7^{n+1} + 1.2f_8^{n+1} - 0.05f_9^{n+1} = 0.05f_7^n + 0.8f_8^n + 0.05f_9^n \tag{10.296}$$

$i = 9$

$$-0.05f_8^{n+1} + 1.2f_9^{n+1} - 0.05f_{10}^{n+1} = 0.05f_8^n + 0.8f_9^n + 0.05f_{10}^n \tag{10.297}$$

$i = 10$

$$-0.05f_9^{n+1} + 1.2f_{10}^{n+1} - 0.05f_{11}^{n+1} = 0.05f_9^n + 0.8f_{10}^n + 0.05f_{11}^n \tag{10.298}$$

Thus the equations for $t = 1.0$ are:

$$-0.05f_1^{n+1} + 1.2f_2^{n+1} - 0.05f_3^{n+1} = 0.05(-10000) + 0.8(-9900) + 0.05(-9600) \tag{10.299}$$

$$-0.05f_2^{n+1} + 1.2f_3^{n+1} - 0.05f_4^{n+1} = 0.05(-9900) + 0.8(-9600) + 0.05(-9100) \tag{10.300}$$

$$-0.05f_3^{n+1} + 1.2f_4^{n+1} - 0.05f_5^{n+1} = 0.05(-9600) + 0.8(-9100) + 0.05(-8400) \tag{10.301}$$

$$-0.05f_4^{n+1} + 1.2f_5^{n+1} - 0.05f_6^{n+1} = 0.05(-9100) + 0.8(-8400) + 0.05(-7500) \tag{10.302}$$

$$-0.05f_5^{n+1} + 1.2f_6^{n+1} - 0.05f_7^{n+1} = 0.05(-8400) + 0.8(-7500) + 0.05(-6400) \tag{10.303}$$

$$-0.05f_6^{n+1} + 1.2f_7^{n+1} - 0.05f_8^{n+1} = 0.05(-7500) + 0.8(-6400) + 0.05(-5100) \tag{10.304}$$

$$-0.05f_7^{n+1} + 1.2f_8^{n+1} - 0.05f_9^{n+1} = 0.05(-6400) + 0.8(-5100) + 0.05(-3600) \tag{10.305}$$

$$-0.05f_8^{n+1} + 1.2f_9^{n+1} - 0.05f_{10}^{n+1} = 0.05(-5100) + 0.8(-3600) + 0.05(-1900) \tag{10.306}$$

$$-0.05f_9^{n+1} + 1.2f_{10}^{n+1} - 0.05f_{11}^{n+1} = 0.05(-3600) + 0.8(-1900) + 0.05(0) \tag{10.307}$$

Thus the equations for $t = 1.0$ are:

$$+1.2f_2^{n+1} - 0.05f_3^{n+1} = -9350 \tag{10.308}$$

$$-0.05f_2^{n+1} + 1.2f_3^{n+1} - 0.05f_4^{n+1} = -8630 \tag{10.309}$$

$$-0.05f_3^{n+1} + 1.2f_4^{n+1} - 0.05f_5^{n+1} = -8180 \tag{10.310}$$

$$-0.05f_4^{n+1} + 1.2f_5^{n+1} - 0.05f_6^{n+1} = -7550 \quad (10.311)$$

$$-0.05f_5^{n+1} + 1.2f_6^{n+1} - 0.05f_7^{n+1} = -6740 \quad (10.312)$$

$$-0.05f_6^{n+1} + 1.2f_7^{n+1} - 0.05f_8^{n+1} = -5750 \quad (10.313)$$

$$-0.05f_7^{n+1} + 1.2f_8^{n+1} - 0.05f_9^{n+1} = -4580 \quad (10.314)$$

$$-0.05f_8^{n+1} + 1.2f_9^{n+1} - 0.05f_{10}^{n+1} = -3230 \quad (10.315)$$

$$-0.05f_9^{n+1} + 1.2f_{10}^{n+1} = -1650 \quad (10.316)$$

Thus the equations for $t = 1.0, 2.0, 3.0$, etc. are:

$$-0.05f_1^{n+1} + 1.2f_2^{n+1} - 0.05f_3^{n+1} = 0.05f_1^n + 0.8f_2^n + 0.05f_3^n \quad (10.317)$$

$$-0.05f_2^{n+1} + 1.2f_3^{n+1} - 0.05f_4^{n+1} = 0.05f_2^n + 0.8f_3^n + 0.05f_4^n \quad (10.318)$$

$$-0.05f_3^{n+1} + 1.2f_4^{n+1} - 0.05f_5^{n+1} = 0.05f_3^n + 0.8f_4^n + 0.05f_5^n \quad (10.319)$$

$$-0.05f_4^{n+1} + 1.2f_5^{n+1} - 0.05f_6^{n+1} = 0.05f_4^n + 0.8f_5^n + 0.05f_6^n \quad (10.320)$$

$$-0.05f_5^{n+1} + 1.2f_6^{n+1} - 0.05f_7^{n+1} = 0.05f_5^n + 0.8f_6^n + 0.05f_7^n \quad (10.321)$$

$$-0.05f_6^{n+1} + 1.2f_7^{n+1} - 0.05f_8^{n+1} = 0.05f_6^n + 0.8f_7^n + 0.05f_8^n \quad (10.322)$$

$$-0.05f_7^{n+1} + 1.2f_8^{n+1} - 0.05f_9^{n+1} = 0.05f_7^n + 0.8f_8^n + 0.05f_9^n \quad (10.323)$$

$$-0.05f_8^{n+1} + 1.2f_9^{n+1} - 0.05f_{10}^{n+1} = 0.05f_8^n + 0.8f_9^n + 0.05f_{10}^n \quad (10.324)$$

$$-0.05f_9^{n+1} + 1.2f_{10}^{n+1} - 0.05f_{11}^{n+1} = 0.05f_9^n + 0.8f_{10}^n + 0.05f_{11}^n \quad (10.325)$$

Thus, solving this set of equations for the values of the function f, we should get a unique value for each node of the computational domain at a specific time (t) as follows:

$$f(x,t) = \begin{array}{l} f_1 \ \text{(node 1)} \\ f_2 \ \text{(node 2)} \\ f_3 \ \text{(node 3)} \\ f_4 \ \text{(node 4)} \\ f_5 \ \text{(node 5)} \\ f_6 \ \text{(node 6)} \\ f_7 \ \text{(node 7)} \\ f_8 \ \text{(node 8)} \\ f_9 \ \text{(node 9)} \\ f_{10} \ \text{(node 10)} \\ f_{11} \ \text{(node 11)} \end{array} \tag{10.326}$$

where the subscripts mean the number of the node in the computational domain; here in this example i varies from 1 to 11.

Therefore, the solutions for this set of equations over the time passing are:

$$t = 0.0\text{:} \ f(x,0) = \begin{array}{l} f_1 = 1000 \ \text{(node 1)} \\ f_2 = -9900 \ \text{(node 2)} \\ f_3 = -9600 \ \text{(node 3)} \\ f_4 = -9100 \ \text{(node 4)} \\ f_5 = -8400 \ \text{(node 5)} \\ f_6 = -7500 \ \text{(node 6)} \\ f_7 = -7400 \ \text{(node 7)} \\ f_8 = -5100 \ \text{(node 8)} \\ f_9 = -3600 \ \text{(node 9)} \\ f_{10} = -1900 \ \text{(node 10)} \\ f_{11} = 0.000 \ \text{(node 11)} \end{array} \tag{10.327}$$

$$t = 1.0\text{:} \ f(x,1) = \begin{array}{l} f_1 = -9000 \ \text{(node 1)} \\ f_2 = -8118.3112806 \ \text{(node 2)} \\ f_3 = -7839.4707332 \ \text{(node 3)} \\ f_4 = -7428.986317 \ \text{(node 4)} \\ f_5 = -6856.2008754 \ \text{(node 5)} \\ f_6 = -6119.8346919 \ \text{(node 6)} \\ f_7 = -5219.8317302 \ \text{(node 7)} \\ f_8 = -4156.1268325 \ \text{(node 8)} \\ f_9 = -2927.2122504 \ \text{(node 9)} \\ f_{10} = -1496.9671771 \ \text{(node 10)} \\ f_{11} = 1000 \ \text{(node 11)} \end{array} \tag{10.328}$$

$t = 2.0$: $f(x, 2) =$

$$\begin{aligned} &f_1 = -8000 \text{ (node 1)} \\ &f_2 = -6714.1062801 \text{ (node 2)} \\ &f_3 = -6406.5507224 \text{ (node 3)} \\ &f_4 = -6065.1110571 \text{ (node 4)} \\ &f_5 = -5596.1146484 \text{ (node 5)} \\ &f_6 = -4993.6405056 \text{ (node 6)} \\ &f_7 = -4257.2574848 \text{ (node 7)} \\ &f_8 = -3386.5391299 \text{ (node 8)} \\ &f_9 = -2373.6816332 \text{ (node 9)} \\ &f_{10} = -1093.8200681 \text{ (node 10)} \\ &f_{11} = 2000 \text{ (node 11)} \end{aligned} \tag{10.329}$$

$t = 3.0$: $f(x, 3) =$

$$\begin{aligned} &f_1 = -7000 \text{ (node 1)} \\ &f_2 = -5586.4421909 \text{ (node 2)} \\ &f_3 = -5242.6125808 \text{ (node 3)} \\ &f_4 = -4952.2597472 \text{ (node 4)} \\ &f_5 = -4567.6213517 \text{ (node 5)} \\ &f_6 = -4074.6526947 \text{ (node 6)} \\ &f_7 = -3472.0433218 \text{ (node 7)} \\ &f_8 = -2758.3870289 \text{ (node 8)} \\ &f_9 = -1913.2453716 \text{ (node 9)} \\ &f_{10} = -699.5018905 \text{ (node 10)} \\ &f_{11} = 3000 \text{ (node 11)} \end{aligned} \tag{10.330}$$

$t = 4.0$: $f(x, 4) =$

$$\begin{aligned} &f_1 = -6000 \text{ (node 1)} \\ &f_2 = -4663.4580585 \text{ (node 2)} \\ &f_3 = -4296.9934031 \text{ (node 3)} \\ &f_4 = -4044.3836159 \text{ (node 4)} \\ &f_5 = -3728.2133776 \text{ (node 5)} \\ &f_6 = -3324.7374465 \text{ (node 6)} \\ &f_7 = -2831.4853378 \text{ (node 7)} \\ &f_8 = -2244.9106612 \text{ (node 8)} \\ &f_9 = -1526.3705319 \text{ (node 9)} \\ &f_{10} = -317.9821055 \text{ (node 10)} \\ &f_{11} = 4000 \text{ (node 11)} \end{aligned} \tag{10.331}$$

$t = 5.0$: $f(x,5) =$

$$\begin{aligned} f_1 &= -5000 \quad (\text{node } 1)\\ f_2 &= -3893.3085961 \quad (\text{node } 2)\\ f_3 &= -3527.4063064 \quad (\text{node } 3)\\ f_4 &= -3304.4427572 \quad (\text{node } 4)\\ f_5 &= -3043.2198665 \quad (\text{node } 5)\\ f_6 &= -2712.834038 \quad (\text{node } 6)\\ f_7 &= -2308.7970444 \quad (\text{node } 7)\\ f_8 &= -1824.2950276 \quad (\text{node } 8)\\ f_9 &= -1198.2836185 \quad (\text{node } 9)\\ f_{10} &= 49.4881826 \quad (\text{node } 10)\\ f_{11} &= 5000 \quad (\text{node } 11) \end{aligned} \tag{10.332}$$

Example 10.10 Numerical modeling of the wave equation

Considering a second order wave equation for $0 \le x \le L$ and $0 \le t$, we intend to develop a numerical solution. The equation, as already mentioned, is:

$$\frac{\partial^2 f}{\partial t^2} = C^2 \frac{\partial^2 f}{\partial x^2} \tag{10.333}$$

Assuming constant wave speed c and Dirichlet boundary condition for $0 \le t$ as well as initial conditions as follows:

$$f(t,0) = g_1(t) \tag{10.334}$$

$$f(t,l) = g_2(t) \tag{10.335}$$

$$f(0,x) = h_1(t) \tag{10.336}$$

$$\frac{\partial f}{\partial t}(0,x) = h_2(x) \tag{10.337}$$

We are going to approximate the equation using a finite difference method with accuracy in the order of $(\Delta t)^2$ and $(\Delta x)^2$.

Assuming fixed width of cells:

$$\Delta x = \frac{L}{n} \tag{10.338}$$

$$t_i = i\Delta t \tag{10.339}$$

$$x_j = j\Delta x \tag{10.340}$$

We can replace the second order derivatives using finite difference approximations:

$$\frac{\partial^2 f}{\partial t^2}(t_i, x_j) \approx \frac{f(t_{i-1}, x_j) - 2f(t_i, x_j) + f(t_{i+1}, x_j)}{(\Delta t)^2} + O((\Delta x)^2) \tag{10.341}$$

$$\frac{\partial^2 f}{\partial x^2}(t_i, x_j) \approx \frac{f(t_i, x_{j-1}) - 2f(t_i, x_j) + f(t_i, x_{j+1})}{(\Delta t)^2} + O((\Delta t)^2) \tag{10.342}$$

where the errors are of the orders of $(\Delta t)^2$ and $(\Delta x)^2$.

Now we can substitute the finite difference approximations into the partial differential equation and rearrange the terms:

$$f_{i+1,j} = -f_{i-1,j} + \left(\frac{c\Delta t}{\Delta x}\right)^2 f_{i,j-1} + 2\left(1 - \left(\frac{c\Delta t}{\Delta x}\right)^2\right) f_{i,j} + \left(\frac{c\Delta t}{\Delta x}\right)^2 f_{i,j+1} \tag{10.343}$$

The boundary conditions require that

$$f_{i,0} = g_1(t_i) \tag{10.344}$$

$$f_{i,n} = g_2(t_i) \tag{10.345}$$

This allows us to rewrite the system in matrix form:

$$f_{i+1} = -f_{i-1} + mf_i + a_i \tag{10.346}$$

where:

$$m = \begin{bmatrix} 2\left[1-\left(\frac{c\Delta t}{\Delta x}\right)^2\right] & \left(\frac{c\Delta t}{\Delta x}\right)^2 & 0 & & 0 & 0 \\ \left(\frac{c\Delta t}{\Delta x}\right)^2 & 2\left[1-\left(\frac{c\Delta t}{\Delta x}\right)^2\right] & \left(\frac{c\Delta t}{\Delta x}\right)^2 & 0 & & 0 \\ 0 & \left(\frac{c\Delta t}{\Delta x}\right)^2 & 2\left[1-\left(\frac{c\Delta t}{\Delta x}\right)^2\right] & \left(\frac{c\Delta t}{\Delta x}\right)^2 & 0 & \\ & 0 & \left(\frac{c\Delta t}{\Delta x}\right)^2 & 2\left[1-\left(\frac{c\Delta t}{\Delta x}\right)^2\right] & \left(\frac{c\Delta t}{\Delta x}\right)^2 & 0 \\ & & & . & . & . \\ 0 & & & & . & . \quad \left(\frac{c\Delta t}{\Delta x}\right)^2 \\ 0 & 0 & & & 0 & \left(\frac{c\Delta t}{\Delta x}\right)^2 \; 2\left[1-\left(\frac{c\Delta t}{\Delta x}\right)^2\right] \end{bmatrix} \tag{10.347}$$

and

$$f_i = \begin{bmatrix} f_{i,1} \\ f_{i,2} \\ f_{i,3} \\ f_{i,4} \\ . \\ . \\ f_{i,n-1} \end{bmatrix} \tag{10.348}$$

and

$$a_i = \begin{bmatrix} \left(\frac{c\Delta t}{\Delta x}\right)^2 g_{(1),i} \\ 0 \\ 0 \\ 0 \\ \cdot \\ 0 \\ \left(\frac{c\Delta t}{\Delta x}\right)^2 g_{(2),i} \end{bmatrix} \tag{10.349}$$

where it can be solved using an iterative method. Thus f_{i+1} can be computed based on f_{i-1} and f_i.

For approximation of f_1 with error on the order of $(\Delta t)^2$, for $i = 1$ or $t = \Delta t$ we find

$$f_{1,j} \cong f(0, x_j) + (\Delta t)\frac{\partial f}{\partial t}(0, x_j) + \frac{1}{2}(c\Delta t)^2 \frac{\partial^2 f}{\partial x^2}(0, x_j) \tag{10.350}$$

and we have

$$f(0, x) = h_{(1)}(t) \tag{10.351}$$

$$\frac{\partial f}{\partial t}(0, x) = h_{(2)}(x) \tag{10.352}$$

hence

$$f_{1,j} \cong h_{(1)}(x_j) + (\Delta t)h_{(2)}(x_j) + \frac{1}{2}(c\Delta t)^2 \frac{\partial^2 h_{(1)}(x_j)}{\partial x^2} \tag{10.353}$$

or

$$f_{1,j} \cong h_{(1),j} + (\Delta t)h_{(2),j} + \frac{1}{2}\left(\frac{c\Delta t}{\Delta x}\right)^2 (h_{(1),j-1} - 2h_{(1),j} + h_{(1),j+1}) \tag{10.354}$$

and

$$f_{1,j} = \frac{1}{2}\left(\frac{c\Delta t}{\Delta x}\right)^2 h_{(1),j+1} + \left(1 - \left(\frac{c\Delta t}{\Delta x}\right)^2\right) h_{(1),j} + \frac{1}{2}\left(\frac{c\Delta t}{\Delta x}\right)^2 h_{(1),j-1} + (\Delta t)h_{(2),j} \tag{10.355}$$

If we want to explain it in matrix form we can write:

$$f_1 = \frac{1}{2}\mathrm{m}f_0 + (\Delta t)h_{(2)} + \frac{1}{2}a_0 \tag{10.356}$$

where

$$f_0 = h_{(1)} \tag{10.357}$$

Example 10.11 Application of different methods to the Poisson equation

Consider the 1D Poisson equation with the following form and boundary conditions:

$$-\frac{\partial^2 f}{\partial x^2}=\varphi \tag{10.358}$$

$$\begin{cases} f(x)=0 & \text{at} \quad x=0 \\ f(x)=0 & \text{at} \quad x=1 \end{cases} \tag{10.359}$$

where f represents the velocity potential.

The finite difference equations are approximated from this differential form using a Taylor expansion. Assuming a one-dimensional domain including N nodes with the same grid steps we have:

$$f_{i-1}\cong f_i-\Delta x\left(\frac{\partial f}{\partial x}\right)_{(i)}+\frac{1}{2}\Delta x^2\left(\frac{\partial^2 f}{\partial x^2}\right)_{(i)}-O(\Delta x^2) \tag{10.360}$$

$$f_{i+1}\cong f_i+\Delta x\left(\frac{\partial f}{\partial x}\right)_{(i)}-\frac{1}{2}\Delta x^2\left(\frac{\partial^2 f}{\partial x^2}\right)_{(i)}+O(\Delta x^2) \tag{10.361}$$

It can be seen that the Taylor series are expanded only up to the quadratic terms and the higher order terms are neglected. Therefore, the accuracy of these approximations is second order in space.

Thus we get:

$$\left(\frac{\partial f}{\partial x}\right)_{(i)}\cong-\left(\frac{f_{i-1}-f_{i+1}}{2\Delta x}\right) \tag{10.362}$$

and

$$\left(\frac{\partial^2 f}{\partial x^2}\right)_{(i)}\cong\left(\frac{f_{i-1}-2f_i+f_{i+1}}{\Delta x^2}\right) \tag{10.363}$$

with the finite-differences formulation as follows:

$$\varphi_{(i)}=-\left(\frac{f_{i-1}-2f_i+f_{i+1}}{\Delta x^2}\right) \quad \mathrm{i}=2,3,\ldots,N-2,N-1 \tag{10.364}$$

For example:

$$f_3\cong f_2+\Delta x\left(\frac{\partial f}{\partial x}\right)_{(2)}-\frac{1}{2}\Delta x^2\left(\frac{\partial^2 f}{\partial x^2}\right)_{(2)}+O(\Delta x^2) \tag{10.365}$$

$$f_1\cong f_2-\Delta x\left(\frac{\partial f}{\partial x}\right)_{(2)}+\frac{1}{2}\Delta x^2\left(\frac{\partial^2 f}{\partial x^2}\right)_{(2)}-O(\Delta x^2) \tag{10.366}$$

Therefore, these formulations can be obtained for all nodes located inside the computational domain.

$$\varphi_{(2)} = -\left(\frac{f_1 - 2f_2 + f_3}{\Delta x^2}\right) \tag{10.367}$$

$$\varphi_{(3)} = -\left(\frac{f_2 - 2f_3 + f_4}{\Delta x^2}\right) \tag{10.368}$$

$$\varphi_{(4)} = -\left(\frac{f_3 - 2f_4 + f_5}{\Delta x^2}\right) \tag{10.369}$$

$$\vdots$$

$$\varphi_{(n-1)} = -\left(\frac{f_{n-2} - 2f_{n-1} + f_n}{\Delta x^2}\right) \tag{10.370}$$

and a tridiagonal matrix can be written:

$$M = \frac{1}{\Delta x^2}\left\{\begin{array}{cccccccccc} 2 & -1 & 0 & & & & & 0 & 0 & 0 \\ -1 & 2 & -1 & 0 & & & & & 0 & 0 \\ 0 & -1 & 2 & -1 & 0 & & & & & 0 \\ & & \cdot & \cdot & \cdot & & & & & \\ & & & \cdot & \cdot & \cdot & & & & \\ & & & & \cdot & \cdot & \cdot & & & \\ & & & & & \cdot & \cdot & \cdot & & \\ 0 & & & & & 0 & -1 & 2 & -1 & 0 \\ 0 & 0 & & & & & 0 & -1 & 2 & -1 \\ 0 & 0 & 0 & & & & & 0 & -1 & 2 \end{array}\right\} \tag{10.371}$$

with the solution vector as:

$$x^T = \left[f_1 \quad f_2 \quad f_3 \quad . \quad . \quad . \quad f_N\right] \tag{10.372}$$

and the right-hand side of the equation as:

$$C^T = \left[\varphi_1 \quad \varphi_2 \quad \varphi_3 \quad . \quad . \quad . \quad \varphi_N\right] \tag{10.373}$$

we obtain a matrix equation:

$$M \cdot X = C \quad \text{for:} \quad i = 2, 3, \ldots, N-1 \tag{10.374}$$

The matrix equation is solved considering the boundary conditions.

As a different solution, remembering the finite volume method described before in the previous chapters (The finite volume method will be also discussed further in detail in the next section):

Similarly we need to approximate the differential equation.

We have:

$$-\frac{\partial}{\partial x}\left(\frac{\partial f}{\partial x}\right) = \varphi \tag{10.375}$$

Integrating the equation over the control volume we have:

$$\left(\frac{\partial f}{\partial x}\right)_1 - \left(\frac{\partial f}{\partial x}\right)_2 = \int_1^2 \varphi dx = 0 \tag{10.376}$$

$$\left(\frac{f_n - f_1}{\Delta x_1}\right) - \left(\frac{f_2 - f_n}{\Delta x_2}\right) = \overline{\varphi}\Delta x \tag{10.377}$$

where $\overline{\varphi}$ is the average of φ over the control volume (see the next section for further details on the finite volume method).

As an additional solution based on the finite element method (for further information and details, readers are referred to the books on the numerical modeling, and solving the PDEs using the finite elements method), we have:

$$\int_{x_{i-1}}^{x_i} w\left(\left(\frac{\partial^2 f}{\partial x^2}\right) + \varphi\right) dx = 0 \quad \text{for } i = 2, 3, 4, \ldots, N \tag{10.378}$$

assuming

$$f = mx + c \tag{10.379}$$

and it can be written as functions of f_1 and f_2

$$\begin{Bmatrix} f_1 \\ f_2 \end{Bmatrix} = \begin{bmatrix} 1 & x_1 \\ 1 & x_2 \end{bmatrix} \begin{Bmatrix} c \\ m \end{Bmatrix} \tag{10.380}$$

Now we can compute the value of the variable f at the location x in element as follows:

$$f = b_1 f_1 + b_2 f_2 \tag{10.381}$$

where b_1 and b_2 are the basis functions:

$$b = \begin{bmatrix} 1 & x \end{bmatrix} \begin{bmatrix} 1 & x_1 \\ 1 & x_2 \end{bmatrix}^{-1} \tag{10.382}$$

Thus considering the boundary conditions, the system of linear equations can be solved for the unknown functions at the nodes.

10.4 FINITE VOLUME METHOD

The finite volume method was introduced and described to some extent in the previous chapters. In this section the method is briefly evaluated and discussed with a few simple examples. Considering a one-dimensional problem of linear convection with the following equation:

$$\frac{\partial \varnothing(\mathrm{x}, \mathrm{t})}{\partial \mathrm{t}} + \mathrm{C}\frac{\partial \varnothing(\mathrm{x}, \mathrm{t})}{\partial \mathrm{t}} = 0 \tag{10.383}$$

we can write

$$\frac{d(\Delta x \overline{\Phi}_j)}{dt} - \left(f_{j-\frac{1}{2}} - f_{j+\frac{1}{2}}\right) = \int_{x_{j-\frac{1}{2}}}^{x_{j+\frac{1}{2}}} s(x, t) dt \tag{10.384}$$

which becomes:

$$\frac{d(\Delta x\overline{\Phi}_j)}{dt} - \left(f_{j-\frac{1}{2}} - f_{j+\frac{1}{2}}\right) = 0 \tag{10.385}$$

where

$$\hat{f}_{j-\frac{1}{2}} = \frac{f^l_{j-\frac{1}{2}} + f^r_{j-\frac{1}{2}}}{2} = \frac{c\overline{\varnothing}_{j-1} + c\overline{\varnothing}_j}{2} \tag{10.386}$$

and

$$\hat{f}_{j+\frac{1}{2}} = \frac{f^l_{j+\frac{1}{2}} + f^r_{j+\frac{1}{2}}}{2} = \frac{c\overline{\varnothing}_j + c\overline{\varnothing}_{j+1}}{2} \tag{10.387}$$

Substitute into integral equation:

$$\frac{d(\Delta x\overline{\Phi}_j)}{dt} - \left(f_{j-\frac{1}{2}} - f_{j+\frac{1}{2}}\right) \approx \frac{d(\Delta x\overline{\varnothing}_j)}{dt} - \left(\hat{f}_{j-\frac{1}{2}} - \hat{f}_{j+\frac{1}{2}}\right) = \frac{d(\Delta x\overline{\varnothing}_j)}{dt} + \frac{c\overline{\varnothing}_j + c\overline{\varnothing}_{j+1}}{2} - \frac{c\overline{\varnothing}_{j-1} + c\overline{\varnothing}_j}{2} \tag{10.388}$$

and

$$\Delta x\frac{d\overline{\varnothing}_j}{dt} + \frac{c\overline{\varnothing}_{j+1} + c\overline{\varnothing}_{j-1}}{2} = 0 \tag{10.389}$$

For further information about the finite volume method and its applications please see: Causon et al., (2011); LeVeque (1990, 2002). Following are a few simple examples of solving steady state diffusion equation in one, two and three dimensional problems with the finite volume method for further exercises:

Example 10.12 One-dimensional problems

Unlike the finite difference method, which is based on the discretization of the differential form of the conservation equations, the finite volume method is based on the discretization of the integral form of the conservation equations.

As a simple example, regarding the one-dimensional steady state diffusion equation, the process is governed by:

$$\frac{d}{dx}\left(\delta\frac{df}{dx}\right) + \vartheta = 0 \tag{10.390}$$

where δ is the diffusion coefficient and ϑ is the source term.

The computational domain needs to be divided into the discrete control volumes. The governing equation should be integrated over the control volume:

$$\int_{\Delta V} \frac{d}{dx}\left(\delta \frac{df}{dx}\right) dV + \int_{\Delta V} \vartheta dV = 0 \tag{10.391}$$

Then the discretized equation will be obtained.

Assuming linear approximation for the diffusion coefficient and source term, we get:

$$\begin{aligned} &\int_{\Delta V} \frac{d}{dx}\left(\delta \frac{df}{dx}\right) dV + \int_{\Delta V} \vartheta dV = \left(\delta A \tfrac{df}{dx}\right)_2 - \left(\delta A \tfrac{df}{dx}\right)_1 + \overline{\vartheta}\Delta V = \\ &= A_2 \left(\frac{\delta_2 + \delta_n}{2}\right)\left(\frac{f_2 - f_n}{\delta x_2}\right) - A_1 \left(\frac{\delta_n + \delta_1}{2}\right)\left(\frac{f_n - f_1}{\delta x_1}\right) + \vartheta_u + \vartheta_n f_n = 0 \end{aligned} \tag{10.392}$$

where A is the cross-sectional area of the control volume face, ΔV is the volume and $\overline{\vartheta}$ is the average value of source ϑ over the control volume.

Thus we get:

$$\left(\frac{\left(\frac{\delta_2 + \delta_n}{2}\right)}{\delta x_2} A_2 + \frac{\left(\frac{\delta_n + \delta_1}{2}\right)}{\delta x_1} A_1 - S_n\right) f_n = \left(\frac{\left(\frac{\delta_n + \delta_1}{2}\right)}{\delta x_1} A_1\right) f_1 + \left(\frac{\left(\frac{\delta_2 + \delta_n}{2}\right)}{\delta x_2} A_2\right) f_2 + S_u \tag{10.393}$$

Then the discretized equations should be written for each node. Boundary conditions need to be considered. Finally, the system of linear algebraic equations is solved. Therefore, the distribution of the parameter f at each node will be determined.

Averaging the diffusion coefficients (δ) and neglecting the source term (ϑ), the discretized equations above can be rewritten in a simplified form. Let $f_{(\mathrm{I})}$ and $f_{(\mathrm{II})}$ be the values of variable f in two boundary nodes and c is constant; A and L are the area of the section and the length of the computational domain.

Dividing the whole length of the computational domain into N equal control volumes gives us

$$\Delta x = \frac{L}{N} \tag{10.394}$$

Discretized equations for internal nodes are:

$$\left(\frac{c_2}{\Delta x_2} A_2 + \frac{c_1}{\Delta x_1} A_1\right) f_n = \left(\frac{c_1}{\Delta x_1} A_1\right) f_1 + \left(\frac{c_2}{\Delta x_2} A_2\right) f_2 \quad \text{for } n = 2, 3, 4, \ldots, N-1 \tag{10.395}$$

For the boundaries we should rewrite the discretized equations as follows:

For the first node, we have:

$$\left(\frac{k}{\Delta x} A + \frac{2k}{\Delta x} A\right) f_1' = \left(\frac{k}{\Delta x} A\right) f_2 + \left(\frac{2k}{\Delta x} A\right) f_{(\mathrm{I})} \tag{10.396}$$

And for the last node, we have:

$$\left(\frac{k}{\Delta x} A + \frac{2k}{\Delta x} A\right) f_N' = \left(\frac{k}{\Delta x} A\right) f_1 + \left(\frac{2k}{\Delta x} A\right) f_{(\mathrm{II})} \tag{10.397}$$

The resulting set of algebraic equations:

$$a_1 f'_1 - b_1 f'_2 = d_1 f_{(I)} \tag{10.398}$$

$$-a_2 f'_1 + b_2 f'_2 - c_2 f'_3 = 0 \tag{10.399}$$

$$-a_3 f'_2 + b_3 f'_3 - c_3 f'_4 = 0 \tag{10.400}$$

$$-a_4 f'_3 + b_4 f'_4 - c_4 f'_5 = 0 \tag{10.401}$$

$$\vdots$$

$$-a_i f'_{i-1} + b_i f'_i - c_i f_{i+1} = 0 \tag{10.402}$$

$$\vdots$$

$$-a_N f'_{N-1} + b_N f'_N = d_N f_{(II)} \tag{10.403}$$

or

$$\begin{bmatrix} a_1 & -b_1 & 0 & & & 0 & 0 \\ -a_2 & b_2 & -c_3 & 0 & & & 0 \\ 0 & -a_3 & b_3 & -c_3 & 0 & & \\ & 0 & -a_4 & b_4 & -c_4 & 0 & \\ & & & . & . & . & \\ 0 & & & & . & . & . \\ 0 & 0 & & & 0 & -a_N & b_N \end{bmatrix} \begin{bmatrix} f'_1 \\ f'_2 \\ f'_3 \\ f'_4 \\ . \\ . \\ f'_N \end{bmatrix} = \begin{bmatrix} d_1 f_{(I)} \\ 0 \\ 0 \\ 0 \\ . \\ . \\ d_N f_{(II)} \end{bmatrix} \tag{10.404}$$

Then the values of known coefficients will be substituted in the matrix and the values of the unknowns ($f'_1, f'_2, f'_3, \ldots, f'_N$) will be determined by solving the matrix.

Example 10.13 Two-dimensional problems

Consider the two-dimensional steady state diffusion equation as follows:

$$\frac{\partial}{\partial x}\left(\delta \frac{\partial f}{\partial x}\right) + \frac{\partial}{\partial y}\left(\delta \frac{\partial f}{\partial y}\right) + \vartheta = 0 \tag{10.405}$$

We integrate the equation over the control volume:

$$\int_{\Delta V} \frac{\partial}{\partial x}\left(\delta \frac{\partial f}{\partial x}\right) dV + \int_{\Delta V} \frac{\partial}{\partial y}\left(\delta \frac{\partial f}{\partial y}\right) dV + \int_{\Delta V} \vartheta dV = 0 \tag{10.406}$$

where for this 2D problem:

$$dV = dx.dy \tag{10.407}$$

Since $A_r = A_l = \Delta y$ and $A_t = A_b = \Delta x$, we have:

$$\left[\delta_r A_r \left(\frac{\partial f}{\partial x}\right)_r - \delta_l A_l \left(\frac{\partial f}{\partial x}\right)_l\right] + \left[\delta_t A_t \left(\frac{\partial f}{\partial y}\right)_t - \delta_b A_b \left(\frac{\partial f}{\partial y}\right)_b\right] + \overline{\vartheta}\Delta V = 0 \tag{10.408}$$

Linear approximation:

$$\delta_l A_l \left(\frac{\partial f}{\partial x}\right)_l = \delta_l A_l \frac{f_n - f_l}{\Delta x_l} \tag{10.409}$$

$$\delta_r A_r \left(\frac{\partial f}{\partial x}\right)_r = \delta_r A_r \frac{f_r - f_n}{\Delta x_r} \tag{10.410}$$

$$\delta_t A_t \left(\frac{\partial f}{\partial y}\right)_t = \delta_t A_t \frac{f_t - f_n}{\Delta y_t} \tag{10.411}$$

$$\delta_b A_b \left(\frac{\partial f}{\partial y}\right)_b = \delta_b A_b \frac{f_n - f_b}{\Delta y_b} \tag{10.412}$$

with the source term:

$$\overline{\vartheta}\Delta V = \vartheta_u + \vartheta_n f_n \tag{10.413}$$

we have:

Substituting and rearranging:

$$\left(\frac{\delta_l A_l}{\Delta x_l} + \frac{\delta_r A_r}{\Delta x_r} + \frac{\delta_b A_b}{\Delta y_b} + \frac{\delta_t A_t}{\Delta y_t} - \vartheta_n\right) f_n = \left(\frac{\delta_l A_l}{\Delta x_l}\right) f_l + \left(\frac{\delta_r A_r}{\Delta x_r}\right) f_r + \left(\frac{\delta_b A_b}{\Delta y_b}\right) f_b + \left(\frac{\delta_t A_t}{\Delta y_t}\right) f_t + \vartheta_u \tag{10.414}$$

Example 10.14 Three-dimensional problems

Consider the three-dimensional steady state diffusion equation:

$$\frac{\partial}{\partial x}\left(\delta \frac{\partial f}{\partial x}\right) + \frac{\partial}{\partial y}\left(\delta \frac{\partial f}{\partial y}\right) + \frac{\partial}{\partial z}\left(\delta \frac{\partial f}{\partial z}\right) + \vartheta = 0 \tag{10.415}$$

When the equation is formally integrated over the control volume:

$$\int_{\Delta V} \frac{\partial}{\partial x}\left(\delta \frac{\partial f}{\partial x}\right) dV + \int_{\Delta V} \frac{\partial}{\partial y}\left(\delta \frac{\partial f}{\partial y}\right) dV + \int_{\Delta V} \frac{\partial}{\partial z}\left(\delta \frac{\partial f}{\partial z}\right) dV + \int_{\Delta V} \vartheta dV = 0 \tag{10.416}$$

or

$$\begin{aligned}&\left[\delta_r A_r \left(\frac{\partial f}{\partial x}\right)_r - \delta_l A_l \left(\frac{\partial f}{\partial x}\right)_l\right] + \left[\delta_t A_t \left(\frac{\partial f}{\partial y}\right)_t - \delta_b A_b \left(\frac{\partial f}{\partial y}\right)_b\right] \\ &+ \left[\delta_f A_f \left(\frac{\partial f}{\partial y}\right)_f - \delta_h A_h \left(\frac{\partial f}{\partial y}\right)_h\right] + \overline{\vartheta}\Delta V = 0\end{aligned} \tag{10.417}$$

Linear approximation:

$$\delta_r A_r \left(\frac{\partial f}{\partial x}\right)_r = \delta_r A_r \frac{f_r - f_n}{\Delta x_r} \tag{10.418}$$

$$\delta_l A_l \left(\frac{\partial f}{\partial x}\right)_l = \delta_l A_l \frac{f_n - f_l}{\Delta x_l} \tag{10.419}$$

$$\delta_t A_t \left(\frac{\partial f}{\partial x}\right)_t = \delta_t A_t \frac{f_t - f_n}{\Delta x_t} \tag{10.420}$$

$$\delta_b A_b \left(\frac{\partial f}{\partial x}\right)_b = \delta_b A_b \frac{f_t - f_b}{\Delta x_b} \tag{10.421}$$

$$\delta_f A_f \left(\frac{\partial f}{\partial x}\right)_f = \delta_f A_f \frac{f_f - f_n}{\Delta x_f} \tag{10.422}$$

$$\delta_h A_h \left(\frac{\partial f}{\partial x}\right)_h = \delta_h A_h \frac{f_n - f_h}{\Delta x_h} \tag{10.423}$$

With the source term:

$$\overline{\vartheta} \Delta V = \vartheta_u + \vartheta_n f_n \tag{10.424}$$

By substituting and rearranging we have:

$$\left(\frac{\delta_l A_l}{\Delta x_l} + \frac{\delta_r A_r}{\Delta x_r} + \frac{\delta_b A_b}{\Delta y_b} + \frac{\delta_t A_t}{\Delta y_t} + \frac{\delta_f A_f}{\Delta y_f} + \frac{\delta_h A_h}{\Delta y_h} - \vartheta_n\right) f_n =$$
$$\left(\frac{\delta_l A_l}{\Delta x_l}\right) f_l + \left(\frac{\delta_r A_r}{\Delta x_r}\right) f_r + \left(\frac{\delta_b A_b}{\Delta y_b}\right) f_b + \left(\frac{\delta_t A_t}{\Delta y_t}\right) f_t + \left(\frac{\delta_f A_f}{\Delta y_f}\right) f_f + \left(\frac{\delta_h A_h}{\Delta y_h}\right) f_h + \vartheta_u \tag{10.425}$$

10.5 SOLUTION OF THE NAVIER-STOKES EQUATIONS

Basically, the main result of the solving process of the Navier–Stokes equations is a velocity field defined at different points in the region of interest. When the velocity field is calculated, other quantities of interest such as pressure can be calculated. An independent pressure equation does not exist in the Navier–Stokes equations and the momentum equations include pressure gradient.

To calculate the pressure using Navier–Stokes for incompressible flows we have (see: Constantin and Foias 1988; Drazin and Riley 2006 and Quartapelle 1993):

$$\frac{\partial \rho u^{\rightarrow}}{\partial t} + \nabla.(\rho \vec{u}\,\vec{u}) + \nabla p - \mu \nabla^2 \vec{u} - \rho \vec{g} = 0 \tag{10.426}$$

Hence it is clear that the Navier–Stokes equations are nonlinear partial differential equations in their general form. The Navier–Stokes equations explain the balance of momentum. However, basically, an additional explanation of the mass conservation or continuity equation is also required to solve a problem:

$$\nabla.\vec{u} = 0 \tag{10.427}$$

Thus for constant viscosity and density, the following equation (which is a Poisson equation) is obtained by combining the conservation equations:

$$\frac{\partial}{\partial x_i}\left(\frac{\partial p}{\partial x_i}\right) = -\frac{\partial}{\partial x_i}\left(\frac{\partial(\rho u_i u_j)}{\partial x_i}\right) \tag{10.428}$$

The derivatives of momentum terms must be approximated in a form consistent with those of the momentum equations. The pressure equation can be derived from discretized momentum/continuity equations.

Therefore, first we need to solve the incompressible momentum equations to calculate the velocity field for a known pressure and then we can solve the Poisson equation to compute the corrected pressure field. The explicit or implicit methods can be applied (see: Constantin and Foias 1988; Drazin and Riley 2006 and Quartapelle 1993).

Consider the equation:

$$\frac{\partial \rho f_i}{\partial t} = -\frac{\partial(\rho f_i f_j)}{\partial x_j} - \frac{\partial p}{\partial x_i} + \frac{\partial \tau_{ij}}{\partial x_j} \tag{10.429}$$

where the variable f is velocity or u.

Using an explicit method, it can be replaced by

$$(\rho f_i)^{n-1} - (\rho f_i)^n = \Delta t\left(-\frac{\partial(\rho f_i f_j)^n}{\partial x_j} + \frac{\partial \tau_{ij}{}^n}{\partial x_j} - \frac{\partial p^n}{\partial x_i}\right) \tag{10.430}$$

In general, the new velocity field we obtain at time $n + 1$ does not satisfy the continuity equation:

$$\frac{\delta(\rho f_i)^{n+1}}{\delta x_i} = 0 \tag{10.431}$$

Taking the numerical divergence of the Navier–Stokes equations, we get:

$$\frac{\partial(\rho f_i)^{n+1}}{\partial x_i} - \frac{\partial(\rho f_i)^n}{\partial x_i} = \Delta t\left[\frac{\partial}{\partial x_i}\left(-\frac{\partial(\rho f_i f_j)^n}{\partial x_j} + \frac{\partial \tau_{ij}^n}{\partial x_j} - \frac{\partial p^n}{\partial x_i}\right)\right] \quad (10.432)$$

The first and second terms need to be zero; thus we obtain:

$$\frac{\delta}{\delta x_i}\left(\frac{\delta p^n}{\delta x_i}\right) = \frac{\delta}{\delta x_i}\left(-\frac{\partial(\rho f_i f_j)^n}{\partial x_j} + \frac{\partial \tau_{ij}^n}{\partial x_j}\right) \quad (10.433)$$

Therefore we have to begin with velocity at time t_n. Then we can solve the Poisson equation for the pressure at time t_n. Finally we compute the velocity field at the new time step using the momentum equation and go to the next time step.

It might be more difficult to use an implicit method for solving the incompressible Navier–Stokes equations.

Recalling the main equation, we have:

$$\frac{\partial \rho f_i}{\partial t} = -\frac{\delta(\rho f_i f_j)}{\delta x_j} - \frac{\delta p}{\delta x_i} + \frac{\delta \tau_{ij}}{\delta x_j} \quad (10.434)$$

Thus using an implicit method we get:

$$(\rho f_i)^{n+1} - (\rho f_i)^n = \Delta t\left(-\frac{\delta(\rho f_i f_j)^{n+1}}{\delta x_j} + \frac{\delta \tau_{ij}^{n+1}}{\delta x_j} - \frac{\delta p^{n+1}}{\delta x_i}\right) \quad (10.435)$$

Taking the numerical divergence of the Navier–Stokes equations we get:

$$\frac{\delta(\rho f_i)^{n+1}}{\delta x_i} - \frac{\delta(\rho f_i)^n}{\delta x_i} = \Delta t\left[\frac{\delta}{\delta x_i}\left(-\frac{\delta(\rho f_i f_j)^{n+1}}{\delta x_j} + \frac{\delta \tau_{ij}^{n+1}}{\delta x_j} - \frac{\delta p^{n+1}}{\delta x_i}\right)\right] \quad (10.436)$$

and similarly we get:

$$\frac{\delta}{\delta x_i}\left(\frac{\delta p^{n+1}}{\delta x_i}\right) = \frac{\delta}{\delta x_i}\left(\frac{\delta(\rho f_i f_j)^{n+1}}{\delta x_j} + \frac{\delta \tau_{ij}^{n+1}}{\delta x_j}\right) \quad (10.437)$$

However, as can be seen, the velocities and pressures are both unknown at time t_{n+1}; therefore, the Poisson equation and momentum equations need to be solved simultaneously. To solve the equation we may apply several methods, such as Newton-Raphson, Secant methods, alternate direction implicit scheme, etc.

For solving the nonlinear algebraic equations, we have

$$(\rho f_i)^{n+1} - (\rho f_i)^n = \Delta t\left(-\frac{\partial(\rho f_i f_j)^{n+1}}{\partial x_j} + \frac{\partial \tau_{ij}^{n+1}}{\partial x_j} - \frac{\partial p^{n+1}}{\partial x_i}\right) \tag{10.438}$$

or

$$\frac{\partial}{\partial x_i}\left(\frac{\partial p^{n+1}}{\partial x_i}\right) = \frac{\partial}{\partial x_i}\left(\frac{\partial(\rho f_i f_j)^{n+1}}{\partial x_j} + \frac{\partial \tau_{ij}^{n+1}}{\partial x_j}\right) \tag{10.439}$$

where

$$f_i^{n+1} = f_i^n + \Delta f_i \tag{10.440}$$

and

$$f_i^{n+1} f_j^{n+1} = f_i^n f_j^n + f_i^n \Delta f_j + f_j^n \Delta f_i + \Delta f_i \Delta f_j \tag{10.441}$$

Thus, the momentum equations are changed as follows:

$$(\rho f_i)^{n+1} - (\rho f_i)^n = \rho \Delta f_i$$
$$= \Delta t\left(-\frac{\delta(\rho f_i f_j)^n}{\delta x_j} - \frac{\delta(\rho f_i^n \Delta f_j)^n}{\delta x_j} - \frac{\delta(\rho \Delta f_i f_j^n)^n}{\delta x_j} + \frac{\delta \tau_{ij}^n}{\delta x_j} + \frac{\delta \Delta \tau_{ij}}{\delta x_j} - \frac{\delta p^n}{\delta x_i} - \frac{\delta \Delta p}{\delta x_i}\right) \tag{10.442}$$

where a large system of nonlinear momentum equations must be solved for the velocity field. It is very difficult to solve the system directly, but we can employ an iterative scheme to solve it.

Using the finite difference method we can approximate first and second derivatives. For example, for the Laplace equation we have:

$$\nabla F_{i,j} \approx \frac{F_{i-1,j} - 2F_{i,j} + F_{i+1,j}}{(\Delta x)^2} + \frac{F_{i,j-1} - 2F_{i,j} + F_{i,j+1}}{(\Delta y)^2} \tag{10.443}$$

A first derivative can be approximated:

$$F_{i,j} \approx \frac{F_{i+1,j} - F_{i-1,j}}{2\Delta x} \tag{10.444}$$

for a point between the points $F_{i+1,j}$ and $F_{i,j}$. The approximation is:

$$F_{i+\frac{1}{2},j} \approx \frac{F_{i+1,j} - F_{i,j}}{\Delta x} \tag{10.445}$$

Please see Constantin and Foias (1988); Drazin and Riley (2006) and Quartapelle (1993) for further information.

REFERENCES

Causon, D.M., Mingham, C.G., 2010. Introductory Finite Difference Methods for PDEs. Ventus Publishing ApS.

Causon, D.M., Mingham, C.G., Qian, L., 2011. Introductory Finite Volume Methods for PDEs. Ventus Publishing ApS.

Constantin, P., Foias, C., 1988. Navier-Stokes Equations. The University of Chicago Press, Chicago and London.

Crank, J., Nicolson, P., 1947. A practical method for numerical evaluation of solutions of partial differential equations of the heat conduction type. Proc. Camb. Phil. Soc. 43, 50–67.

Crank, J., Nicolson, P., 1996. A practical method for numerical evaluation of solutions of partial differential equations of the heat conduction type. Adv. Comput. Math. 6 (1), 207–226.

Davis, T.A., 2006. Direct Methods for Sparse Linear Systems, Vol. 2. Siam.

Drazin, P.G., Riley, N., 2006. The Navier-Stokes Equations: A Classification of Flows and Exact Solutions, London Mathematical Society, Vol. 13. Cambridge University Press, Lecture Note Series 334.

Farlow, S.J., 1993. Partial Differential Equations for Scientists and Engineers. Dover Publications, Inc., New York.

Greenbaum, A., 1997. Iterative Methods for Solving Linear Systems. SIAM, Philadelphia.

Kelley, C.T., 2003. Solving Nonlinear Equations with Newton's Method number 1 in Fundamental Algorithms for Numerical Calculations. SIAM, Philadelphia.

Kelley, C.T., 1995. Iterative Methods for Linear and Nonlinear Equations, number 16 in Frontiers in Applied Mathematics. SIAM, Philadelphia.

LeVeque, R.J., 2007. Finite Difference Methods for Ordinary and Partial Differential Equations, Steady State and Time Dependent Problems. SIAM.

LeVeque, R.J., 2002. Finite volume methods for hyperbolic problems. Cambridge university press.

LeVeque, R.J., 1990. Numerical Methods for Conservation Laws, Lectures in Mathematics, ETH-Zurich. Birkhauser-Verlag, Basel.

Logan, J.D., 2004. Applied Partial Differential Equations. Springer-Verlag, New York, LLC.

Olshanskii, M.A., Tyrtyshnikov, E.E., 2014. Iterative Methods for Linear Systems: Theory and Applications. SIAM, / approx. xiv + 244 / ISBN 978-1-611973-45-7.

Quartapelle, L., 1993. Numerical Solution of the Incompressible Navier-Stokes Equations, Birkhauser-Verlag. Basel.

Saad, Y., 2003. Iterative Methods for Sparse Linear Systems, second ed. SIAM, Philadelpha, PA.

Smith, G.D., 1985. Numerical Solution of Partial Differential Equations: Finite Difference Methods. Clarendon Press, Oxford.

Thomas, L.H., 1949. Elliptic Problems in Linear Differential Equations Over a Network. Columbia University, New York, Watson Sci. Comput. Lab Report.

Thomas, J.W., 1995. Numerical Partial Differential Equations: Finite Difference Methods. Springer-Verlag, New York, Inc.

Zachmanoglou, E.C., Thoe, D.W., 1986. Introduction to Partial Differential Equations with Applications. Dover Publications, Inc., New York.

CHAPTER 11

Case Studies and Worked-Out Examples

11.1 INTRODUCTION

This chapter describes the case studies, worked-out examples and analysis of the results. The examples are outlined in the following sections in detail.

A series of two-dimensional (2D) and three-dimensional (3D) laboratory experiments has been carried out in three wave tanks with different scales. The broad range of laboratory tests undertaken is to examine the wave pattern behind submerged breakwaters caused by 2D/3D wave transformation due to overtopping and diffraction. Experiments have been carried out with the objective of providing a more thorough understanding of the hydrodynamics behavior in the vicinity of 2D and particularly 3D submerged breakwaters, to prepare the required data for artificial neural network (ANN) modeling. The objective of the first sections is to describe in detail the experimental set-up, equipment and facilities, and the method of data sampling throughout the tests. Then the results and analysis of numerical models including ANN (artificial neural networks) models and SWE (shallow water equations) are presented. Discussions on the results of prediction models are provided with example. A simplified 3D analysis tool and initial prediction approach for calculating wave field around 3D submerged breakwaters is also represented in this chapter.

11.2 WORKED-OUT EXAMPLES

11.2.1 Two-Dimensional Experiments

Tests in 0.45 m Wide Flume

The tests were carried out using small scale models in a wave-current flume 14 m by 0.45 m by 0.75 m at University College London (UCL) (Figure 11.1). The horizontal bottom of the flume was constructed with aluminum while the side walls of the flume were constructed with glass to provide clear visualization panels to monitor wave transformation behind the breakwater during the tests. Measurements were made at different positions along the flume including in front of the breakwater and

Numerical Models for Submerged Breakwaters.
© 2016 Elsevier Ltd.
All rights reserved.

Figure 11.1 14 m by 0.45 m by 0.75 m wave-current flume used for two-dimensional tests.

wave maker, over the side slopes, over the crest and at several points behind the breakwater. Offshore water depth was maintained at 400 mm for all experiments.

Two piston type paddles (each 0.45 m wide) were in the wave flume, one at either end, one to generate waves and the other as a wave absorber (as indicated in Figure 11.1). The wave paddle in the flume was used to generate monochromatic waves. A second wave absorption system was also employed to minimize wave reflection in the flume. This consisted of a horizontal plate at the end of the flume and in front of the wave absorbing paddle. The beach was made of a perforated stainless steel plate fixed by metal bars to the top of the flume. The length of the beach was about 2 m and its width was 0.45 m. The false beach was kept submerged for all tests. Depending on the generated wave height and length, the beach was positioned at different depths below the still water surface level to ensure the maximum wave energy absorption. The effectiveness of the wave absorbing system was investigated for all test conditions in the flume without the breakwater.

Several breakwaters with different trapezoidal body sections were constructed and tested in the wave flume. The breakwaters were constructed with smooth material and without permeability. The breakwaters were constructed with horizontal top and side slopes of 1V:1H in both directions (Figure 2.1).

The breakwaters were fixed on the bottom of the flume so that the shoreward toes of all the breakwaters were located at a distance of almost

7.00 m onshore of the wave paddle. Figure 2.2 shows an experiment with a breakwater in the flume where wave breaking occurs over the breakwater crest.

Wave height and water level changes along the flume were measured by 13 wave probes. The wave probes measure the changes in resistance between two wires as the air–water interface fluctuates with the passage of waves. The voltage outputs of the wave probes are converted to water levels using calibration factors for each wave probe. To calibrate the wave probes, several different water levels (five or more) were measured against the corresponding output voltages from the probes. The voltage conversion factor was determined as the slope of the best fit linear curve to the height-voltage diagram. Examining the variation of calibration factor for each probe with time indicated that calibration of the probes should be done every week to achieve errors less than ± 1.0 mm.

The voltage is transmitted to the multichannel data logger system and is displayed on the computer screen at the same time using the software Measure Foundry. The sampling rate of water surface elevation was set to 100 samples per second. All the tests were monitored with a video camera installed at the side of the flume with the observations being recorded on digital camera for further analysis. The location of the probes and their distances from each other in the flume depend on the breakwater crest width employed in the test. Figures 11.2a to 11.2d indicate the location of the probes in each test condition (with and without breakwater).

A total of 12 test series were conducted in this wave flume with the same wave conditions but different breakwater geometries. Table 11.1 summarizes the range of data for this data set. A total of 180 tests were performed under different combinations of 15 monochromatic wave climates including 5 wave heights and 3 wave periods with 2 different submergence depths. Fifteen tests were also carried out in the flume with the same water depth and almost similar wave conditions without a breakwater in the flume to examine the generated wave heights and beach reflection. The range of tested variations in water depth, wave period and wave length, target wave height, and submergence depth are presented in Table 11.1.

Tests in 1.2 m Wide Flume

These tests were carried out in a wave flume 20 m by 1.2 m wide by 1.3 m high (Figure 11.3). The horizontal bottom of the flume and the side walls were constructed with glass to provide clear visualization panels to monitor wave transformation around the breakwater during the tests.

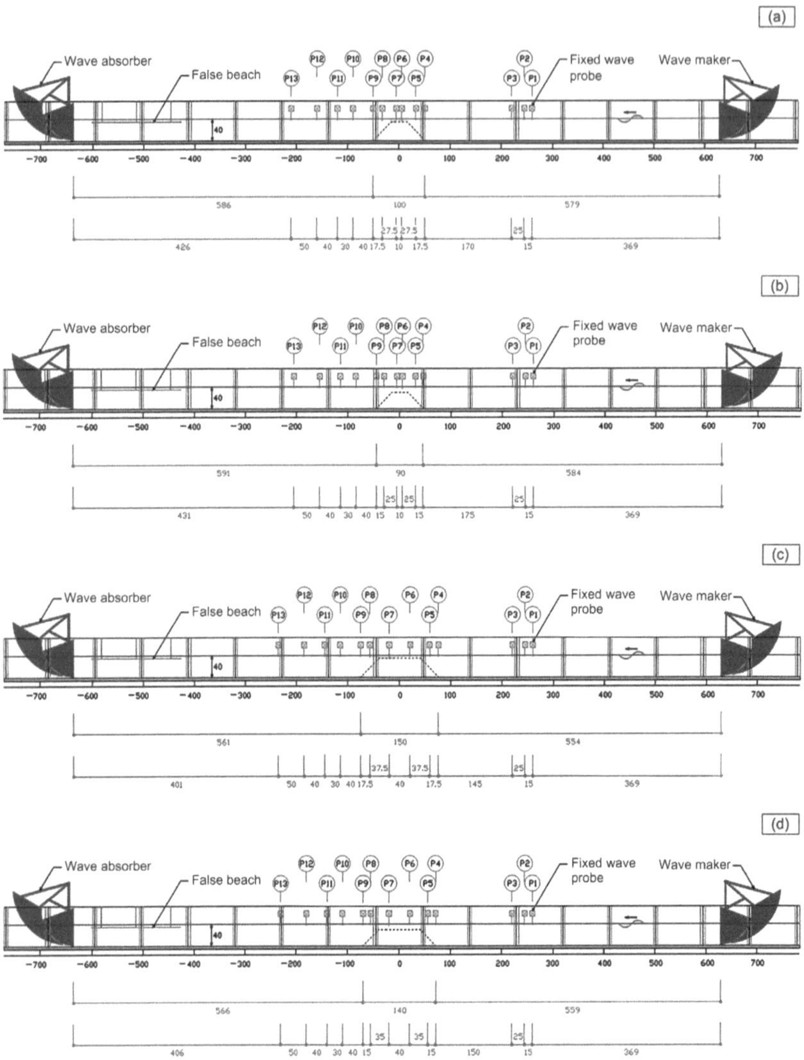

Figure 11.2 Wave probe positions along the wave-current flume for $h_s = 50$ mm and 100 mm and $B = 300$ and 800 mm: (a): $h_s = 50$ mm and $B = 300$ mm; (b): $h_s = 100$ mm and $B = 300$ mm; (c): $h_s = 50$ mm and $B = 800$ mm; and (d): $h_s = 100$ mm and $B = 800$ mm.

Measurements were made in different locations around the breakwater, including in front of the breakwater and wave maker, over the side slopes, over the crest and at several points behind the breakwater. Offshore water depth was maintained at 300 mm for all experiments. One 1.2 m wide

Table 11.1 Range of physical parameters considered in the 2D experiments (0.45 m wide flume)

Parameters	Range
Wave period (s)	1.00-1.11-1.25
Offshore wave length (mm)	1464-1729-2052
Target wave height (mm)	20-40-60-80-100
Submergence depth (mm)	50–100
Breakwater crest width (mm)	30-40-50-60-70-80
Water depth (mm)	400

Figure 11.3 20 m by 1.2 m by 1.3 m wave flume with piston type wave generator.

piston type wave maker was employed to generate monochromatic waves (as indicated in Figure 11.3). A 1V:15H sloping artificial stainless steel beach at the end of the flume was used to absorb the wave energy to reduce wave reflection in the flume. The length of the beach from the toe to the end was 5.5 m. The effectiveness of the wave absorbing system was investigated for all test conditions in the flume without breakwater.

A 2D breakwater with trapezoidal section was constructed and tested in the wave flume. The breakwater was constructed with smooth marine plywood material and with horizontal top and side slopes of 1V:1H in

Figure 11.4 Wave transformation over the 2D small scale submerged breakwater in 1.2 m wide flume.

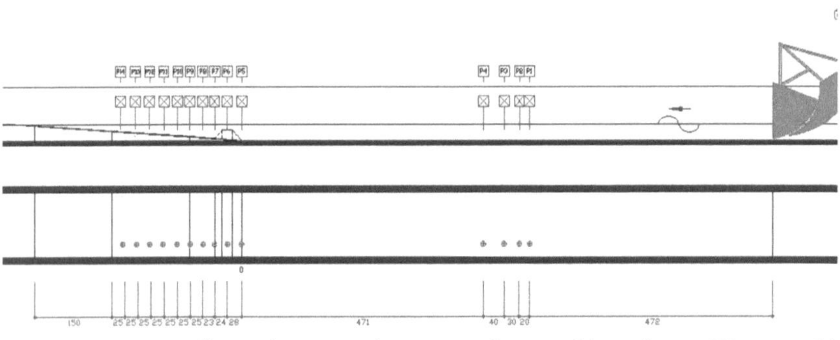

Figure 11.5 Wave flume layout and wave probes positions for a 2D test with 190 mm breakwater crest width, 300 mm water depth at the toe of the breakwater and 112 mm submergence depth. (All measurements in cm.)

both directions (Figure 11.4). The breakwater was fixed on the bottom of the flume so that the shoreward toe of the breakwater was located at a distance of 10.00 m onshore of the wave maker.

Wave height and water level changes along the flume were measured by 14 wave gauges. All the tests were monitored with a video camera installed on the top and one at the side of the flume, with the observations being recorded on video tape and digital camera for further analysis. Figure 11.5 indicates the location of the probes throughout the tests

Table 11.2 Range of physical parameters considered in the experiments in 1.2 m wide flume

Parameters	Range
Wave period (s)	0.80-1.00-1.133-2.00
Target wave height (mm)	20-30-40-50-60-70-80-90-100-110-120
Submergence depth (mm)	0-25-60-110
Breakwater crest width (mm)	190
Water depth (mm)	190-215-250-300

(with and without breakwater). The signals were transmitted to the multichannel data logger system and were displayed on the computer screen at the same time for further analysis during the tests. The sampling rate of water surface elevation was set to 100 Hz.

A series of experiments was conducted in the wave flume under various wave conditions and submergence depths. Table 11.2 summarizes the test conditions for these experiments. A total of 176 tests were performed under different combinations of 44 monochromatic wave climates including 11 wave heights, 4 wave periods and 4 water depths. Forty-four tests were also carried out with the same water depth and similar wave conditions without the breakwater in the flume to examine the generated wave heights and beach reflection.

11.2.2 Three-Dimensional Experiments

Small Scale Model

These tests were carried out in the 1.2 m wide flume and described in detail in the previous section (see Figure 11.6). A small scale breakwater roundhead and an adjacent trapezoidal trunk section were constructed and tested in the wave flume. The breakwater was constructed with smooth marine plywood material and with horizontal top and side slopes of 1V:1H in all directions (Figure 11.6).

The breakwater was fixed on the bottom of the flume so that the shoreward toe of the breakwater along the beach was located at a minimum distance of 10.00 m onshore of the wave maker. Wave height and water level changes along the flume were measured by 24 wave gauges. All the tests were monitored with a video camera installed on the top and one at the side of the flume with the observations being recorded on video tape and digital camera for further analysis.

Figure 11.6 The 3D small scale submerged breakwater in the 20 × 1.2 × 1.3 m wave flume.

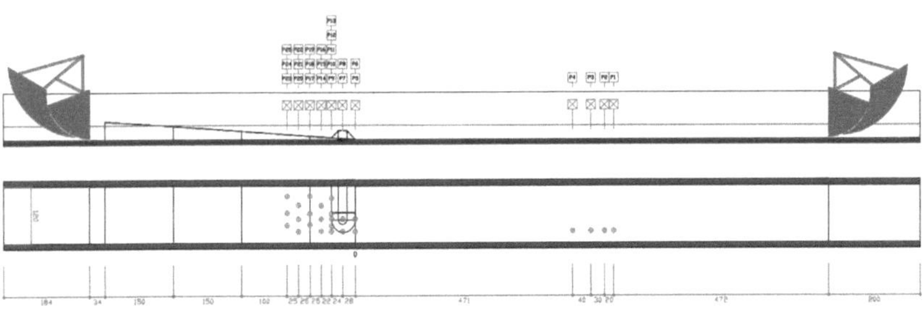

Figure 11.7 Wave flume layout and wave probes positions for a 3D test with 190 mm breakwater crest width, 300 mm water depth at the toe of the breakwater and 112 mm submergence depth (all measurements in centimeters).

Figure 11.7 indicates the position of the probes throughout the experiments (with and without breakwater). The sampling rate of water surface elevation was set at 100 samples per second. The signals were transmitted to the multichannel data logger system and recorded.

A series of experiments was conducted in the wave flume under various wave conditions and submergence depths. Table 11.2 summarizes the conditions in these 3D tests and the range of physical parameters, which are similar to the 2D data set collected in the 1.2 m wide flume as mentioned earlier. A total of 176 tests were performed under different combinations of 44 monochromatic wave climates, including 11 wave heights and 4 wave period conditions with 4 different water depths.

Large Scale Model

These tests were carried out using a large scale model in a 20 m by 2.5 m wide by 1.3 m high wave tank. The horizontal bottom of the flume was constructed with concrete (relatively smooth) while the side walls of the flume were constructed with glass to provide clear visualization panels to monitor wave transformation around the breakwater during the tests. Measurements were made in different locations around the breakwater. Offshore water depth was maintained at 1000 mm for all experiments. Seven flap type paddles (each 0.5 m wide) were in the wave flume to generate monochromatic waves (as indicated in Figure 11.8). A parabolic artificial stainless steel beach (with a length of 2 m) at the end of the tank was constructed to absorb wave energy and reduce the wave reflection.

Figure 11.8 20 m by 2.5 m wide by 1.3 m high wave tank with seven flap type paddles and a 3D large scale model.

The effectiveness of the wave absorbing system was investigated for all test conditions in the flume without breakwater.

A large scale breakwater roundhead and the adjacent trapezoidal trunk section was constructed and tested in the wave flume. The breakwater was constructed with smooth marine plywood material and with horizontal top and side slopes of 1V:1H in all directions. The breakwater was fixed on the bottom of the flume so that the shoreward toe of the breakwater was located at the distance of 10.00 m onshore of the wave paddles (Figure 11.8).

Wave height and water level changes along the flume were measured by 31 wave gauges. All the tests were monitored with video cameras installed on the top and side of the flume with the observations being recorded on video tape and digital camera for further analysis. The sampling rate of water surface elevation was set to 100 samples per second. Figure 11.9 shows the facilities employed in these experiments. Figure 11.10 indicates the position of the probes throughout the experiments (with and without breakwater).

A broad range of 3D tests was conducted in this wave flume with various wave conditions. Table 11.3 summarizes these test conditions. A total of 83 tests were performed under different combinations of monochromatic wave climates including 9 wave heights and 9 wave periods with a constant water depth of one meter. Eighteen tests were also carried out with the same water depth and similar wave conditions without the breakwater in the flume to examine the generated wave heights and beach reflection. The range of wave parameters (wave periods and target wave heights) in this data set is presented in Table 11.3.

11.3 DATA PROCESSING

The details of the experimental set-up and measurement were discussed in the previous section. However, an important part of the experimental studies is processing of the collected data to enhance its interpretation.

11.3.1 Data Acquisition

As mentioned before, the output signals of the wave probes were sent to the multichannel data logger system and stored (in volts) in MATLAB® MAT files, Excel® spreadsheets and TXT files for analysis and processing using different software. Calibration was undertaken in MATLAB to convert voltage data to the water surface level in millimeters. This was

Figure 11.9 Laboratory facilities and equipment utilized for water surface elevation measurement.

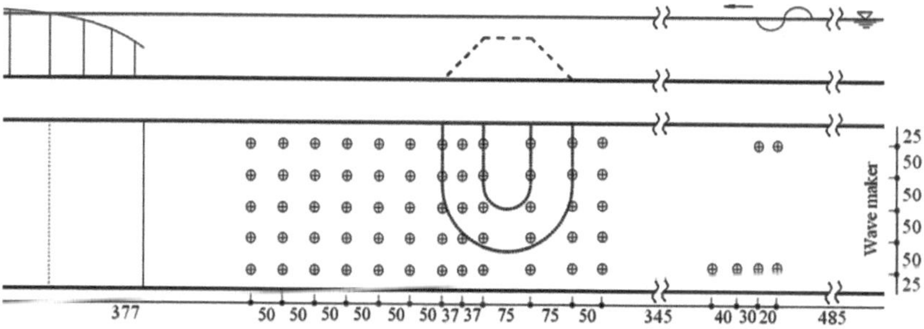

Figure 11.10 Large wave tank layout and wave probes positions for a 3D test with 250 mm breakwater crest width, 1000 mm water depth and 250 mm submergence depth (all measurements in centimeters).

Table 11.3 Range of physical parameters considered in the 3D experiments (Large scale model; 2.5 m wide flume)

Parameters	Range
Wave period (s)	0.66-0.71-0.77-0.83-0.91-1.00-1.11-1.25-1.43-1.66-2.00
Target wave amplitude (mm)	10-20-30-40-50-60-70-80-90-100
Submergence depth (mm)	250
Breakwater crest width (mm)	750
Water depth (mm)	1000

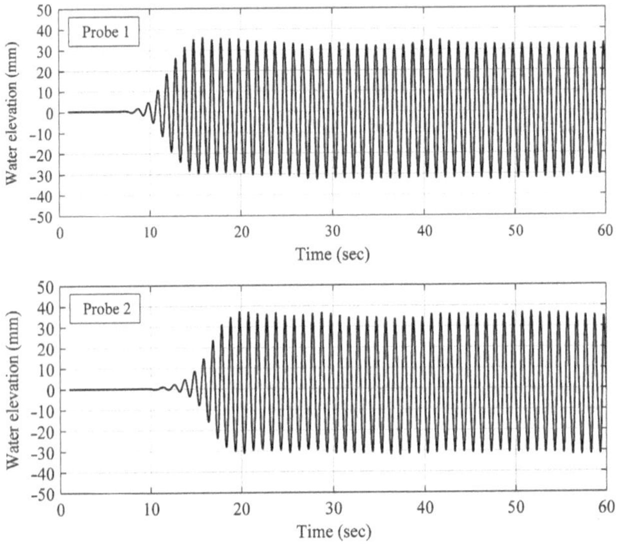

Figure 11.11 Two time series of water surface level collected by two wave probes in the small wave flume.

achieved using the calibration data collected in the laboratory for each probe individually and by measuring the still water surface level (SWL) in a calm flume for 10 different wave probe depths. Figure 11.11 illustrates two time series of water surface in the flume obtained after raw data calibration. These two time series are related to two different measurements, one close to the wave paddle (Figure 11.11a) and the other one at the far end of the flume (Figure 11.11b).

As can be deduced from the figure, the form of the propagated waves usually varies along the flume depending on the wave steepness and relative water depth. Free secondary harmonic waves exist when waves are generated by a sinusoidal moving wave paddle (Tajziehchi, 2006) (see Biésel and Suquet (1951) for further information). In addition, higher harmonic waves generated by wave propagation over a beach slope or particularly a submerged breakwater also affect the general form of observed waves in the flume (Goda et al., 1999). Therefore, studying the form of wave propagation along the flume is required to obtain a reliable wave data analysis.

11.3.2 Experiments Without Breakwater

Water surface elevation data collected by wave probes in the flume without a breakwater were examined to give a better understanding of the characteristics of the generated waves. In the case of monochromatic long waves, a free second harmonic wave always exists when h/L is less than 0.1 (Flick and Guza, 1980). These secondary waves travel with a smaller speed and amplitude than primary waves (Biésel and Suquet, 1951). Therefore, looking at the water surface profile, the existence of the secondary waves may be identified. The existence of the secondary waves may be the cause of a number of troubles in laboratory experiments for long water waves (Tajziehchi, 2006). Figure 11.12 shows a sample of wave profiles and water levels at different stations in the wave flume without breakwater. The water surface variations were measured at the same time. As indicated in the figure, depending on the distance from the wave maker, water surface levels may vary. For waves of small amplitude, the secondary waves are relatively small and may be negligible in most cases (Tajziehchi, 2006).

Monitoring of water surface elevation in the flume without breakwater was carried out by 14 fixed probes, as shown in Figure 11.7. Spectral analysis of the recorded signals from the probes, based on the fast Fourier transformation (FFT), was also carried out as shown in Figure 11.13 for a selected test condition in the flume without breakwater. The figure shows the shape of the wave amplitude spectrum and results from eight probes located at different positions along the flume. This reveals the existence of higher harmonic waves at some specific points; for instance, secondary harmonics are clearly most evident for probes close to the beach.

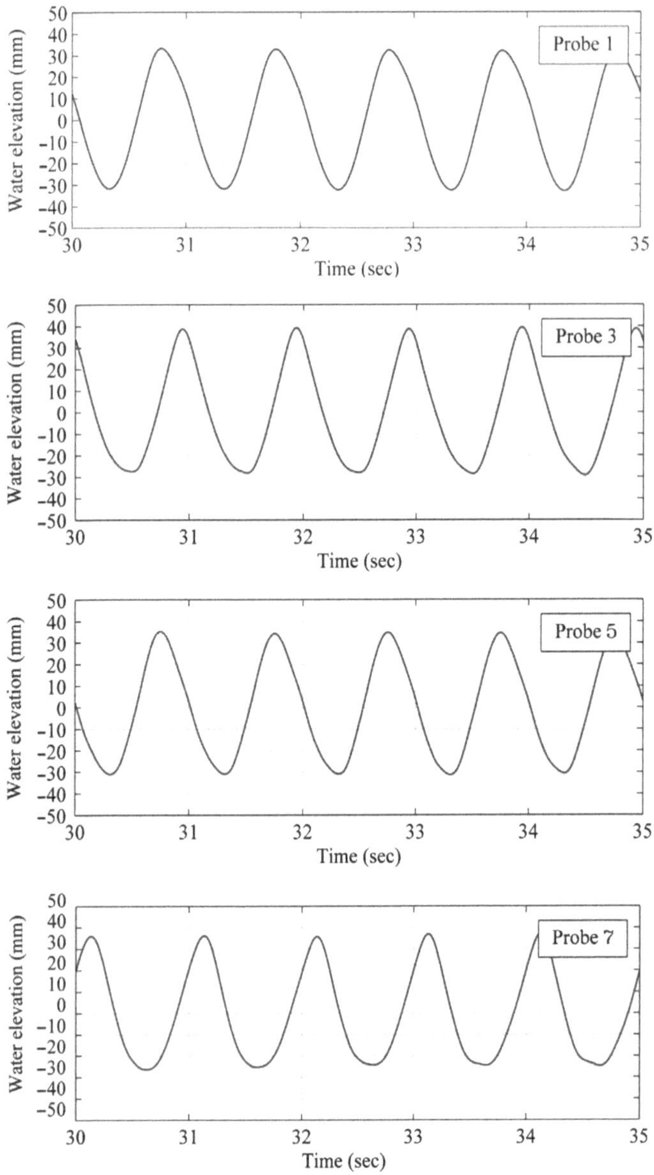

Figure 11.12 Wave profile variation at different stations along the wave flume (without breakwater, $h = 300$ mm, $H_i = 64$ mm, $T = 1.00$ s). Probe locations are shown in Figure 11.2.

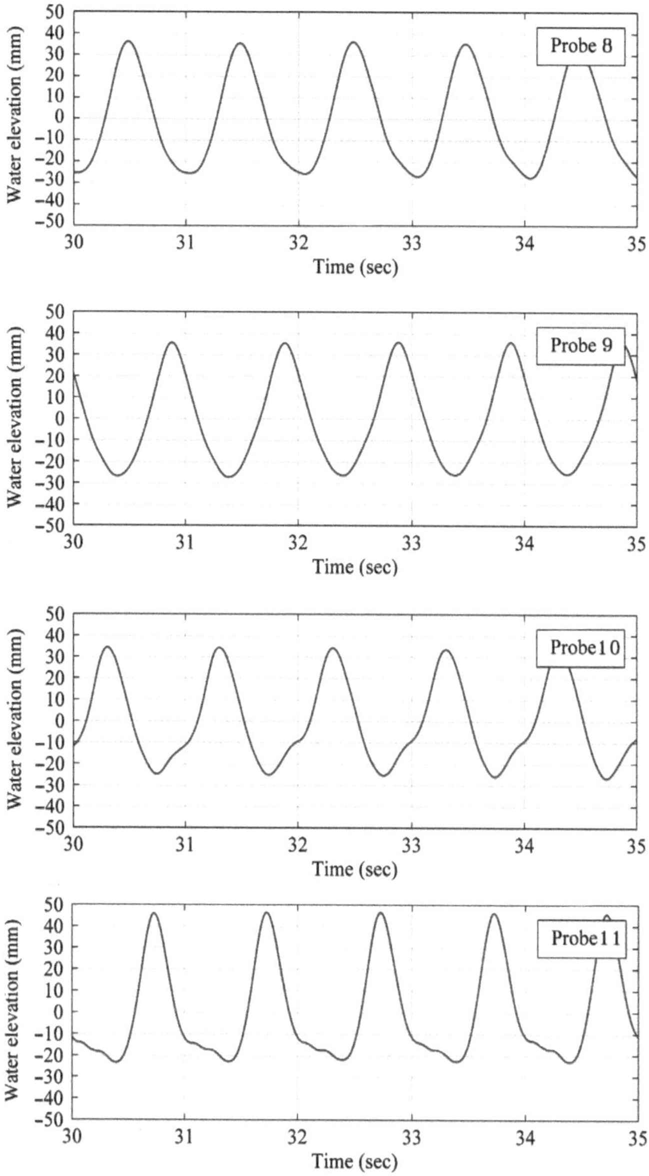

Figure 11.12 (Continued)

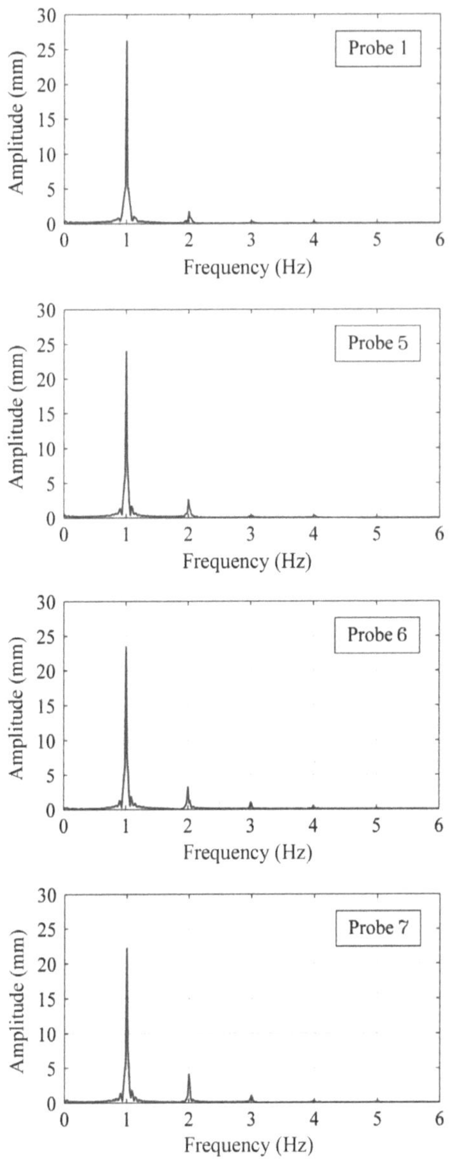

Figure 11.13 Wave amplitude spectrum at different stations along the wave flume (without breakwater, $h = 300$ mm, $H_i = 51$ mm, $T = 1.00$ s). Probe locations are shown in Figure 11.2.

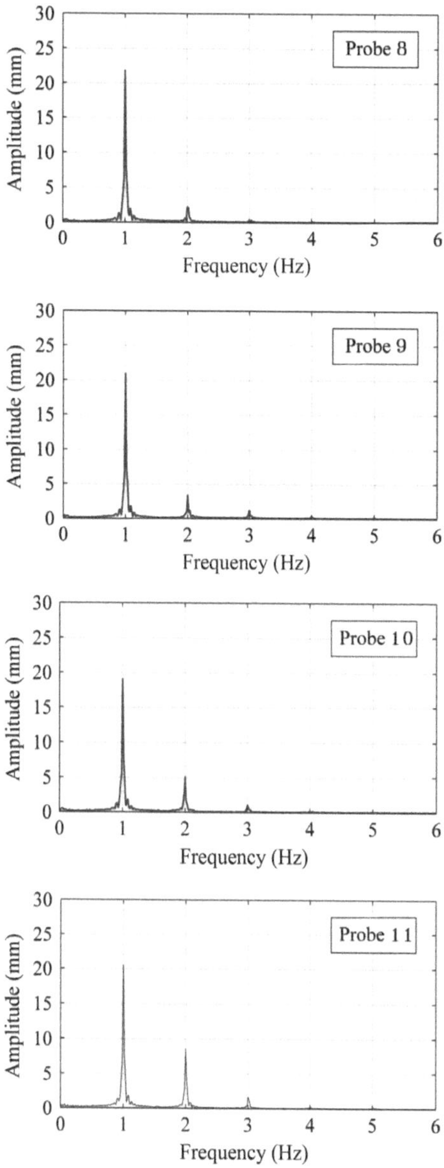

Figure 11.13 (Continued)

Regarding the reflected waves in the flumes as mentioned before, the wave energy absorber systems were employed in the flumes to minimize wave reflection. As discussed, a number of experiments were carried out in the flumes without breakwater but with the same water depth and almost similar wave conditions to examine the generated wave heights and beach reflection. When a primary incident wave travels toward the beach, a primary reflected wave travels from the beach toward the wave maker. Then a secondary incident wave travels from the wave maker toward the shore. This process is repeated till the steady state is attained after a few reflections (Ursell et al., 1960). In the presence of a highly dissipative beach or wave energy absorber system, the secondary reflected wave is usually negligible (Tajziehchi, 2006). The calculated reflection coefficient for the tests without breakwater in the present experimental study revealed that the reflection coefficient of the flume varies between 0.03 and 0.11.

11.3.3 Experiments with Breakwater

Several experimental and numerical studies have been conducted to investigate higher harmonics generated over submerged breakwaters (Goda et al., 1999; Longuet-Higgins, 1977; Losada et al., 1997; Massel, 1983; Tajziehchi, 2006). When waves approach a submerged obstacle, part of the wave energy is reflected by the barrier, while above the obstacle higher harmonics are generated and transmitted to the deeper water behind the barrier as free waves (Tajziehchi, 2006). The amplitude spectrum at different measurement points around the head and behind the breakwater for a selected test is shown in Figure 11.14. The figure corresponds to a test with breaking wave (Series 3, No. 17). As can be seen, although the form of wave spectra offshore of the breakwater shows similar trends to those in the flume without breakwater, higher harmonics are observed seaward of the breakwater.

The method of Mansard and Funke (1980) was implemented to decompose the incident and reflected wave heights using simultaneous measurements from three fixed wave probes. A MATLAB (Matlab, R2010b) script was written to read wave data and compute the incident and reflected wave heights using the Mansard and Funke (1980) method. The reflection coefficient K_r is not of primary concern in this analysis.

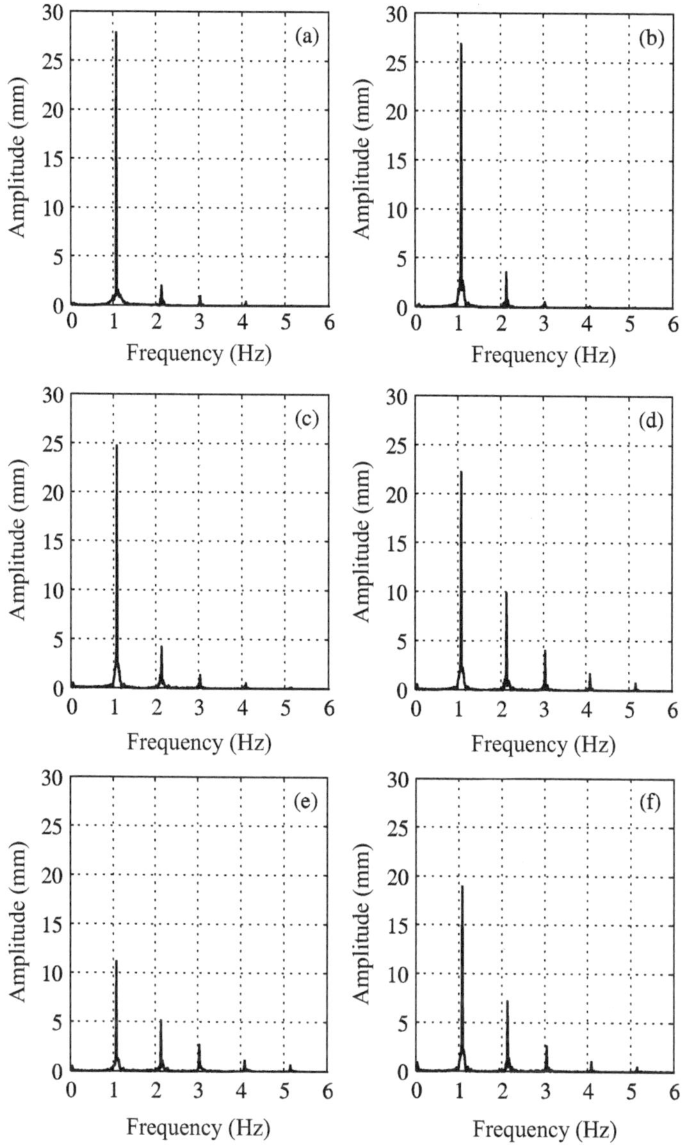

Figure 11.14 Wave amplitude spectrum at different stations around and behind breakwater in the wave flume ($h = 300$ mm, $H_i = 51$ mm, $T = 1.00$ s). Probe locations are shown in Figure 11.2. (a): Probe No. P5; (b): Probe No. P6; (c): Probe No. P9; (d): Probe No. P11; (e): Probe No. P13; (f): Probe No. P15; (g): Probe No. P16; (h): Probe No. P18; (i): Probe No. P20; and (j): Probe No. P22.

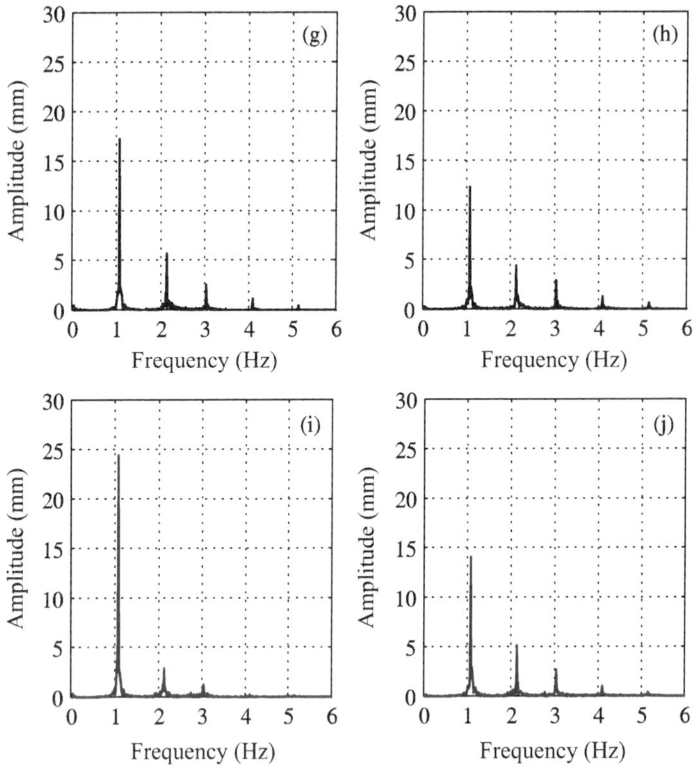

Figure 11.14 (Continued)

11.4 ANALYSIS AND DISCUSSION OF RESULTS

Processed wave data is analyzed in detail and discussed in this section. Waves transmitted onshore of the submerged breakwater are a combination of wave overtopping, which includes higher harmonic free waves due to wave breaking over the breakwater crest, and wave diffraction. Hence, transmitted wave heights vary spatially and depend on the number of significant harmonics generated and transmitted over the breakwater as well as wave diffraction patterns around the breakwater head. This phenomenon was clearly observed throughout the laboratory experimental tests and is discussed in detail in this section. Wave height transmission patterns behind 2D and 3D submerged breakwaters are analyzed and the effectiveness of several parameters (dimensional and dimensionless) are examined.

11.4.1 The Root-Mean-Square Transmitted Wave Height ($H_{t,rms}$)

In order to obtain a better understanding of the wave pattern behind the breakwater, a large number of measurement points were considered to give a better interpolation. $H_{t,\mathrm{rms}}$ for each wave probe behind the structure was calculated with Equation 11.1 and was adopted as being representative of transmitted wave height at that specific point:

$$H_{t,rms} = \sqrt{\frac{\sum_{i=1}^{N} H_{t,n}^2}{N}} \tag{11.1}$$

with $H_{t,n}$ denoting the individual transmitted wave heights in a certain time series with a total number of N waves.

As the total wave spectrum was taken into account, the computed transmitted wave height ($H_{t,\mathrm{rms}}$) is an average and includes the fundamental and higher harmonic waves.

Figure 11.15 presents contour maps of the average transmitted wave height ($H_{t,\mathrm{rms}}$) for measurement points over, around and behind the breakwater for a number of selected tests. The tests selected from series 1 correspond to the experiments with $h = 300$ mm and $h_s = 110$ mm. Test numbers 14, 17, 21, 35, 38 and 43 were selected (see Table 11.28 for details of tests and further information).

Graphs on the left side correspond to the experiments with wave periods of 1.00 s and three different incident wave heights. It can be deduced from these cases that for a small wave height ($H_s = 36$ mm) the smaller transmitted wave heights are located behind the gap while the larger transmitted wave heights are located right behind the leeside toe of the breakwater. Although wave heights onshore of the breakwater are lower than offshore, quite high values for $H_{t,\mathrm{rms}}$ (about 0.022 m and 0.030 m for the areas behind the gap and breakwater crown, respectively) mean low wave dissipation by the breakwater. Considering the second graph (for $H_i = 62$ mm), $H_{t,\mathrm{rms}}$ about 0.038 m and 0.048 m were observed for smaller and larger transmitted wave heights, respectively. Comparing this contour map with the previous graph, it can be deduced that area with smaller transmitted wave height is moving slightly to the right, while the area with larger transmitted wave height is over the trunk and head of the breakwater. Regarding the third graph ($H_\mathrm{i} = 88$ mm), $H_{t,\mathrm{rms}}$ values are 0.044 m and 0.064 m for smaller and larger transmitted wave heights respectively. The area with smaller $H_{t,\mathrm{rms}}$ moves completely to the right

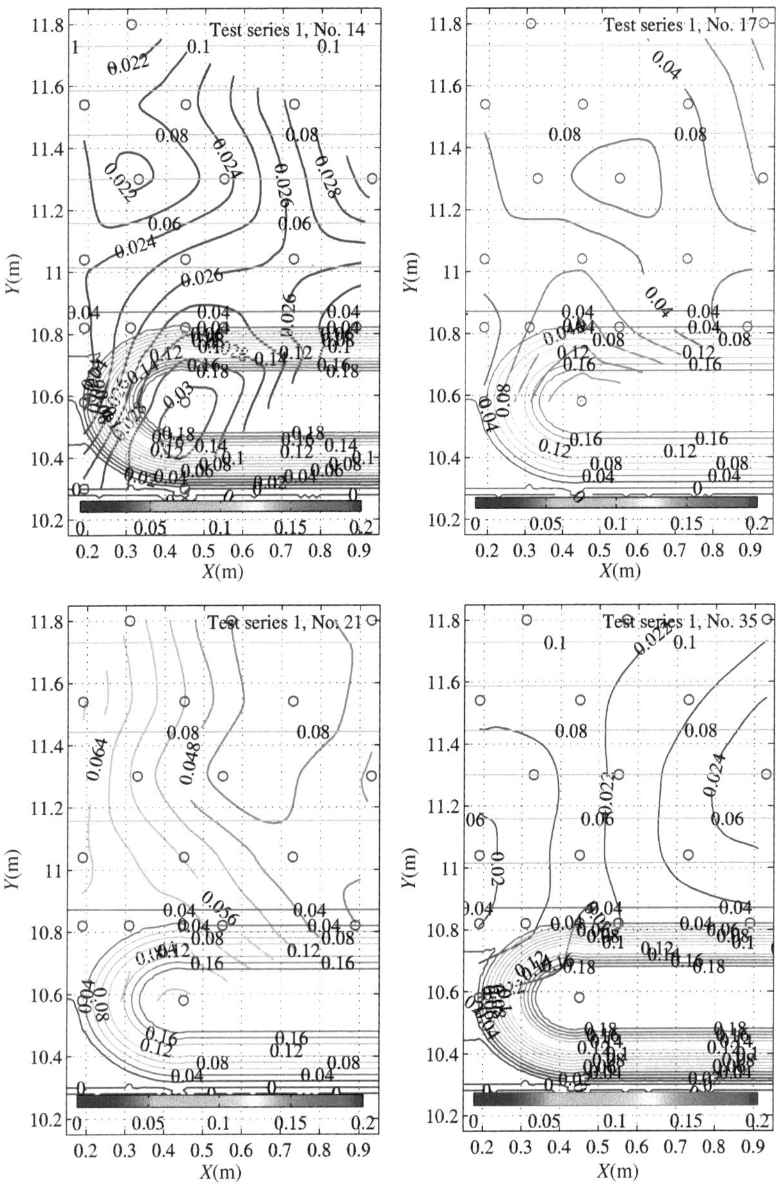

Figure 11.15 $H_{t,rms}$ contour maps (m) for test series 1 ($h_s = 112$ mm) for two wave periods and three incident wave heights. See Table 11.28 for conditions.

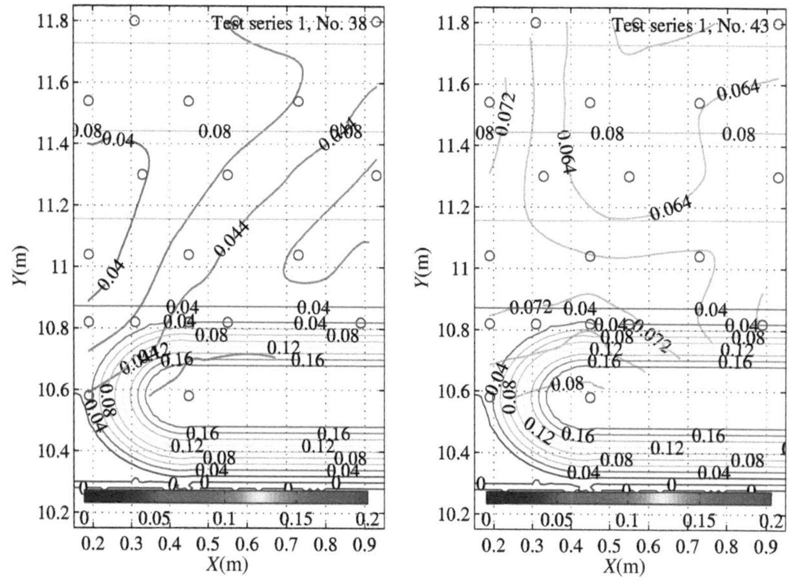

Figure 11.15 (Continued)

side and is located right behind the breakwater trunk. However, the area with larger $H_{t,\text{rms}}$ moves to the left side in line with the gap. A similar trend can be observed in the right hand set of graphs corresponding to the wave period of 2 s. However, the area with smaller $H_{t,\text{rms}}$ in this case is slightly bigger and closer to the breakwater trunk.

Figure 11.16 presents contour maps for a specific wave condition but with four different water depths. The experiments correspond to an averaged incident wave height of 36 mm and a wave period of 1 s. As can be seen in the graphs, area with smaller $H_{t,\text{rms}}$ behind the breakwater reduces significantly by decreasing the submergence depth from 22 mm for Test series 1 ($h_s = 11$ mm) to 7 mm for Test series 6 ($h_s = 0$ mm). The area with smaller $H_{t,\text{rms}}$ also moves from left (behind the gap) to right (behind the breakwater crown) by decreasing the submergence depth, while the area with larger $H_{t,\text{rms}}$ moves in the opposite direction (see Table 11.28 for details of tests and further information).

11.4.2 Wave Transmission Coefficient (K_t)

The wave transmission coefficient K_t is calculated from the average transmitted wave height onshore of the breakwater of each measurement point

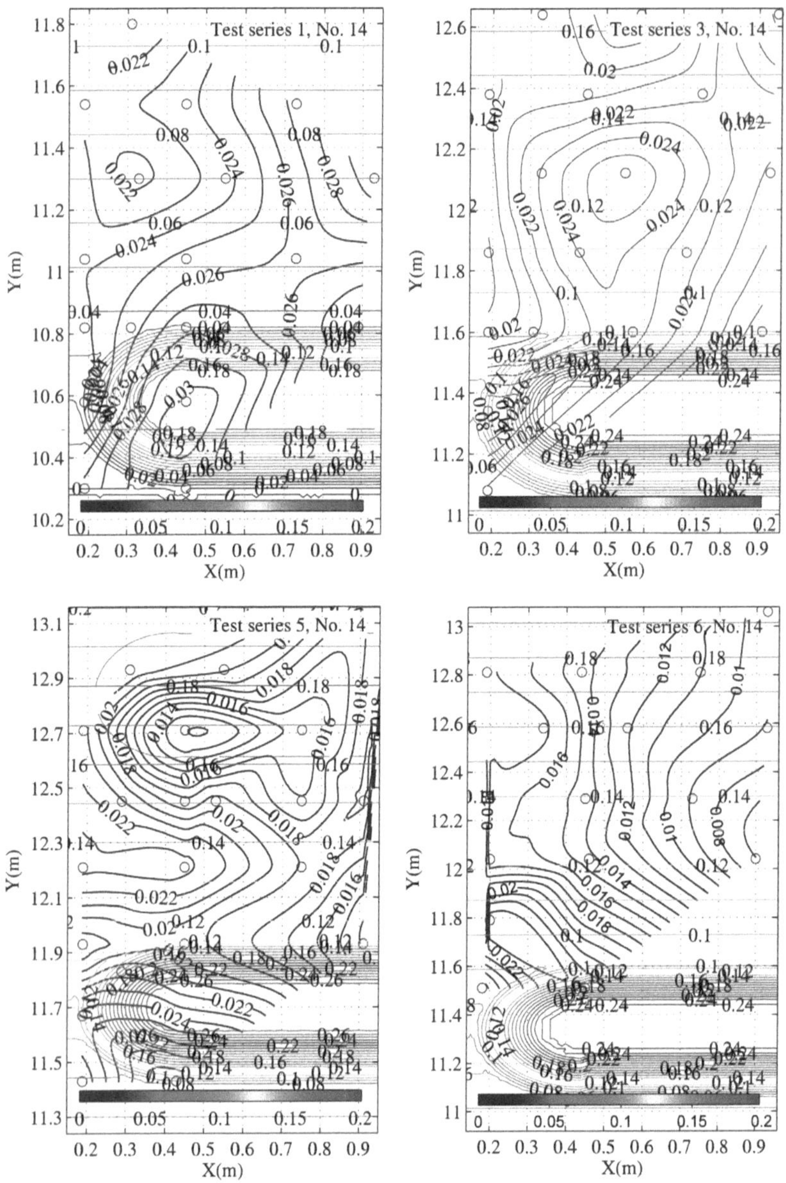

Figure 11.16 $H_{t,\mathrm{rms}}$ contour maps for Test No. 14 for one wave condition with four submergence depths. See Table 11.28 for conditions.

and the incident wave height calculated in the previous sections using wave probes located offshore of the breakwater. As mentioned before, K_t is defined as the ratio of average transmitted wave height to the incident wave height (H_i) as follows:

$$K_t = \frac{H_{t,rms}}{H_i} \tag{11.2}$$

where $H_{t,\mathrm{rms}}$ is calculated using Equation 11.3.

In the case of two-dimensional experiments, transmitted wave height H_t is calculated based on the average of root mean squared wave height ($\hat{H}_{t,rms}$) from the wave probes behind the structure using the following method:

$$\hat{H}_{t,rms} = \frac{\sum_{i=1}^{n} H_{t,rms}}{n} \tag{11.3}$$

$H_{t,rms}$ is the standard deviation of the water surface level for the ith wave probe behind the breakwater, and n is the number of wave probes behind the submerged breakwater. In this study, $\hat{H}_{t,rms}$ was obtained from three wave probes behind the breakwater and this average value was adopted as being representative of transmitted wave height. These three probes are probe numbers 10, 11 and 12 as shown in Figure 11.2 for experiments in the 0.45 m wide flume and probe numbers 8, 9 and 10 as shown in Figure 11.5 for experiments in the 1.2 m wide flume.

Analysis of effective parameters (dimensional and nondimensional) from K_t is examined in the following sections. The purpose of this analysis is to define the dimensional and nondimensional parameters that can most effectively predict the wave transmission coefficient. Dimensional and nondimensional parameters are also used in the numerical model for calculating K_t in Chapter 5.

Dimensional Analysis

The effect of different dimensional parameters including wave characteristics, water depth and breakwater geometry on the 2D wave transmission coefficient has been thoroughly studied (Ahrens, 1987; Seabrook, 1997; Seelig, 1980; Tajziehchi, 2006; van der Meer, 1991).

In the present analysis, graphical interpretation was used to identify trends in the graphs to determine the significance of each parameter. In addition, statistical analysis of variables was performed for more investigation of effectiveness of each dimensional parameter on the transmission coefficient. The variation of the correlation coefficient between K_t and other dimensional parameters was used as an indicator. Five main dimensional variables were selected as the most effective parameters in 2D analysis:

$$K_{t,2D} = f(H_i, L_o, h_s, h, B) \tag{11.4}$$

where H_i is incident wave height, L_o is offshore wave length, B is breakwater crest width, h_s and h are respectively water depth over the crest and at the toe of the breakwater.

The 2D dimensional analysis of the collected laboratory data on these parameters demonstrated that K_t decreases with incident wave height and breakwater crest width while offshore wave length and submergence depth have an inverse effect on K_t, which was consistent with previous studies and confirmed the past research (for instance see: Tajziehchi, 2006).

The wave energy dissipation coefficient is derived easily from K_t and can be used as an indicator to describe the phenomena of wave energy dissipation over the barriers. K_{ed} is defined as follows:

$$K_{ed} = \sqrt{1 - K_t^2 - K_r^2} \tag{11.5}$$

where K_r is the wave reflection coefficient calculated as follows:

$$K_r = \frac{H_r}{H_i} \tag{11.6}$$

where H_r and H_i respectively are reflected wave height and incident wave height which were computed in the earlier sections using the decomposition method of Mansard and Funke (1980).

A sample scatter plot is presented in Figure 11.17 to show the magnitude of the wave energy dissipation based on the variation of incident wave height for two submergence depths. It can be concluded from Figure 11.17 that increasing the incident wave height leads to higher wave energy dissipation. The higher wave energy dissipation is observed for lower submergence depths while the incident wave height is constant. The figure clearly shows the significance of the two parameters H_i and h_s on K_{ed} and consequently on K_t. Similar analyses were conducted for other dimensional parameters and interesting trends were observed. The outcomes were discussed earlier in this section.

However, in the case of three-dimensional analysis, two additional dimensional parameters x and y (corresponding to the Cartesian

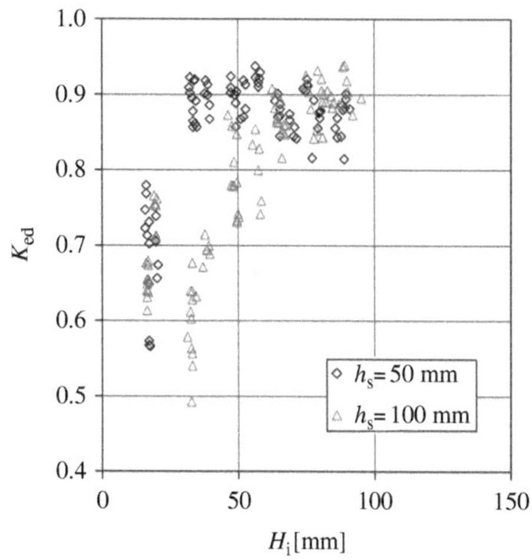

Figure 11.17 2D wave dissipation coefficient against incident wave height for different submergence depths (Set A1).

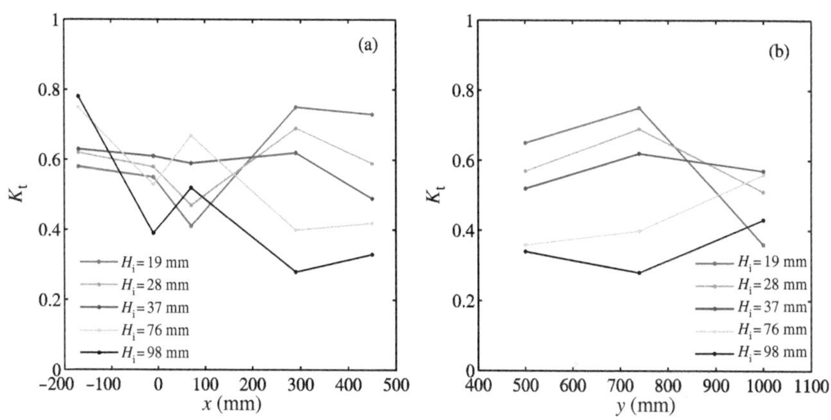

Figure 11.18 Variation of 3D wave transmission coefficient with x and y: (a): $y = 740$ mm; (b): $x = 290$ mm (Set C1: $L_o = 1427$ mm, $h_s = 27$ mm and $h = 222$ mm).

coordinate system) are also considered. Therefore, an improved form of dependency of wave transmission coefficient on dimensional variables is defined and applied in this study (Sharif Ahmadian and Simons, 2014):

$$K_{t,3D} = f\{H_i, L_o, h_s, h, B, \mathrm{x}, \mathrm{y}\} \tag{11.7}$$

The variation of wave transmission coefficient with dimensional parameters and specifically x and y in the 3D case is investigated and analyzed below. $K_{t,3D}$ is plotted in Figure 11.18 as a function of x and y for

Set C1: $L_o = 1427$ mm, $h_s = 27$ mm and $h = 222$ mm. As can be seen in Figure 11.18a (related to $y = 740$ mm), the wave transmission coefficient is decreased significantly by x for larger incident wave heights while it is increased by x for smaller values of H_i. It can be deduced from Figure 11.18b (related to $x = 290$ mm) that K_t is generally increased by y but after a specific point it is decreased significantly for smaller incident wave heights and increased for larger values of H_i.

The set of variables presented in Equations 11.4 and 11.7 will be analyzed further.

Nondimensional Analysis

The results of dimensional analysis are of great worth for the geometric design of submerged breakwaters. The guidance provided from these analyses is recommended for engineering design when wave climates are available. Nevertheless, nondimensional parameters can provide a more productive correlation of the dependent parameters (Tajziehchi, 2006).

In order to represent the combined effect of the wave steepness ($s_o = H_i/L_o$) and the seaward slope of the breakwater α, the surf similarity parameter or Iribarren number ξ_o is introduced as follows:

$$\xi_o = \frac{\tan(\alpha)}{\sqrt{\frac{H_i}{L_o}}} \tag{11.8}$$

Therefore, on the basis of the 2D dimensional analysis, the following 2D dimensionless parameters were considered for further analysis:

$$K_{t,2D} = f(\xi_o, H_i/h, h_s/H_i, B/H_i) \tag{11.9}$$

Several sample plots of the variation of transmission coefficient as a function of the selected nondimensional parameters (Tests in 1.2 m wide flume) are presented in Figure 11.19. Graphical analysis of the dimensionless parameters in Figure 11.19 demonstrates that the transmission coefficient increases with submergence ratio h_s/H_i, surf similarity parameter ξ_o and dimensionless crest width ratio B/H_i and decreases with wave steepness H_i/L_o and inverse water depth ratio H_i/h. K_t generally doesn't change with breakwater crest width to offshore wave length ratio B/L_o within this range ($B/L_o < 0.55$).

It should be noted that wave breaking is the primary cause for wave energy dissipation over the 2D submerged breakwaters. The plots in Figure 11.20

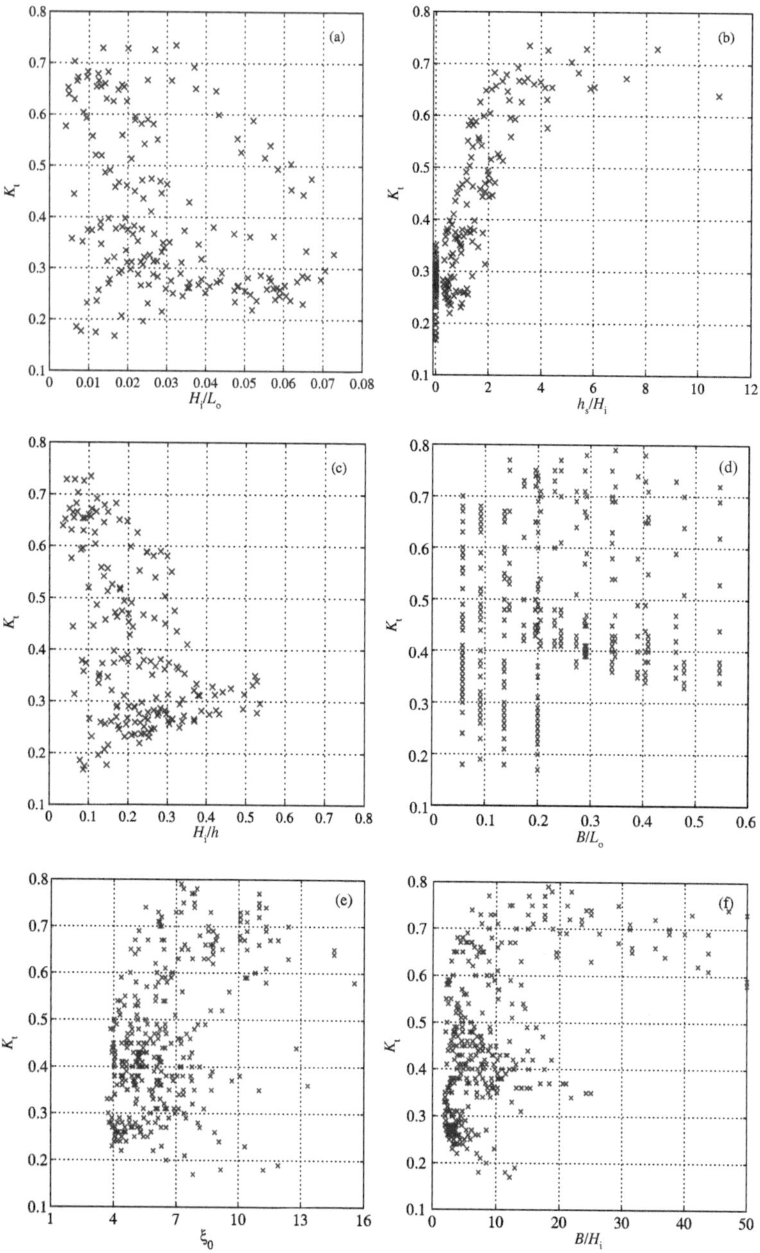

Figure 11.19 Variation of 2D wave transmission coefficient with wave steepness, submergence ratio and nondimensional wave height ratio (tests in 1.2 m wide flume).

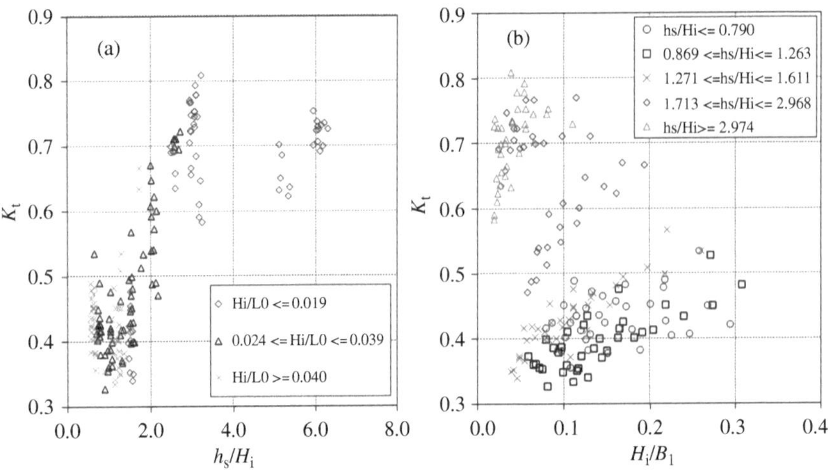

Figure 11.20 Variation of 2D wave transmission coefficient with submergence ratio and incident wave height over breakwater crest width ratio (tests in 0.45 m wide flume).

confirm this fact as lower wave transmission coefficients are observed for higher wave steepness and lower submergence ratio and inverse water depth ratio (for example see: Tajziehchi, 2006). All these three nondimensional parameters are related to the breaking process over the breakwater crest. Wave steepness higher than a certain value and water depth lower than a specific ratio to wave height both lead to wave breaking. The width of the breakwater crest also affects the breaking process by providing a longer shallow water area and consequently enough time to develop the breaking. Friction loss can also logically be a secondary factor in 2D wave energy dissipation over these sorts of breakwaters. However, as breakwaters with a smooth surface were employed in the tests, it is assumed that the friction loss factor is negligible.

More information is presented in Figure 11.20 regarding the effects of the three main nondimensional parameters in the 2D case including wave steepness, submergence ratio and inverse breakwater crest ratio using another set of data with different scale (Tests in 0.45 m wide flume). These analyses have been conducted to check the earlier results in more detail and using a more comprehensive data set.

It can be seen from Figure 11.20 that the general trend is absolutely the same as the earlier graphs obtained from data set B. According to this graphical analysis, the effect of the dimensionless parameters on wave transmission coefficient can be summarized: an increment in wave steepness (H_i/L_o) causes more wave energy dissipation due to stronger wave breaking over the breakwater (Figure 11.20a). The submergence ratio (h_s/H_i) has a direct impact on the wave transmission coefficient (Figure 11.20a and b).

Greater submergence ratio allows coming waves to pass over the breakwater with a smaller degree of breaking and as a result, larger wave energy transfers toward the shore (Figure 11.20). An increment in the breakwater crest ratio also increases the wave energy attenuation by allowing the wave breaking process grows further over the crest (Figure 11.20b). These results are consistent with previous 2D studies on submerged breakwaters by other researchers and confirms the past researches (see: Tajziehchi, 2006).

Considering the three-dimensional analysis, it is assumed the 3D wave transmission coefficient can be defined as a function of the following nondimensional parameters (Sharif Ahmadian and Simons, 2014):

$$K_{t,3D} = f\{\xi_o, H_i/h, h_s/H_i, B/H_i, x/L_o, y/L_o\} \tag{11.10}$$

Regarding the 3D effects of the nondimensional parameters, a particular analysis has been made on the impact of these variables on the wave transmission coefficient. Figure 11.21 shows the graphical analysis for two selected nondimensional parameters (3D small scale tests in 1.2 m wide flume). It can be deduced from Figure 11.21a that variation of K_t with incident wave steepness follows the same general trend as the 2D case, while Figure 11.21b for submergence ratio illustrates a slightly different pattern comparing with the 2D case. Although the general trend is similar, large values of K_t are observed for low submergence ratio values less than 2.00. The reason is the 3D effect of wave transformation and particularly diffraction on the wave pattern behind the structure, so that some specific cases

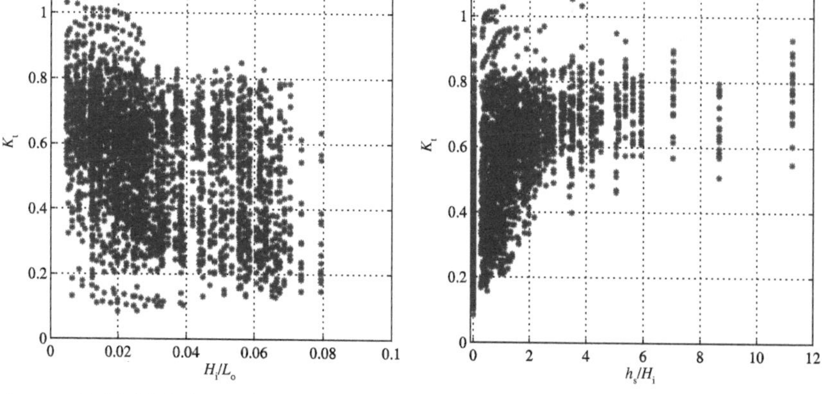

Figure 11.21 Variation of 3D wave transmission coefficient with nondimensional parameters: Wave steepness and inverse incident wave height ratio (3D small scale tests in 1.2 m wide flume).

result in higher transmitted wave height even in low submergence depths. In low submergence depths, wave energy is significantly dissipated over the breakwater crest (2D transmission), while incident waves pass through the gap without considerable changes. Therefore, the difference between the wave height passing over the crest and wave height passing through the gap results in diffraction. This process completely disturbs the wave pattern behind the structure compared with the 2D case. Therefore, the existence of such high K_t in Figure 11.21b for low submergence ratio (even at some points greater than 1.00) is due to the 3D effect.

Figures 11.22 and 11.23 respectively for low ($h_s = 27$ mm) and high ($h_s = 60$ mm) submergence depths show a detailed graphical analysis for

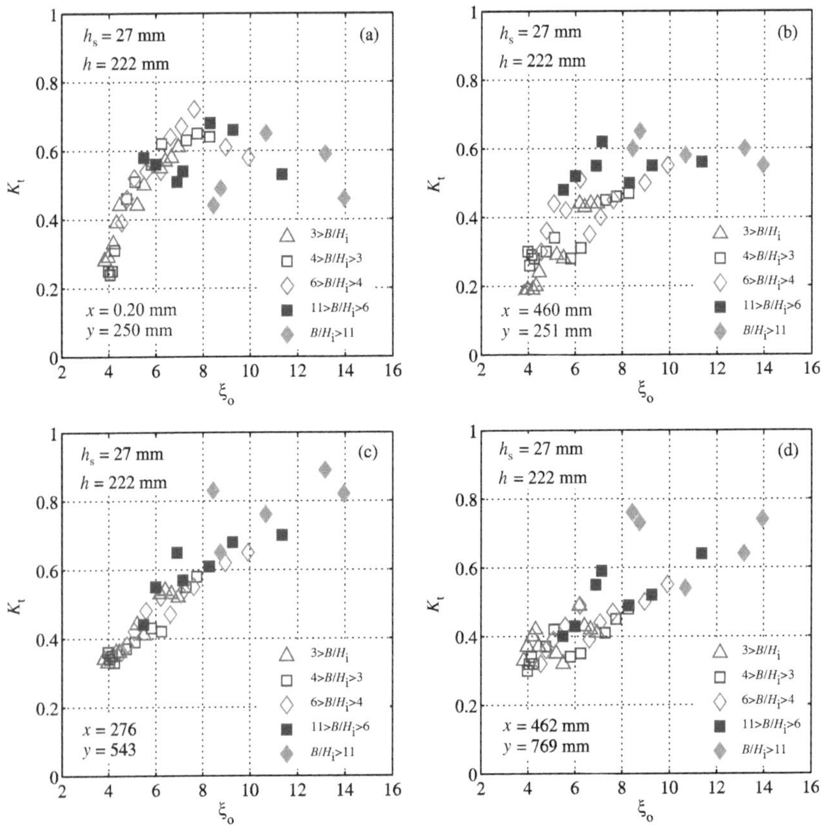

Figure 11.22 Variation of 3D wave transmission coefficient with nondimensional parameters: Surf similarity parameter and dimensionless breakwater crest ratio for four different points behind the 3D structures (3D small scale tests in 1.2 m wide flume; $h_s = 27$ mm and $h = 222$ mm).

four selected nondimensional parameters. Each graph illustrates the variation of K_t with surf similarity parameter (Iribarren Number) for one specific point. It can be deduced from Figure 11.22a that variation of K_t with surf similarity parameter is strongly dependent on the dimensionless breakwater crest width ratio, with different values of K_t observed for different B/H_i values. It can be seen that the variation of K_t is also strongly dependent on submergence depth.

In addition, the values of K_t vary significantly from point to point due to the 3D effects. Figure 11.22a (11.23a) is related to the points behind the breakwater roundhead with $x = 0.20$ mm and $y = 250$ mm ($x = 120$ mm

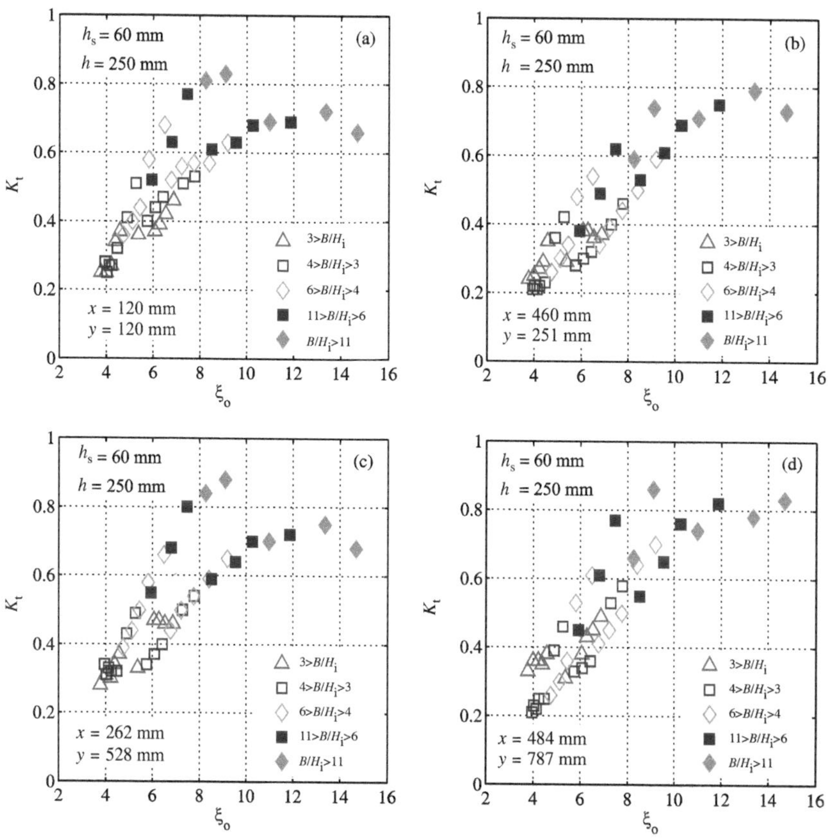

Figure 11.23 Variation of 3D wave transmission coefficient with nondimensional parameters: Surf similarity parameter and dimensionless breakwater crest ratio for four different points behind the 3D structures (3D small scale tests in 1.2 m wide flume; $h_s = 60$ mm and $h = 250$ mm).

and $y = 269$ mm). As expected, K_t increases with surf similarity parameter for points behind the breakwater trunk. However, lower values for K_t are observed for point b; located right behind the breakwater crown with $x = 460$ mm and $y = 251$ mm ($x = 460$ mm and $y = 269$ mm); see Figures 11.22b and 11.23b. This is because of wave dissipation by breaking over the crest (2D transmitted waves only; Figure 11.22a). However, a slight increase in K_t can be observed at the points c and d (combined 2D transmitted and diffracted waves) compared with the point right behind the breakwater trunk (point b) due to the effect of wave diffraction.

These 3D analyses confirm the significance and importance of 3D studies of submerged breakwaters. To conclude, the surf similarity parameter ξ_o, the submergence ratio h_s/H_i, the breakwater crest width ratio B/H_i and relative wave height H_i/h are, in order, the parameters most affecting the wave transmission coefficient in 2D, while x/L_o, y/L_o, the surf similarity parameter ξ_o, the submergence ratio h_s/H_i and the relative wave height H_i/h are, in order, the parameters most affecting the transmission coefficient in 3D cases.

11.4.3 Accuracy Analysis

The comparison of the ANN model with experimental data is done by comparing the five statistical parameters: the root mean squared error (RMSE), the squared multiple correlation coefficient R^2, bias or distortion (β), Wilmott Index I_w and error function ε as defined previously. The values of the five statistical parameters for ANN3D and shallow water equation (SWE) models are provided and graphically compared. The proposed ANN model is verified against new experimental data measured in the wave tanks and not used during the training process. Accuracy analysis is performed under three distinct applications of the model, namely interpolation, extrapolation, and larger scale predictions.

To evaluate the performance of the trained ANN model, it was applied over three ranges. These included interpolation, extrapolation, and prediction at larger scale well beyond that used in training the model. To demonstrate the influence of the input variables on the ANN's predictions, a sensitivity analysis was conducted. The next section considers these analyses. Interpolation predictions were done within the range of input parameters used for training the model as defined in Table 11.1 (Set 1). The neural networks are basically trained within a specific range of inputs and are expected to return acceptable results for new values of input parameters within the same range (Set 2). However, a trained model is likely to be employed to predict wave transmission for input values which are not necessarily within the range of the training data set. Detailed analysis of extrapolation predictions was conducted for values of input parameters outside the range of the training data set (Set 3). This refers both to the wave conditions and to the spatial location of the observation. A separate set of conditions was identified to investigate predictions at a far larger scale, as it is important to have an understanding about the performance of the model when the scales increase significantly. This might be critical if the neural network model is being trained with laboratory data and is going to be used to predict real sea states. The data used in this case were those in Set 4. Finally, a sensitivity analysis was conducted to study the effect of each input parameter on model performance.

Comparison of the proposed model with the experimental data was investigated by comparing five statistical parameters including the root mean squared error (RMSE) and the squared multiple correlation coefficient R^2 as defined in the previous section, and three additional parameters: bias or distortion β, Wilmott index I_w (Wilmott, 1981) and error function ε (Haller et al., 2002). Perfect agreement is achieved if R^2 and the Wilmott index are 1.0, the error function is zero, the value of distortion approaches 1.0 and RMSE is small. R^2 is considered unsatisfactory as a measure of the goodness of fit for a multivariate regression relationship (Draper, 1984) and thus the validity of the model will be assessed mostly by the other parameters. The values of the five statistical parameters are tabulated and compared graphically in the following subsection. The proposed ANN model was validated with new experimental data not used during the training process. Accuracy analysis was performed for the three different scales of application referred to previously and explained in the following.

Interpolation

A statistical comparison was made of wave transmission coefficients around the submerged breakwater for cases involving interpolation. Conditions are as set out in Tables 11.4 and 11.5. Figure 11.24 shows the correlation between predicted and measured values within different regions around the breakwater for a case with input parameters within the range of the training data set. A specific wave period of 1 s and total water depth and submergence depth of 250 mm and 62 mm, respectively,

Table 11.4 The range of training and testing data (dimensional datasets: Set 1 and Set 2)

Parameter	Mean	Standard deviation	Min	Max
H_i (mm)	47	21	12	79
L_o (mm)	1812	348	1427	2126
h_s (mm)	42	16	27	62
h (mm)	235	14	222	250
θ (rad)	1.60	0.59	0	3.14
r (mm)	816	363	10	1341

Table 11.5 The range of training and testing data (nondimensional datasets: Set 1 and Set 2)

Parameter	Mean	Standard deviation	Min	Max
H_i/L_o	0.03	0.01	0.01	0.05
h_s/H_i	1.20	0.96	0.34	5.03
H_i/h	0.20	0.09	0.05	0.36
h_s/h	0.17	0.06	0.12	0.24
θ	1.60	0.59	0	3.14
r/L_o	0.47	0.23	0	0.94

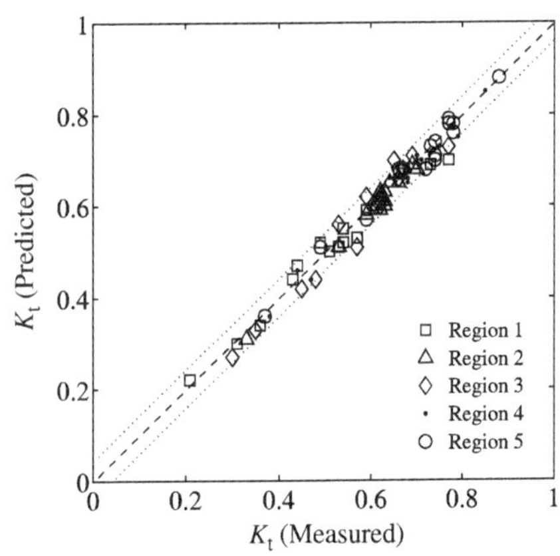

Figure 11.24 Correlation of RBF and experimental values for interpolation cases (dataset: Set 2; test with: H_i = 22 mm, 41 mm, 60 mm, 81 mm, T = 1 s, h_s = 62 mm, h = 250 mm).

were employed in this test. Incident wave height varied from 22 mm up to 81 mm. The values of the statistical indicators and the graphical comparison both indicate that the ANN model is able to calculate wave transmission coefficient with a high accuracy. The predicted values gave high R^2 (97%) and Wilmott number I_w (74%) and low values of ε (0.23) and RMSE (0.04), as well as a value of distortion close to 1.0, indicating very good accuracy (Table 11.6).

Figure 11.25 shows an error contour map for the data plotted in Figure 11.24; this indicates the error in different regions around the submerged breakwater. As can be seen, the nondimensional radial basis function (RBF) model gives low error (mostly less than 2.5%) in most regions behind the breakwater. Higher values of error up to 5% can be seen in some small regions (i.e. far end of regions 1 and 2 and around the head).

Additional analysis was also carried out to examine the ability of the proposed model to predict K_t in various regions behind the breakwater as spatial prediction of K_t is the primary interest of this study. Figure 11.26 presents the results of this analysis for four different points behind the structure. Each graph illustrates the variation of K_t for different submergence ratios at one specific point. The general trend of K_t computed by

Table 11.6 Statistical evaluation of nondimensional RBF model (dataset: Set 2)

Statistical index	I_w	ε	R^2	β	*RMSE*
Nondimensional RBF model	0.74	0.23	0.97	0.99	0.04

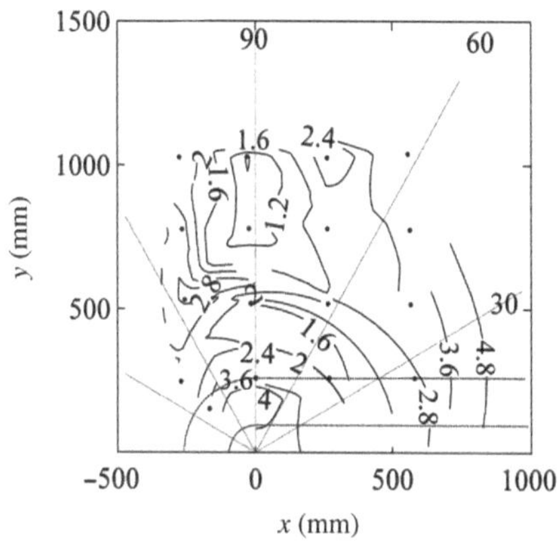

Figure 11.25 Error contour map of RBF for interpolation cases (dataset: Set 2; test with: $H_i = 22$ mm, 41 mm, 60 mm, 81 mm, $T = 1$ mm, $h_s = 62$ mm, $h = 250$ mm).

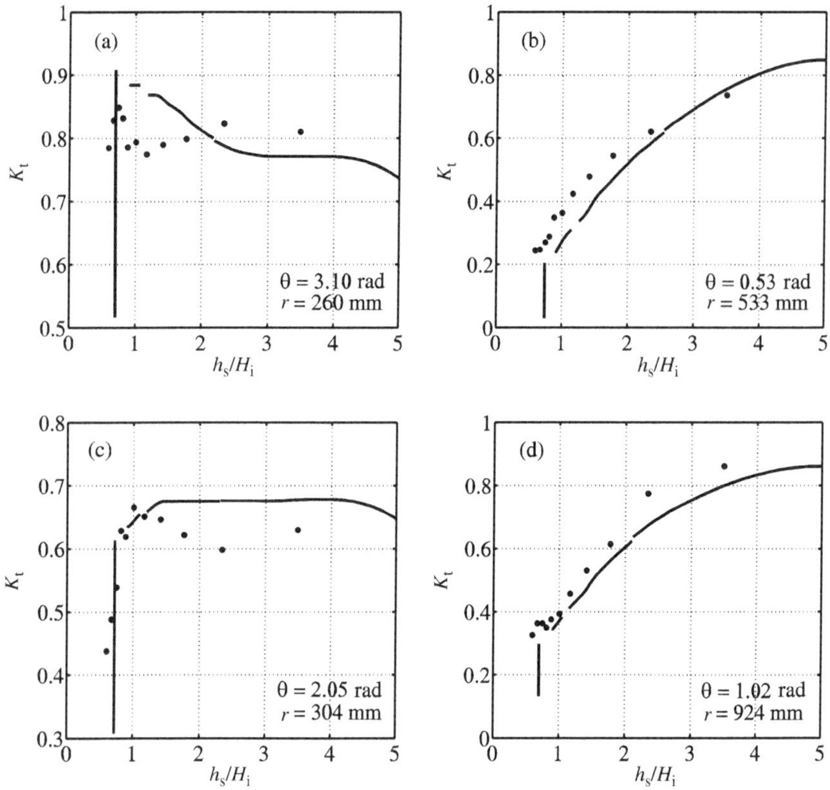

Figure 11.26 Spatial wave transmission coefficient K_t versus submergence ratio h_s/H_i at different locations: measured values (dots), calculated values by proposed ANN model (Solid line).

the ANN model in each graph follows the pattern of measured values reasonably. Figure 11.26a with θ = 3.10 rad and r = 260 mm can be considered as the reference. As expected, K_t increases with submergence ratio for points behind the breakwater (Figure 11.26b and d). K_t decreases at points right behind the breakwater trunk (2D transmitted waves only; Figure 11.26b: θ = 0.53 rad and r = 533 mm) compared with points located in the gap (diffracted waves only; Figure 11.26c: θ = 2.05 rad and r = 304 mm) because of wave dissipation by breaking over the crest. However, an increase in K_t can be observed in the point with θ = 1.02 rad and r = 924 mm (combined 2D transmitted and diffracted waves; Figure 11.26d) compared with a point right behind the breakwater trunk (Figure 11.26b) due to the effect of wave diffraction. It can be also deduced from the plots that the model produces reasonable responses under different conditions and locations behind the breakwater, particularly at points located in the shadow region of the structure, with errors up to a highest value of 10%.

Extrapolation

Data set 3 was employed to test the proposed model in cases involving extrapolation. The parametric range of data is presented in Tables 11.7 and 11.8. Figure 11.27 shows the correlation between predicted and measured values within different regions around the breakwater for a case with input parameters (wave period and water depth over and in front of the structure) within the range of the training data but incident wave heights outside the range of the training data. A specific wave period of 1 s and total water depth and submergence depth of 250 mm and 62 mm, respectively, are employed in this test. Incident wave height varied from 90 mm up to 100 mm. The values of the statistical indicators and graphical comparison indicate that the ANN model is able to predict wave transmission coefficients with high accuracy. The values calculated by the ANN provide acceptable values of R^2 (greater than 90%) and Wilmott

Table 11.7 The range of evaluation data in dimensional form (extrapolation dataset: Set 3)

Parameter	Mean	Standard deviation	Min	Max
H_i (mm)	52	24	10	100
L_o (mm)	2311	808	1427	3380
h_s (mm)	67	36	27	112
h (mm)	258	33	222	300
θ (rad)	1.56	0.56	0	3.14
r (mm)	904	454	10	2099

Table 11.8 The range of nondimensional form of the evaluation data (extrapolation dataset: Set 3)

Parameter	Mean	Standard deviation	Min	Max
H_i/L_o	0.03	0.02	0.01	0.07
h_s/H_i	1.74	1.68	0.27	11.25
H_i/h	0.21	0.10	0.03	0.44
h_s/h	0.25	0.10	0.12	0.37
θ	1.56	0.56	0.00	3.14
r/L_o	0.44	0.28	0.00	1.47

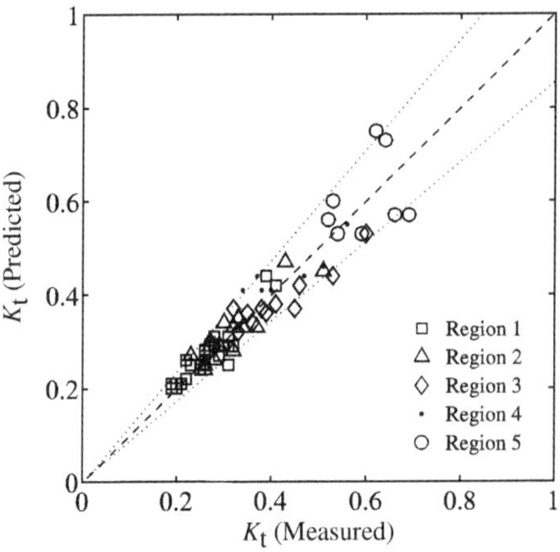

Figure 11.27 Correlation of RBF model results and experiments in the case of extrapolation (dataset Set 3; test with: $H_i = 90$ mm, 93 mm, 96 mm, 100 mm, $T = 1$ s, $h_s = 62$ mm, $h = 250$ mm).

Table 11.9 Statistical evaluation of nondimensional RBF models (extrapolation; Set 3)

Statistical index	I_w	ε	R^2	β	$RMSE$
RBF model	0.74	0.35	0.90	1.00	0.10

number I_w (greater than 70%) and low value of ε (0.35) and RMSE (0.1), as well as a distortion value of 1.0 that indicates good accuracy (Table 11.9). The graph shows very good agreement in most regions with higher correlation in regions behind the breakwater, reducing closer to the region around the end of the breakwater.

Figure 11.28 is an error contour map of the same wave transmission calculated by the RBF model as plotted in Figure 11.27. This shows the error in different regions around the submerged breakwater. As can be seen, the RBF model performs with low error (mostly less than 8%) in most regions behind the breakwater. Less accurate results (errors higher than 8%) were found in some areas in regions 1, 2 and 3 and a narrow band at the far end of the breakwater had a maximum error of 10%.

Additional analysis is also carried out in this section to examine the ability of the proposed model to predict K_t including extrapolation to various regions behind the breakwater. Figure 11.29 presents the results of this analysis for four different points behind the structure. Each graph illustrates the variation of K_t with wave steepness for one specific point. The general

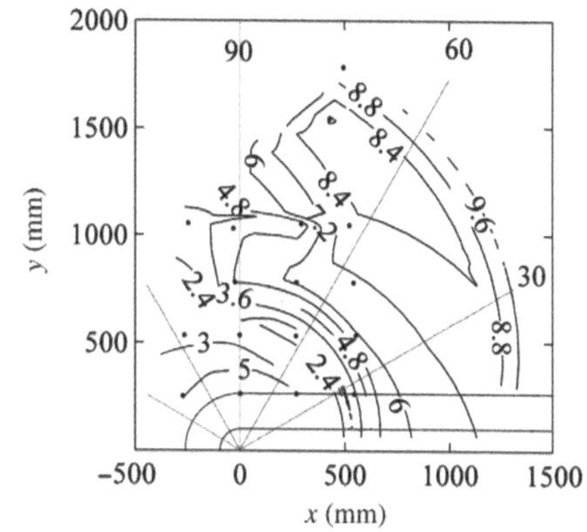

Figure 11.28 Error contour map of RBF model results and experiments for cases of extrapolation (dataset: Set 3; test with: $H_i = 90$ mm, 93 mm, 96 mm, 100 mm, $T = 1$ s, $h_s = 62$ mm, $h = 250$ mm).

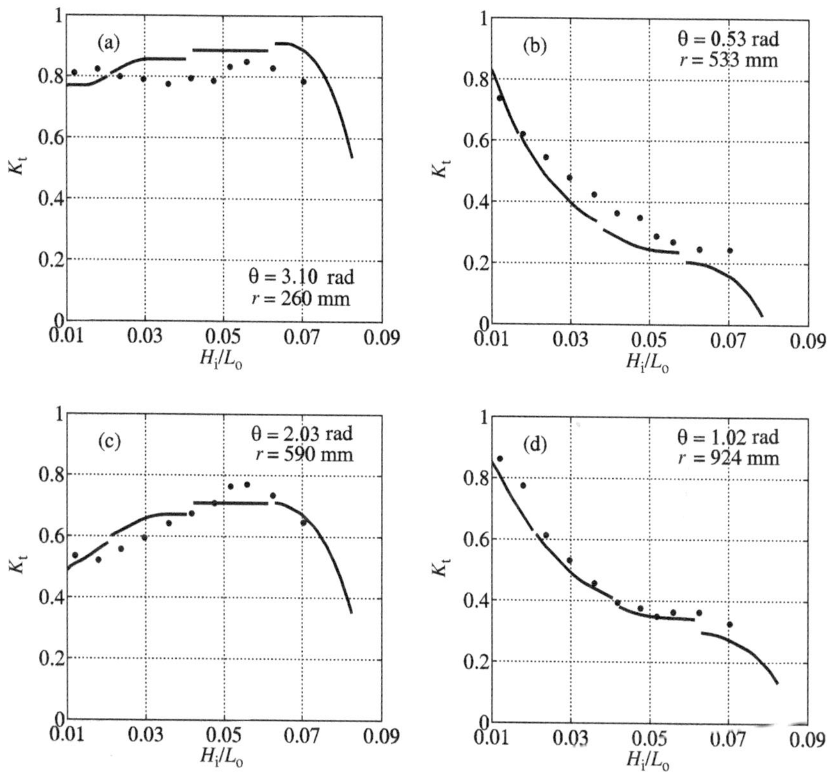

Figure 11.29 Spatial wave transmission coefficient K_t versus wave steepness H_i/L_o at different locations: measured values (dots), calculated values by proposed ANN model (Solid line).

trend of K_t computed by the model follows the experimental pattern reasonably well. Figure 11.29a with $\theta = 3.10$ rad and $r = 260$ mm can be considered as the reference. As expected, K_t decreases with wave steepness for points behind the breakwater (Figure 11.29b and d). As in Figure 11.26, K_t decreases at points right behind the breakwater trunk (2D transmitted waves only; Figure 11.29b: $\theta = 0.53$ rad and $r = 533$ mm) compared with points located in the gap (diffracted waves only; Figure 11.29c: $\theta = 2.05$ rad and $r = 304$ mm) because of wave dissipation by breaking over the crest. In a similar way to Figure 11.8, a slight increase in K_t can be observed at the point $\theta = 1.02$ rad and $r = 924$ mm (combined 2D transmitted and diffracted waves; Figure 11.29d) compared with the point right behind the breakwater trunk (Figure 11.29b) due to effect of wave diffraction. It can also be deduced from the plots that the model produces reasonable responses under different conditions and locations behind the breakwater, particularly at points located in the shadow region of the structure, with errors up to a highest value of 10%. These graphs include the extrapolation cases where the input data are not within the range of the training data. Referring to Table 11.2 and Table 11.7, values of wave steepness larger than 0.05 are considered as extrapolation cases. For the point right behind the breakwater trunk the error between predicted and measured values in both interpolation and extrapolation is almost the same, while at point $\theta = 1.02$ rad and $r = 924$ mm (Figure 11.29d), this error slightly increases for the extrapolation cases ($H_i/L_o > 0.05$). However, even in the extrapolation case the predictions are still acceptable ($e < 10\%$).

Larger Scale

A set of experimental data from the large flume were employed to assess the performance of the ANN model for cases involving input parameters all lying well outside that used for training (Tables 11.10 and 11.11) (see Table 11.29 for details of tests and further information). The evaluation data for these larger scale tests involved a scale factor of 4. The statistical comparison is presented in Table 11.12 and Figure 11.30 shows the correlation between predicted and measured values for these tests. The values of the statistical indicators and the graphical comparison indicate that the ANN model is able to calculate wave transmission coefficients with good accuracy. The values predicted by the ANN model provide fairly good R^2 (84%) and Wilmott number I_w (88%). A low value of ε (0.07) and a convincing value of RMSE (0.11) were also obtained. A value of distortion close to 1.0 also points to the good accuracy of the predictions.

Table 11.10 The range of evaluation data in dimensional form (Larger scale dataset: Set 4) (see Table 11.29 for details of tests and further information)

Parameter	Mean	Standard deviation	Min	Max
H_i (mm)	140	66	40	240
L_o (mm)	3568	871	2352	4175
h_s (mm)	250	0	250	250
h (mm)	1000	0	1000	1000
θ (rad)	1.53	0.39	0.79	2.35
r (mm)	2043	977	530	3644

Table 11.11 The range of nondimensional form of the evaluation data (Larger scale dataset: Set 3) (see Table 11.29 for details of tests and further information)

Parameter	Mean	Standard deviation	Min	Max
H_i/L_o	0.05	0.02	0.01	0.1
h_s/H_i	2.39	1.59	1.04	6.25
H_i/h	0.14	0.07	0.04	0.24
h_s/h	0.25	0.00	0.25	0.25
θ	1.53	0.39	0.79	2.35
r/L_o	0.44	0.28	0.00	1.47

Table 11.12 Statistical evaluation of the nondimensional RBF model in larger scale model with a scale factor of 4

Statistical index	I_w	ε	R^2	β	$RMSE$
Upper scale model	0.73	0.39	0.85	1.05	0.19

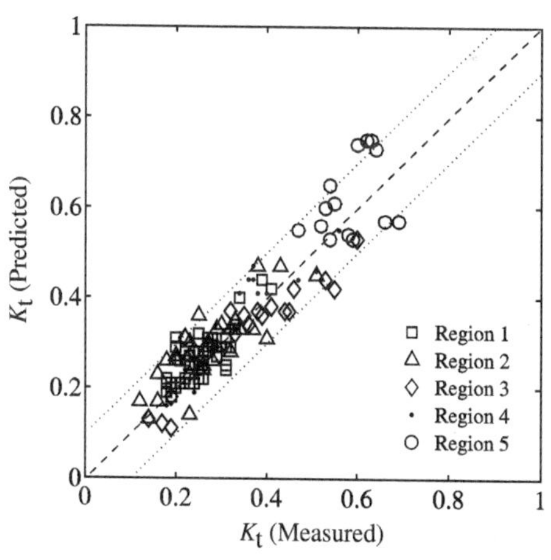

Figure 11.30 Graphical comparison of the nondimensional RBF model tested at larger scale (dataset Set 4; test with: $H_i = 88$ mm, 135 mm, 180 mm, 210 mm, 240 mm, $T = 2$ s, $h_s = 250$ mm, $h = 1000$ mm).

11.4.4 Sensitivity Analysis

All parameters relevant to the wave field around a 3D submerged breakwater have been included in the ANN models described previously. In order to detect the importance of each parameter to the performance of the model, networks were designed and trained using the same training data sets but with each parameter omitted in turn, in order that the most influential parameters could be identified. Assessments were made on the performance of the networks by comparison with the network results using all the statistical parameters. This investigation illustrated that the polar coordinate θ was the most significant parameter in the 3D wave field around the breakwater, with wave height and energy varying spatially around the structure. This confirms the importance of 3D effects on wave height prediction and highlights the inadequacy of 2D models that are unable to deal with spatial variation of wave height behind the breakwater.

From Table 11.13, it can be seen that the parameters H_i/L_o and r/L_o have the next lowest R^2 correlation coefficients (0.77 and 0.76, respectively) and the highest error (RMSE = 0.14 and 0.18); they also have a considerable influence on the results of the networks. It should be noted that the division into training data, validation data and test data for the sensitivity analysis was carried out in the same way as for the main ANN model. The R^2 correlation coefficient and RMSE values for the main ANN model are 0.95 and 0.07, respectively.

A detailed sensitivity analysis is also presented in Figure 11.31 comprising different sensitivity graphs of individual nondimensional input parameters. The sensitivity figures contain several lines. The dashed lines show the band of the 95% confidence interval; one line for the lower boundary (quartile 2.5%) and the other for the upper boundary (quartile 97.5%). The solid line is the ANN prediction (mean value). Figure 11.31a reveals that the ANN model gives a relatively high reliability in the range of $0.003 < H_i/L_o < 0.07$. Outside this interval the model tends to low reliability. Based on Figure 11.31b, the model is reliable in the range $h_s/H_i < 7$. For larger submergence depth the ANN model is

Table 11.13 Sensitivity analysis of RBF model

Omitted parameter	H_i/L_o	h_s/H_i	H_i/h	h_s/h	θ	r/L_o
R^2	0.77	0.90	0.87	0.85	0.50	0.76
RMSE	0.14	0.10	0.11	0.11	0.21	0.18

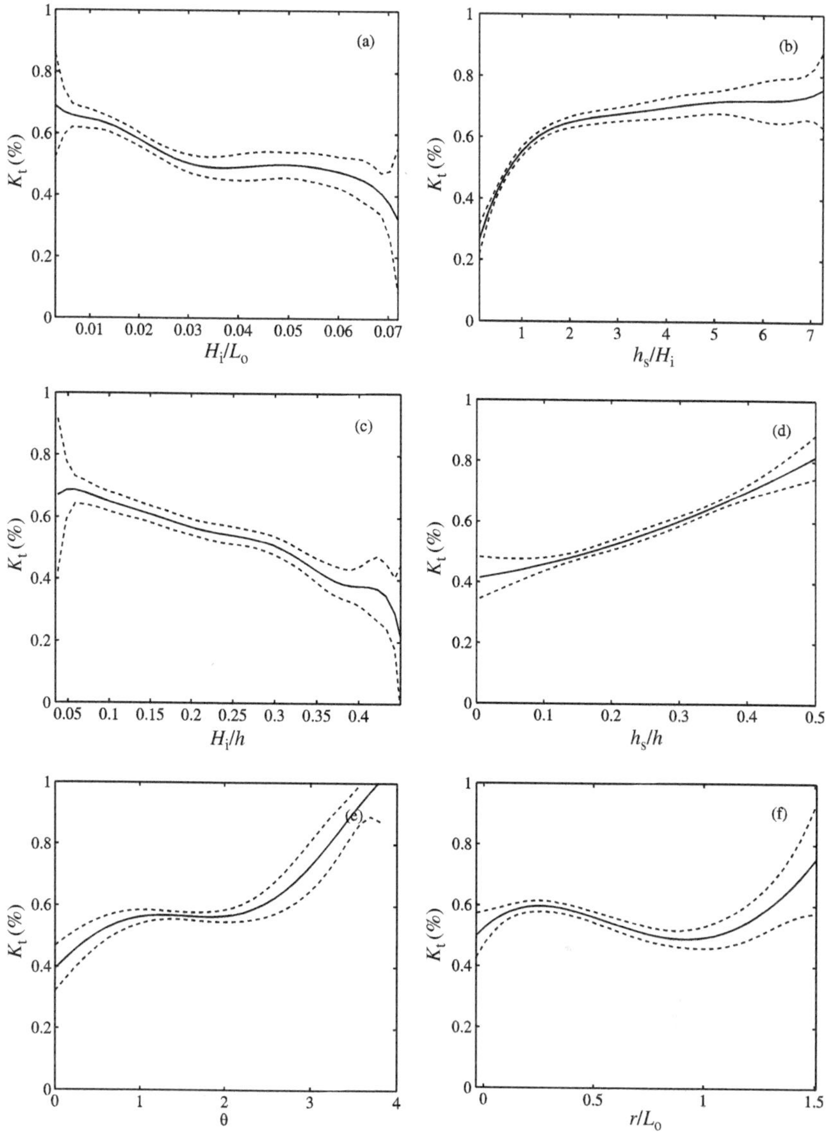

Figure 11.31 Sensitivity of the ANN model to the nondimensional input parameters.

rather unreliable. According to this plot, K_t increases dramatically with submergence ratio from a value of $K_t = 0.28$ at zero submergence to a value of $K_t = 0.63$ at $h_s/H_i = 2$. Beyond this point, no significant influence of the submergence ratio is seen. Figure 11.31c shows that the model is also reliable in the range $0.05 < H_i/h < 0.4$. There is a narrow confidence band with high reliability in Figure 11.31d for $0.12 < h_s/h < 0.37$, and with acceptable reliability outside this range. It can be seen in Figure 11.31e that the confidence band is narrow between values 0 and 3.14 rad, which suggests that the model is reliable across the entire range of θ. The model only seems to be reliable for r/L_o values up to 1.3; further from the breakwater, the model has a rather low reliability (Figure 11.31f).

A summary of the results has also been presented in Table 11.14.

Table 11.14 Summary of the results

No.	H_i	L_o	K_t,2D ANN	K_t,2D M	K_t,2D D	K_t,2D VDM	K_t,2D BC	K_t,2D GA
1-1	52	2126	0.28	0.29	0.51	0.48	0.34	0.42
1-2								
1-3								
1-4								
1-5								
1-6								
1-7								
1-8								
1-9								
1-10								
1-11								
1-12								
1-13								
1-14								

No.	X (mm)	Y (mm)	K_d PP	K_t,3D V	K_t,3D ANN	K_t,3D M
1-1	449	282	0.48	0.64	0.23	0.26
1-2	481	819	0.36	0.56	0.22	0.25
1-3	282	528	0.45	0.62	0.31	0.32
1-4	476	1302	0.30	0.52	0.13	0.10
1-5	303	1049	0.34	0.54	0.26	0.25
1-6	108	820	0.39	0.57	0.37	0.31
1-7	9	280	0.67	0.79	0.40	0.39
1-8	0	530	0.48	0.64	0.43	0.42
1-9	−10	1050	0.34	0.55	0.38	0.37
1-10	−107	820	0.39	0.57	0.45	0.44
1-11	−256	1051	0.34	0.54	0.46	0.53
1-12	−249	530	0.46	0.63	0.52	0.57
1-13	−250	280	0.57	0.71	0.53	0.50
1-14	−250	31	0.70	0.82	0.68	0.77

11.5 ANALYSIS AND DISCUSSION OF SHALLOW WATERS EQUATIONS MODEL RESULTS

Wave propagation around a 3D submerged breakwater is investigated numerically using a model based on the shallow water equations (Figure 11.32).

Simulations for both flumes (coastal flume and ocean tank) are made using the model and the results are presented here.

Coastal Flume Model

The results of simulating the wave field behind the submerged breakwaters in a small scale model are presented here in the forms mentioned previously. The length of the flume is 20 m and the width is 1.2 m. The uniform water depth is 0.3 m. Waves are generated at one line at the end of the flume. Thus, all other model boundaries are closed, i.e. considered as land points. At the generation line ($j = 190$) a time series of waves is imposed. The wave height is $H_i = 0.03$ m and the wave period is $T = 1.00$ s. The grid spacing is selected to be 0.02 m and the time step is 0.01 s. The simulation duration is 90 s (9000 time steps). A 190-point thick sponge layer is used in this model. Similar wave modeling for other wave conditions and water depths was conducted. Time series of the simulated surface elevation extracted at different points, as well as the spatial variation of the transmitted wave height behind the breakwater, is presented.

Figure 11.32 represents a time series of water surface elevation illustrating variation of surface level with time for a specific point located offshore. The figure clearly shows how the surface level increases after a specific period as reflected waves reach the measurement point. This time series matches well the time series obtained from the measurements. The total simulation time

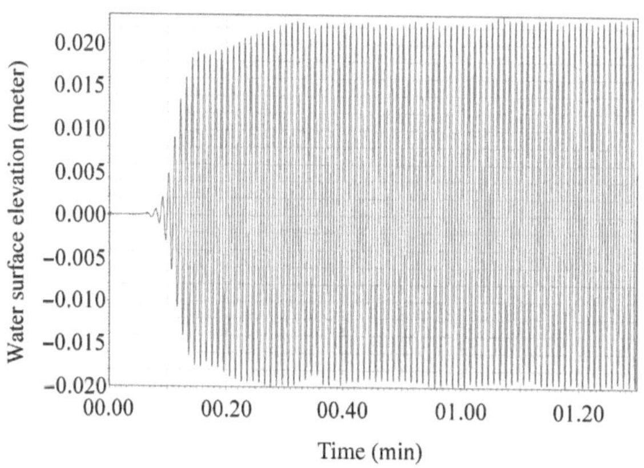

Figure 11.32 Time series of water surface elevation produced by SWE at one point offshore the breakwater.

should allow waves to reach all model areas for calculation of statistical parameters such as H_s and H_{m0}. Figure 11.32 shows the surface elevation time series used in modeling (Simulation time = 90s). The beginning period is not included in calculation of the statistics as no waves have reached the model. The selected time step in simulations should be sufficiently small to be able to resolve properly the shortest individual wave periods in the wave record and additionally to keep away from the instability.

Figure 11.33 presents the two dimensional variation of water surface level along the flume in the presence of the breakwater for a specific wave condition. As can be seen the water surface level decreases gradually over and behind the breakwater and over the beach slope.

Figure 11.34 illustrates a comparison of measured and computed time series by SWE for water surface elevation for some specific tests.

Figure 11.35 presents four time series of water surface elevation calculated by model plotted with measured time series. Figure 11.35a is related to offshore (grid coordinates $x = 31$, $y = 165$) and is used as a reference point for other plots. Perfect agreement with measured time series is observed in this graph. However, less accurate results can be seen at the other points. The accuracy of predictions varies point to point. According to the graph in Figure 11.35b (grid coordinates of $x = 51$, $y = 565$; a point right behind the breakwater trunk), a relatively good agreement is obtained, although a slight discrepancy is observed. For points c ($x = 31$, $y = 595$) and d ($x = 51$, $y = 615$), the disagreement with experimental measurements increases and the model produces underestimated predictions (Figure 11.35c and d).

Ocean Tank Model

The results of simulation of the wave field behind submerged breakwaters in a large scale model are presented here. The length of the flume is 20 m and the width is 2.5 m. The uniform water depth is 1.0 m. Waves are generated at one line at the end of the tank. Thus, all other model boundaries are closed, i.e. considered as land points. At the generation line ($j = 80$) a

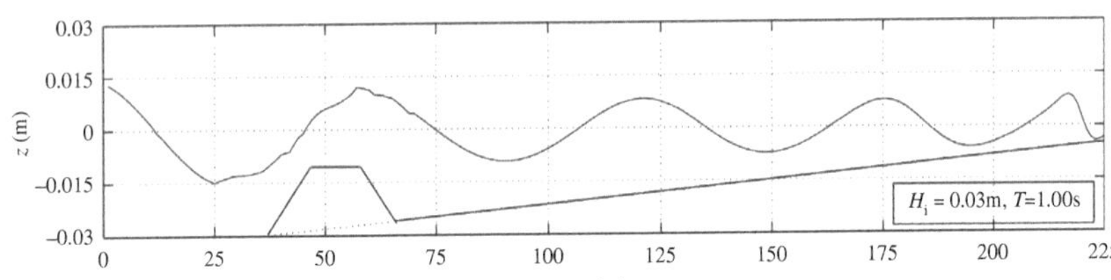

Figure 11.33 2D variation of water surface elevation in a specific line along the flume (X direction Grid = 31 ($x = 600$ mm); $t = 19.6$ s) for SWE ($h_s = 110$ mm).

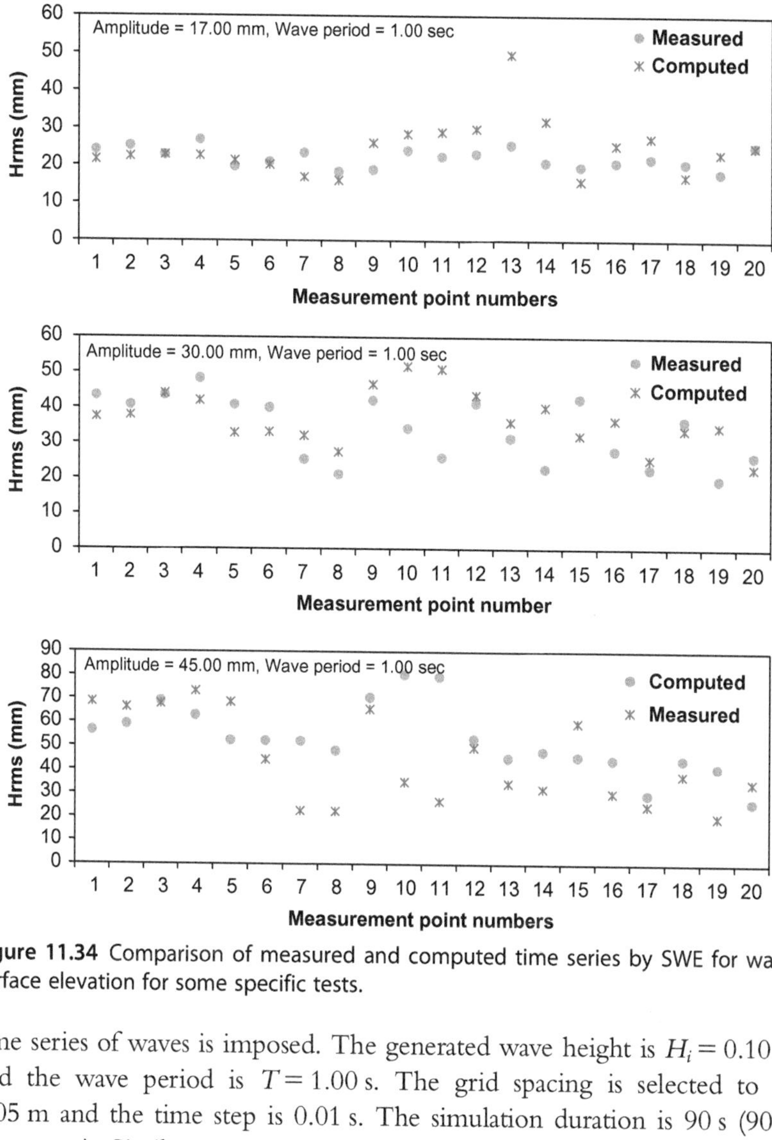

Figure 11.34 Comparison of measured and computed time series by SWE for water surface elevation for some specific tests.

time series of waves is imposed. The generated wave height is $H_i = 0.10$ m and the wave period is $T = 1.00$ s. The grid spacing is selected to be 0.05 m and the time step is 0.01 s. The simulation duration is 90 s (9000 time steps). Similar wave modeling for other wave conditions and water depths was conducted. Time series of the simulated surface elevation extracted at different points, as well as the spatial variation of the transmitted wave height behind the breakwater, is presented.

The variation of water surface elevation along and across the flume in the presence of the submerged breakwater is illustrated in Figure 11.36. The graph clearly shows the variation of water surface elevation at different points behind the breakwater at a specific time.

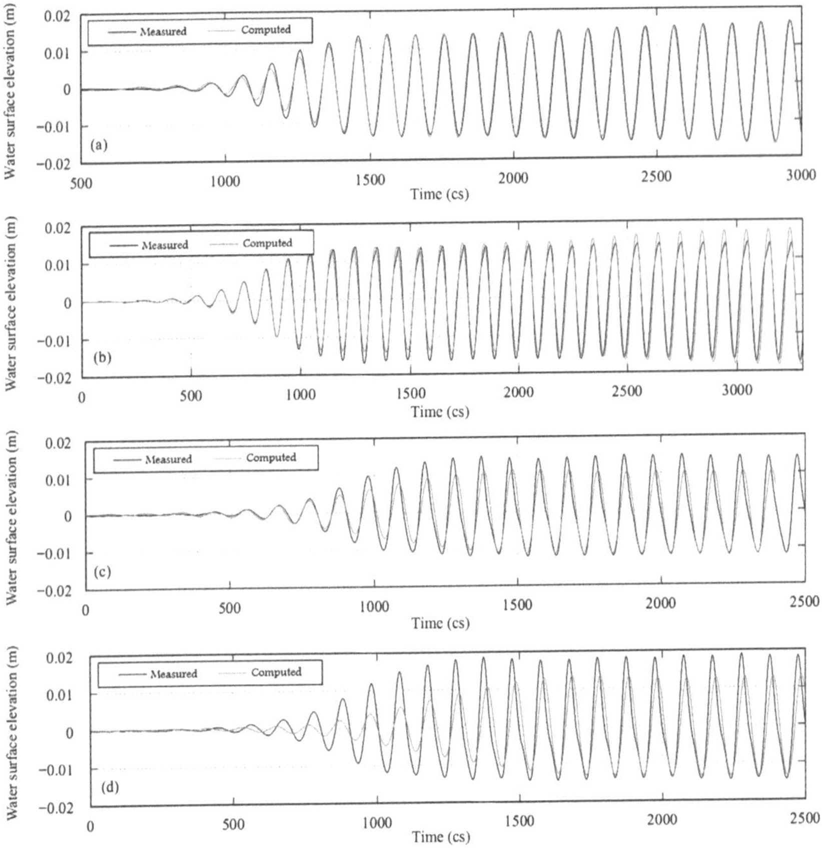

Figure 11.35 Comparison of measured and computed time series by SWE for water surface elevation: (a) $x=31$, $y=165$; (b) $x=51$, $y=565$; (c) $x=31$, $y=595$; and (d) $x=51$, $y=615$. The unit on the x-axis is centisecond (cs).

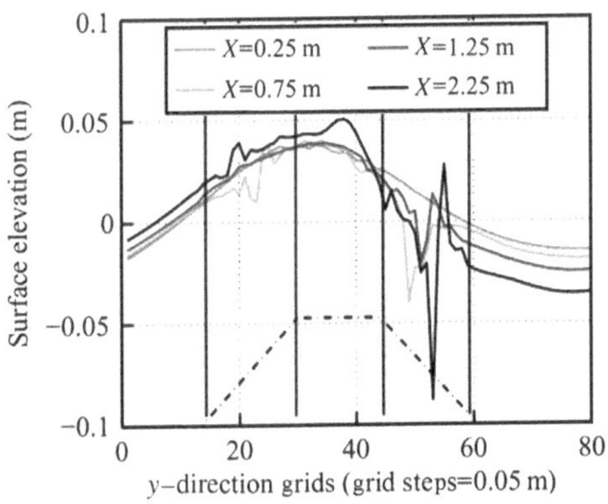

Figure 11.36 2D variation of wave height in four lines along the flume computed by SWE ($t=22$ s).

Figure 11.37 presents several time series of water surface elevations calculated by model compared with experimental measurements. Figure 11.37a is related to offshore (grid coordinates of $x = 25$, $y = 238$) and is used as a reference point for other plots. Perfect agreement with the measured time series is observed in this graph. However, less accurate results are obtained at other points. The accuracy of predictions varies from point to point. From the graph in Figure 11.37b (grid coordinates of $x = 45$, $y = 238$; a point right behind the breakwater trunk), relatively

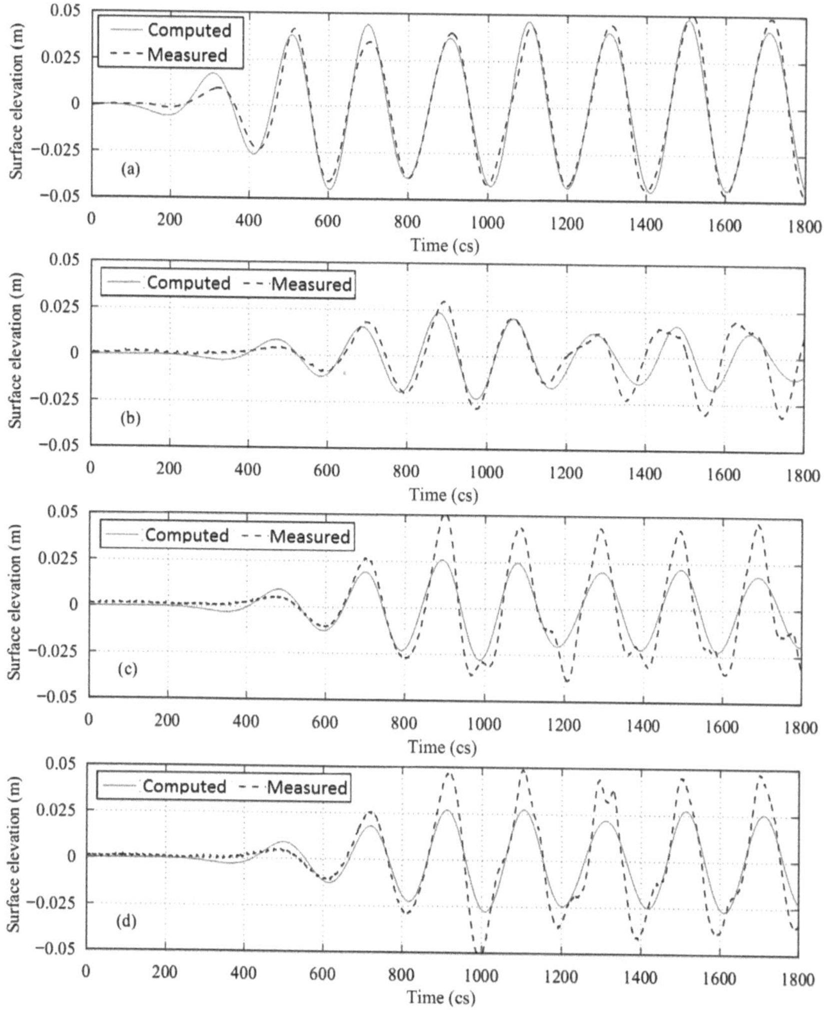

Figure 11.37 Measured and computed time series of water surface elevation by SWE for four different points. The unit on the *x*-axis is centisecond (cs).

good results are obtained, although a slight discrepancy is observed. The disagreement with experimental measurements increases at points c ($x = 25$, $y = 248$) and d ($x = 45$, $y = 248$), as the model produces underestimated predictions (Figure 11.37c and d).

Bathymetric Maps

To start the simulations using the SWE model, an appropriate bathymetry or a two dimensional matrix of water depths of the computational domain is required. The other parameters or features of the model for the area of interest should be specified, such as grid spacing, location and type of boundaries, etc. Basically, the bathymetry determines which wave conditions can be modeled. A 2D grid therefore is used. Figures 11.38 and 11.39 represent the final bathymetry used in the SWE model for a small scale model (coastal flume). Figure 11.40 also represents the final bathymetry used in the SWE model for the large wave tank (ocean tank).

Results

In the contour maps in Figure 11.41, x and y are the width and length of the flume. Therefore, the vertical axis or (z axis) illustrates the values of H_{rms}. The reference point of the applied coordinate system is located on the left corner of the wave paddles, at the one end of the flume. In these graphs, wave trains are propagating from the bottom of the picture to the top. However, for analysis purposes, we only consider the shadow zone behind the breakwater. A contour map of H_{rms} for waves with amplitudes about 5, 10 and 20 cm and frequency of 1 Hz is represented (Figure 11.41; top, middle and bottom plots, respectively). For the 5 cm amplitude (top plot), a region with higher values of wave height is

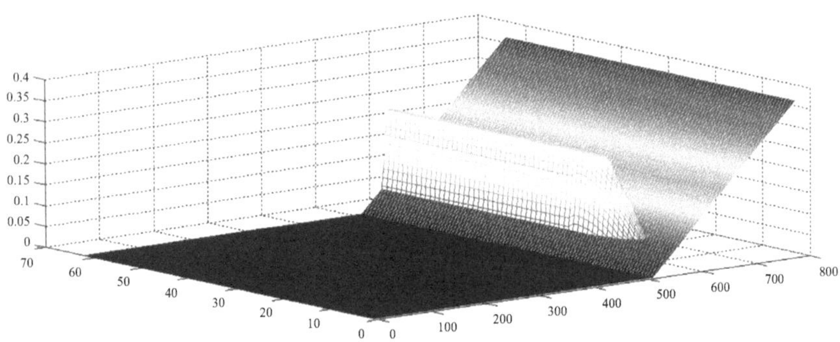

Figure 11.38 Bathymetry of the 20 m by 1.2 m by 1.3 m wave flume (Coastal flume) in SWE model.

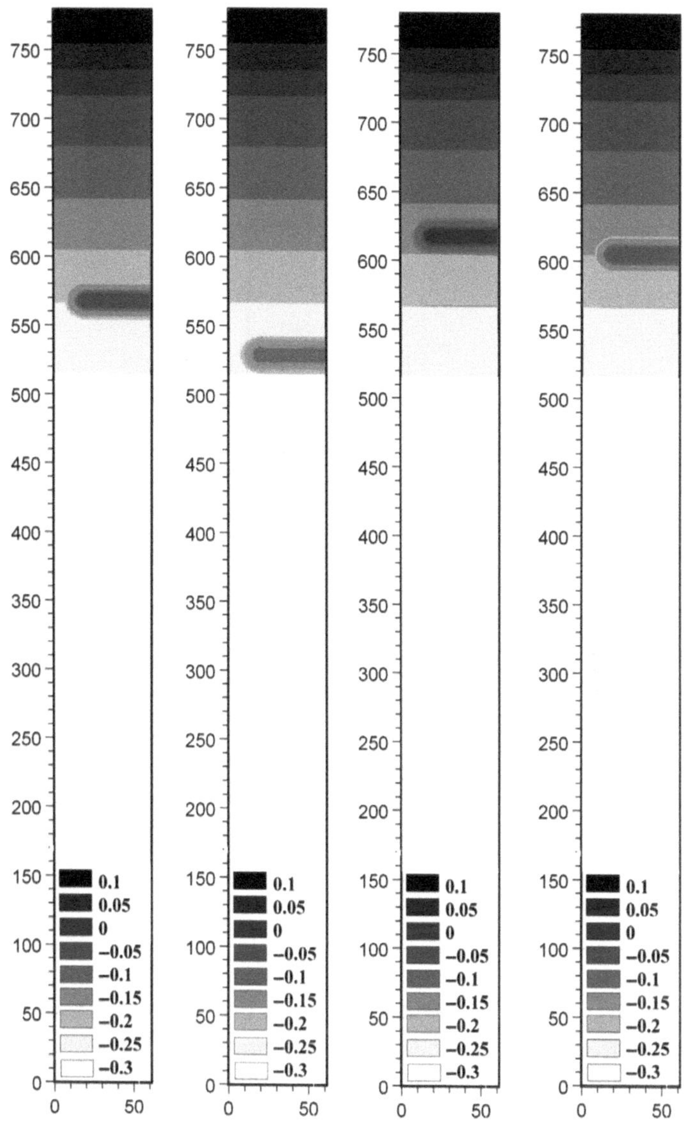

Figure 11.39 Bathymetry of the 20 m by 1.2 m by 1.3 m wave flume (Coastal flume) in SWE model.

observed on the onshore side of the structure's crown, while a region with lower wave height is located right behind the gap. The graph also clearly shows the shoaling effect. Wave height increases over the crest as water depth decreases and, after that, wave height reduces.

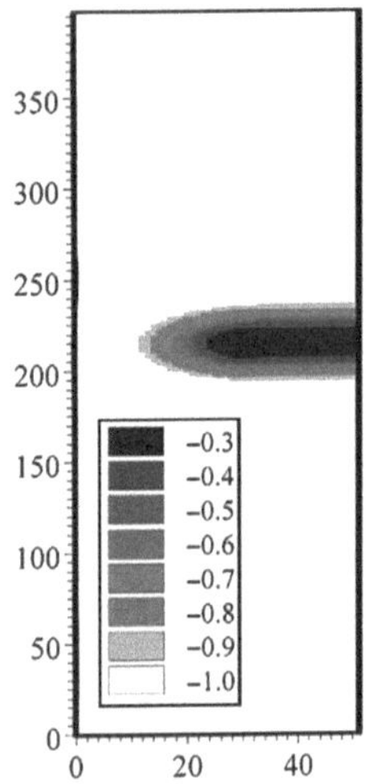

Figure 11.40 Bathymetry of the 20 m by 2.5 m by 1.3 m wave flume (Ocean tank) in SWE.

A contour map of H_{rms} for a wave height with amplitude about 20 cm and frequency of 1 Hz has been presented in the third plot in Figure 11.41. The region with low wave height is located right behind the breakwater crown while offshore wave height is diffracted through the gap behind the breakwater and varies, especially on the shelter area. Wave breaking occurs over the crest and behind the breakwater and their energy is reduced, leading to a decrease in wave height. Therefore, a region with low wave height is observed behind the breakwater. Wave energy propagates from waves with higher energy into the smaller ones. Thus the effect of water wave diffraction can be clearly seen in the graph. For the second graph in Figure 11.41, the variation is between the previous plots. Similar comparison has been made for the last two tests in Figure 11.42 for a wider area, including the top of the breakwater.

The following three contour maps (Figure 11.43) represent the variation of the wave transmission coefficient K_t in the flume behind the breakwater. A contour map of K_t for wave amplitude of 5, 10 and 20 cm with a wave frequency of 1 Hz is presented in Figure 11.43. As we have seen in previous graphs, for the first plot, wave height increases over the crest of the breakwater. K_t is close to 1 behind the breakwater, which means wave energy has not been reduced.

A contour map of K_t for wave amplitude of 20 cm and a frequency of 1 Hz is also presented. This figure shows the effect of wave breaking on the K_t coefficient. When waves are breaking, they lose significant energy, which reduces the wave height. The diffraction can be seen.

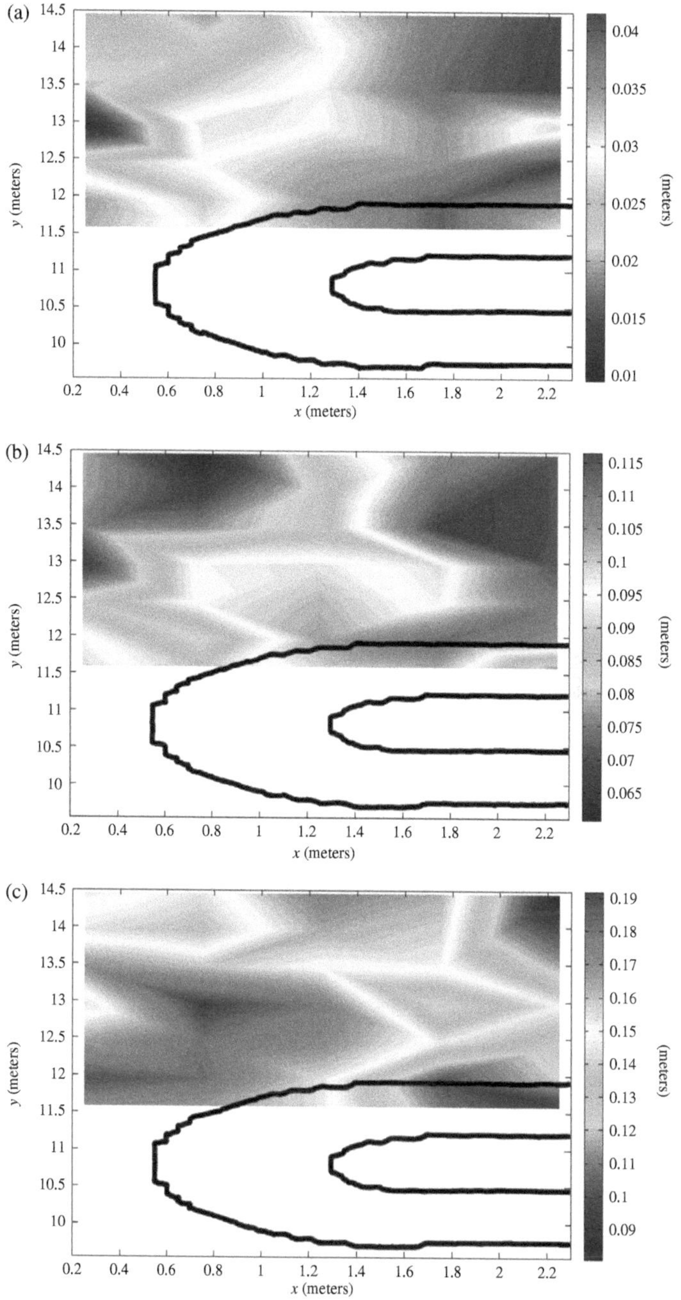

Figure 11.41 H_{rms} contour maps: Top: Test 37; Middle: Test 40; Bottom: Test 84.

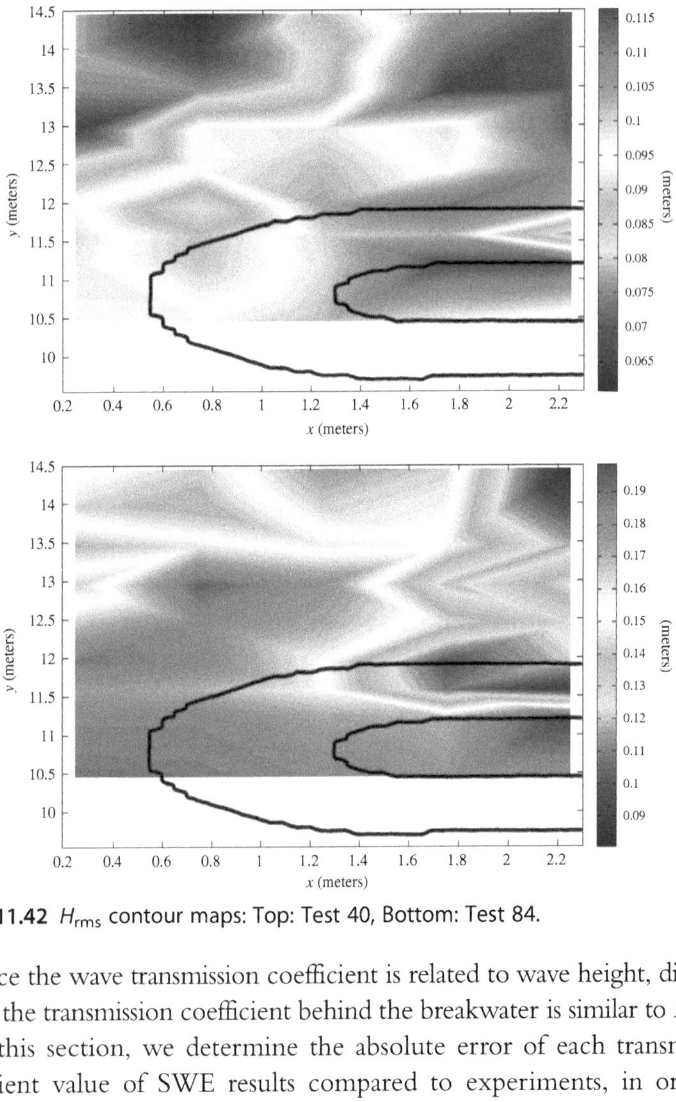

Figure 11.42 H_{rms} contour maps: Top: Test 40, Bottom: Test 84.

Since the wave transmission coefficient is related to wave height, distribution of the transmission coefficient behind the breakwater is similar to H_{rms}.

In this section, we determine the absolute error of each transmission coefficient value of SWE results compared to experiments, in order to observe the accuracy of the SWE results and to evaluate the regions where the model results are not acceptable.

In Figure 11.44 a contour map of absolute error of SWE is compared to experimental values of K_t. The regions where the absolute error is high can be observed on the graph.

The higher error value is located behind the breakwater close to the right boundary and in the gap between the breakwater and the left boundary. In this area, the measured wave height is greater than SWE results.

For a wave amplitude of 5 cm (frequency of 1 Hz), the shoaling effect is the main physical process. A higher wave height region is located on

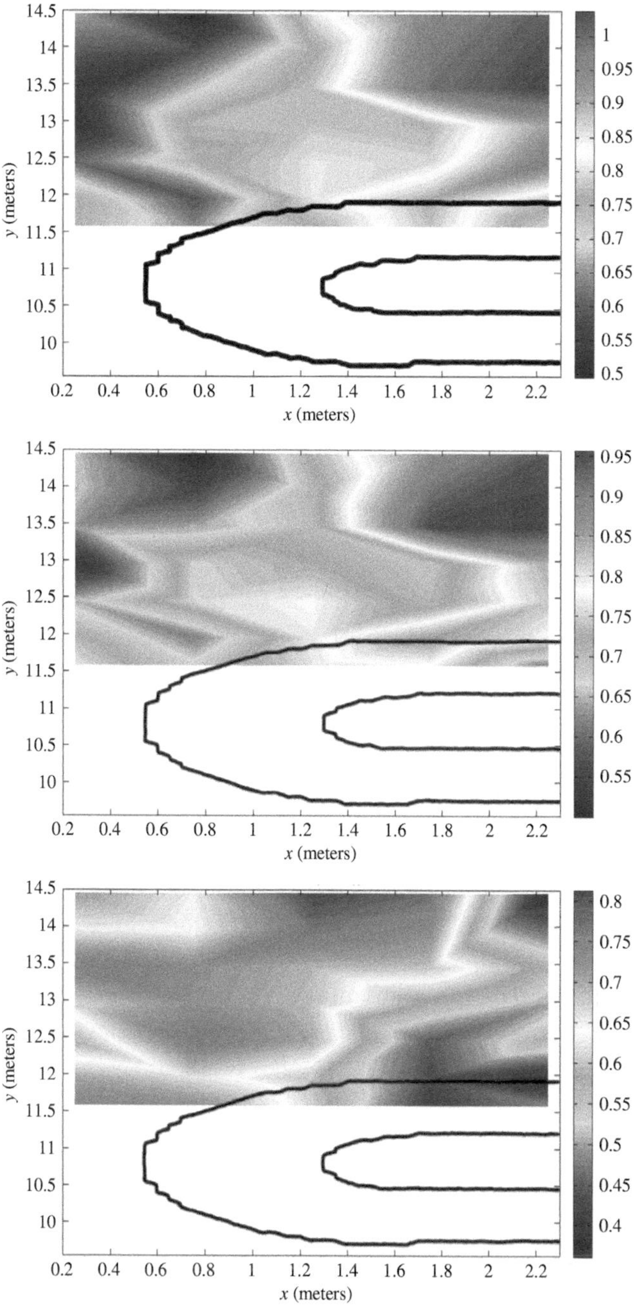

Figure 11.43 K_t contour maps: Top: Test 37, Middle: Test 40, Bottom: Test 84.

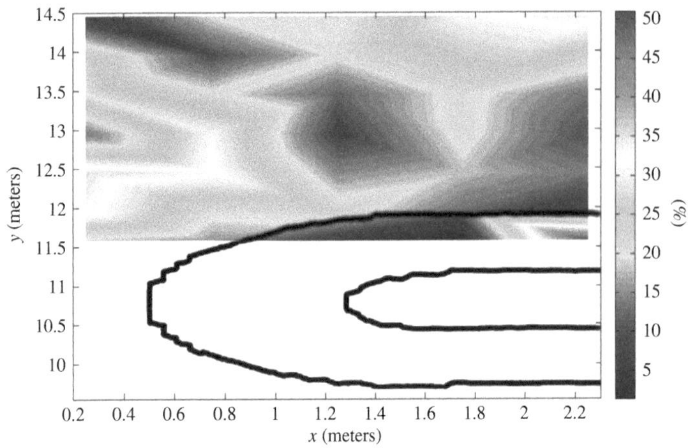

Figure 11.44 Error contour map.

the onshore side of the breakwater. The transmission coefficient is almost 1 behind the breakwater, which means that incident wave energy is not reduced. For the wave amplitude of 10 cm and frequency of 1 Hz, the type of wave breaking is plunging in shallow water conditions. The region with lower wave height is located behind the crest of the breakwater. Waves are diffracted through the gap behind the breakwater and their wave height varies spatially in the shelter area. The wave transmission coefficient is relatively low for this wave condition.

Figure 11.45 also presents a comparison of two wave amplitude spectrums offshore and onshore the breakwater. Figures 11.46–11.48 present contour maps of H_{mo}, K_t and H_s for a specific wave condition ($H_i = 0.10$ m and $T = 1.00$ s) calculated by the SWE model for the small scale model. Figures 11.49–11.53 present contour maps of wave disturbance coefficients including H_{mo}, H_s, K_t, $H_{\max}$ and water surface level, respectively, for a specific test condition ($H_i = 0.03$ m and $T = 1.00$ s), calculated by the SWE for the large scale model. The graphs clearly illustrate the variation of the coefficients.

11.6 COMPARISON OF MODELS

The shallow water equations model is applied for comparison with the ANN3D model results, with particular focus on the 3D wave field behind

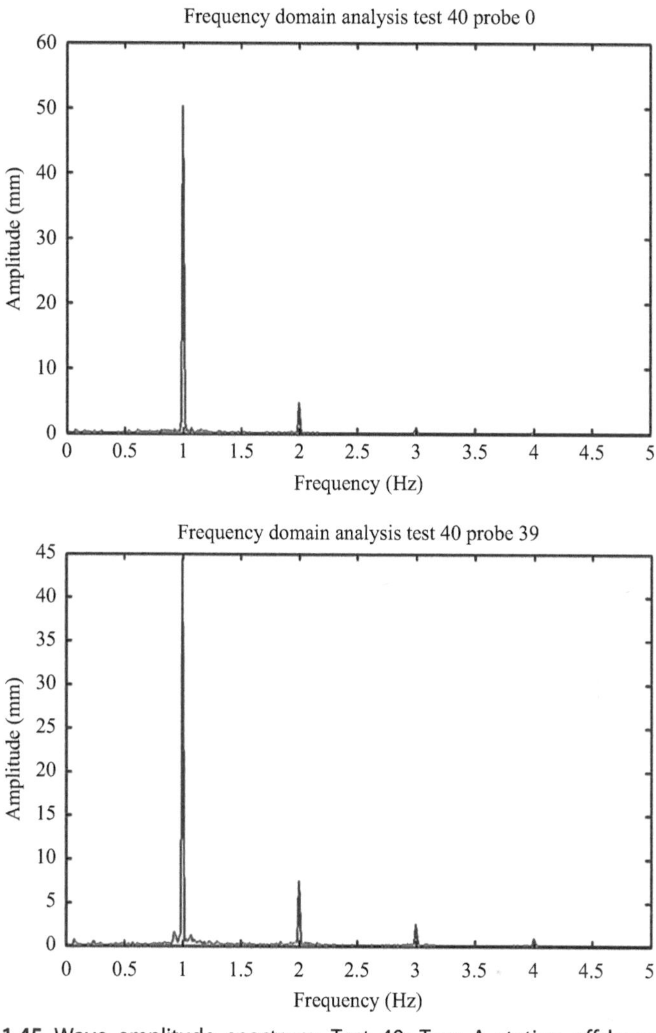

Figure 11.45 Wave amplitude spectrum: Test 40; Top: A station offshore, Bottom: A station onshore.

the submerged breakwater in both small and large scale models. This section therefore provides a report on the capability of two numerical models (ANN vs. SWE) to simulate the wave field around submerged breakwaters.

11.6.1 Small Scale Model

Experimental data used to evaluate the model are listed in previous tables. The tables present the range of physical parameters for each individual experiment.

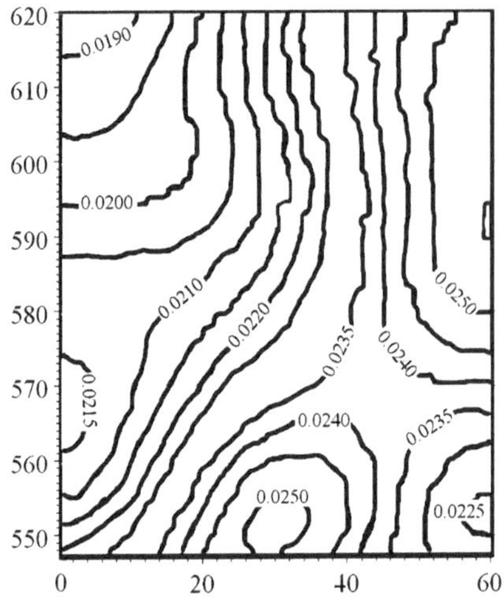

Figure 11.46 Contour map of wave disturbance parameter H_{mo} calculated by SWE model for small scale model.

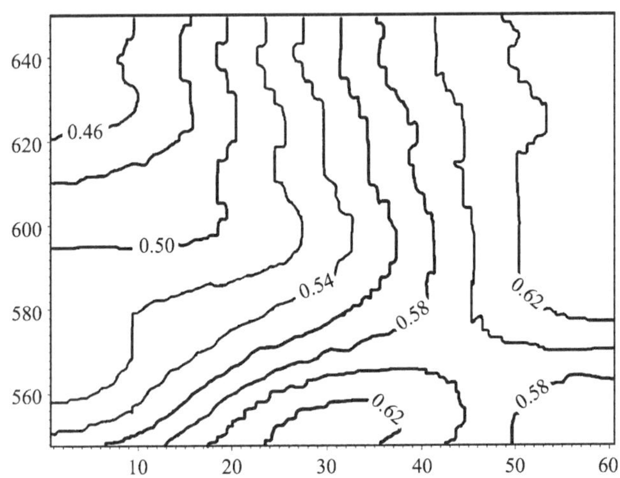

Figure 11.47 Contour map of wave disturbance coefficients K_t calculated by SWE model for small scale model.

Figure 11.54 illustrates the contour maps of the wave transmission coefficient K_t respectively for experimental, ANN3D and SWE models for a specific test condition ($H_i = 27$ mm and $T = 1.00$ s). The submergence depth in this test is 112 mm and the total water depth at the toe of the structure is 300 mm (Test No. 13).

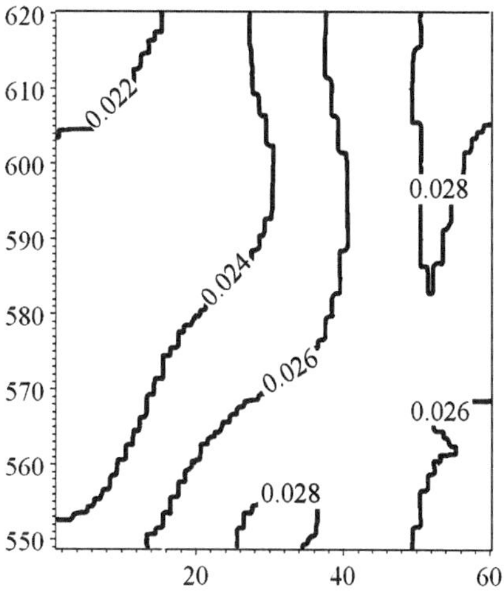

Figure 11.48 Contour map of wave disturbance parameter H_s calculated by SWE model for small scale model.

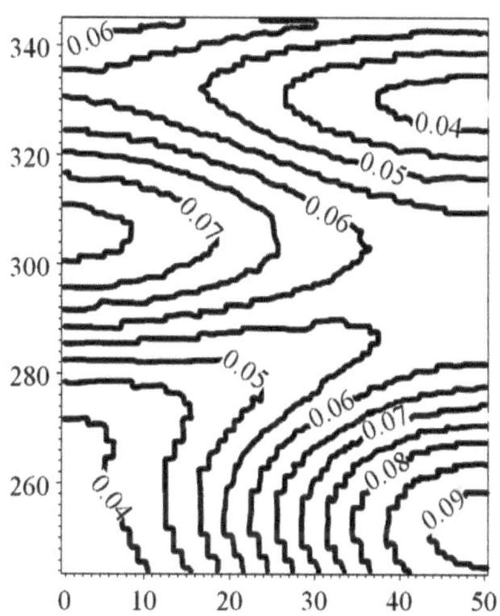

Figure 11.49 2D contour map of wave disturbance parameter H_{mo} calculated by SWE for the large scale model.

Figure 11.55 shows the contour maps of absolute error for the two models. Both models calculate K_t in relatively good agreement with measured values in most areas behind the breakwater. The averaged error for the SWE model is lower than 10% in most areas and up to a maximum of 15% around the head, while the maximum error for the ANN3D model is less than 10%.

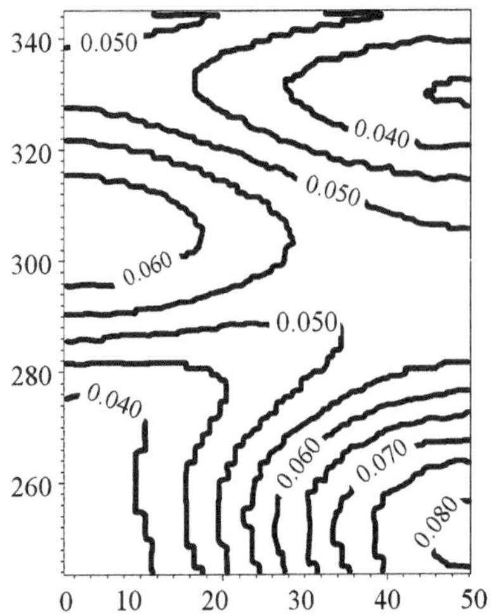

Figure 11.50 2D contour map of wave disturbance parameter H_s calculated by SWE model for large acale model.

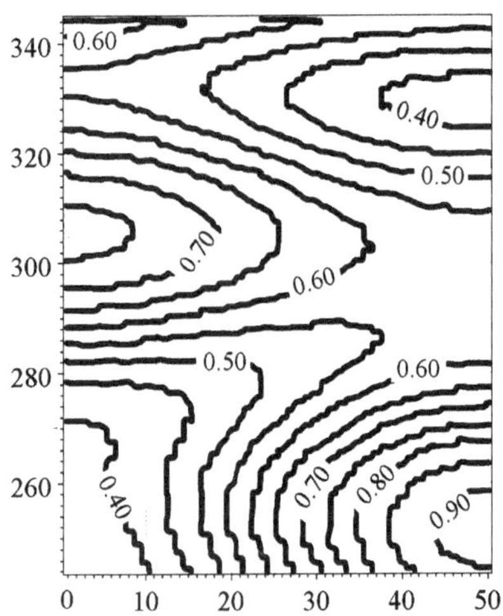

Figure 11.51 2D contour map of wave disturbance coefficient K_t, calculated by SWE for large scale model.

Figure 11.56 shows the correlation between predicted and measured values within different regions around the breakwater for the same tests. Comparison of the models with the experimental data was also investigated by comparing five statistical parameters. The values of the statistical indicators and graphical comparison both indicate that the ANN3D model is able to calculate the wave transmission coefficient with a higher accuracy than SWE.

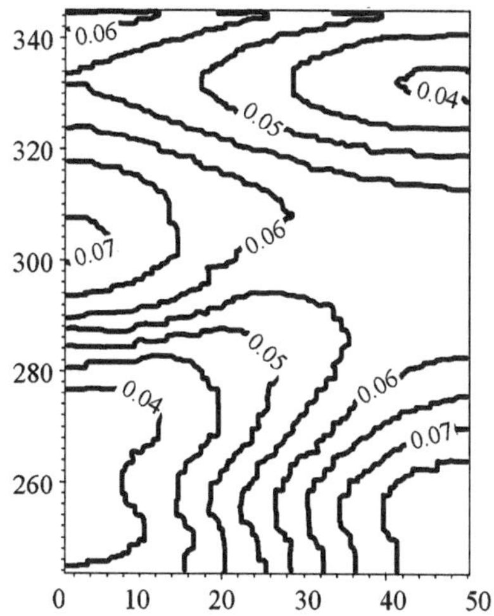

Figure 11.52 2D contour map of wave disturbance parameter H_{max}, calculated by SWE for large scale model.

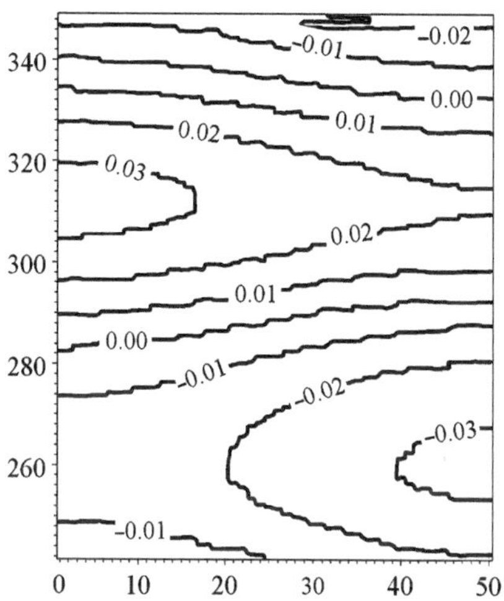

Figure 11.53 2D contour map of disturbance of water surface elevation calculated by SWE for large scale model.

The error distribution is shown in Figure 11.57, using box-and-whisker plots indicating the degree of dispersion (spread) and skewness in K_t, as well as identifying outliers for the two models.

Figure 11.58 shows how accurately the models predict measured K_t using scatter plots that illustrate the correlation between calculated K_t and

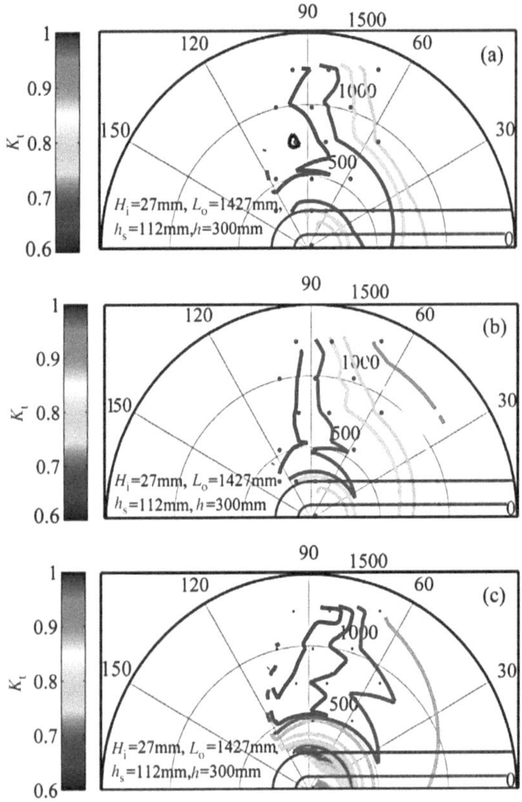

Figure 11.54 Contour maps of wave transmission coefficient: (a) Measured, (b) ANN3D and (c) SWE.

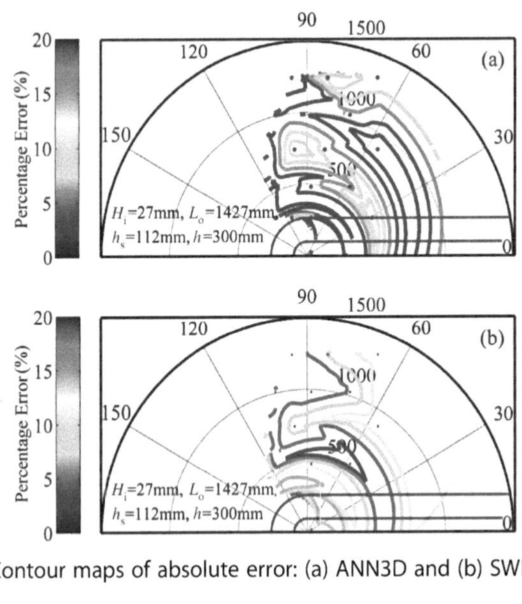

Figure 11.55 Contour maps of absolute error: (a) ANN3D and (b) SWE.

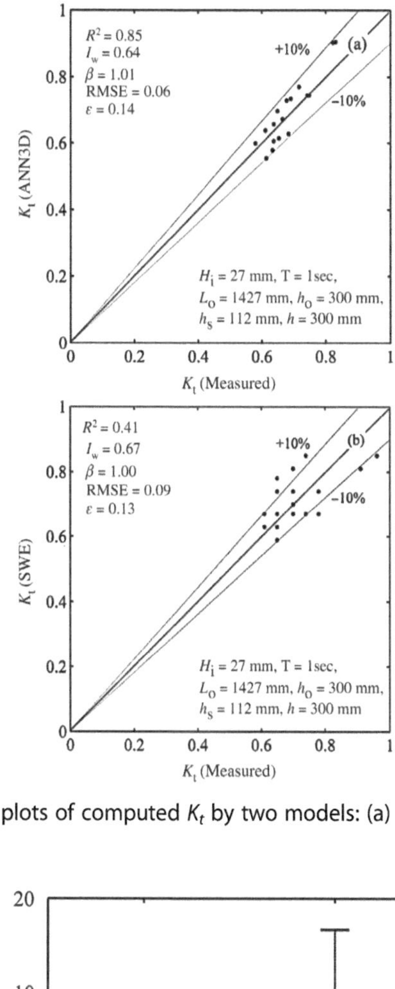

Figure 11.56 Scatter plots of computed K_t by two models: (a) ANN3D and (b) SWE.

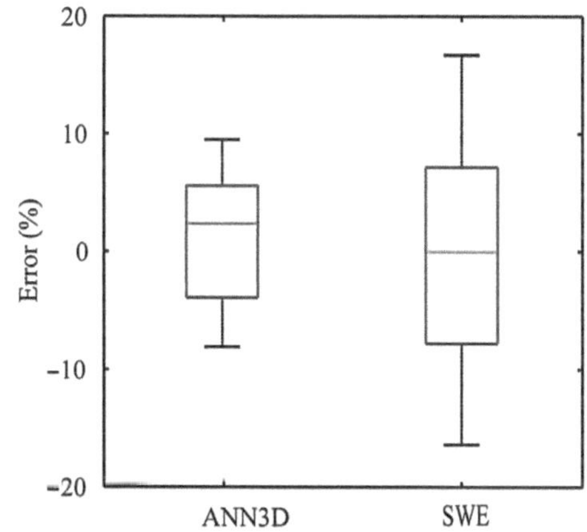

Figure 11.57 Descriptive comparison between predictions from ANN3D and SWE models, by the distribution of percentage error between the estimated and the observed values of K_t using box-and-whisker plots.

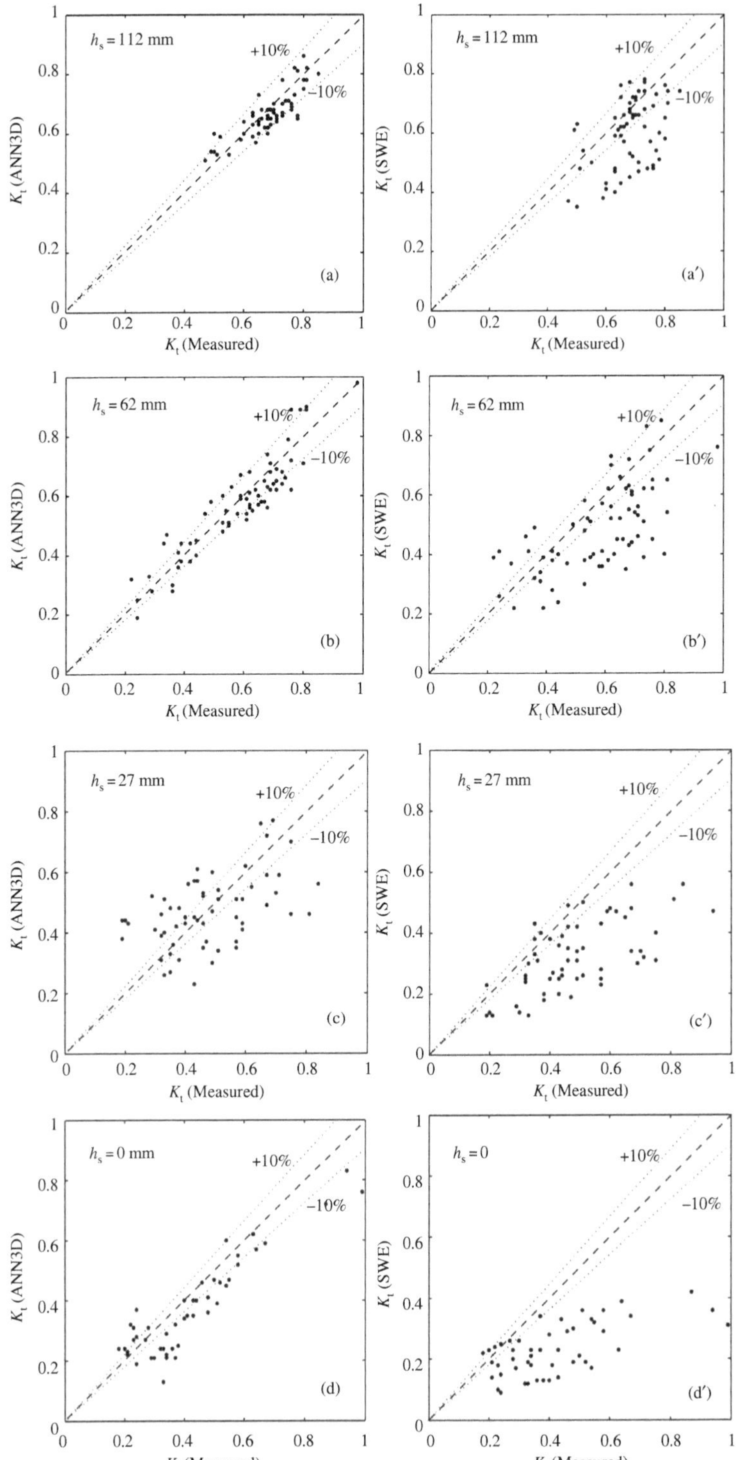

Figure 11.58 Scatter plots of measured and computed K_t, (a, b, c, d): ANN3D and (a′, b′, c′, d′): SWE models in four distinct submergence depths correspond respectively to $h_s = 112$, 62, 27 and 0 mm.

measured data for different values of h_s. It can be deduced from the graphs that the ANN3D model performs more accurately than SWE in all four categories. The ANN3D model results are in very good agreement with the measured data for different conditions (error mostly lower than 10%). Referring to the SWE results, as submergence ratio decreases, a significant increase in error is observed. This is probably due to the overestimated wave energy dissipation by the SWE model, which increases for steeper waves and shallower depths.

The error distribution is shown in Figure 11.59 using box-and-whisker plots indicating the degree of dispersion and skewness in K_t, as well as identifying outliers for ANN3D and SWE models for different values of h_s. Figures 11.59a,b,c and d correspond respectively to $h_s = 112$, 62, 27 and 0 mm. As can be seen in the graphs, decreasing h_s results in a significant increase in error. The ANN model presents more acceptable results for K_t compared to those obtained by SWE, particularly for higher submergence ratio ($h_s > 62$ mm).

Figure 11.60 shows error contour maps for the data plotted in Figure 11.58; this indicates the error in different regions around the

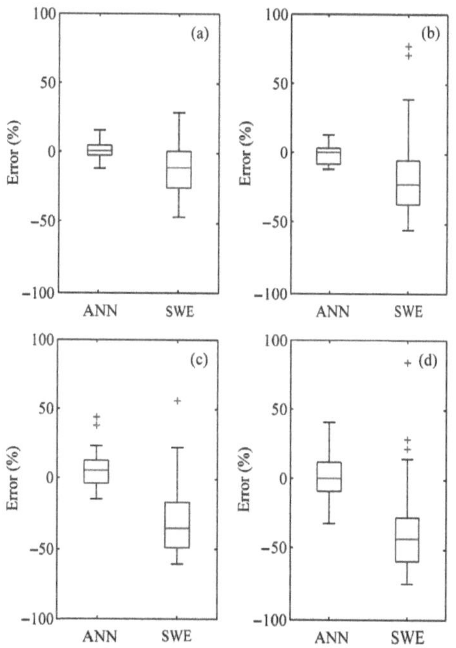

Figure 11.59 Descriptive comparison between predictions from ANN3D and SWE models, by the distribution of percentage error between the estimated and the observed values of K_t using box-and-whisker plots for different values of submergence depth over water depth ratio.

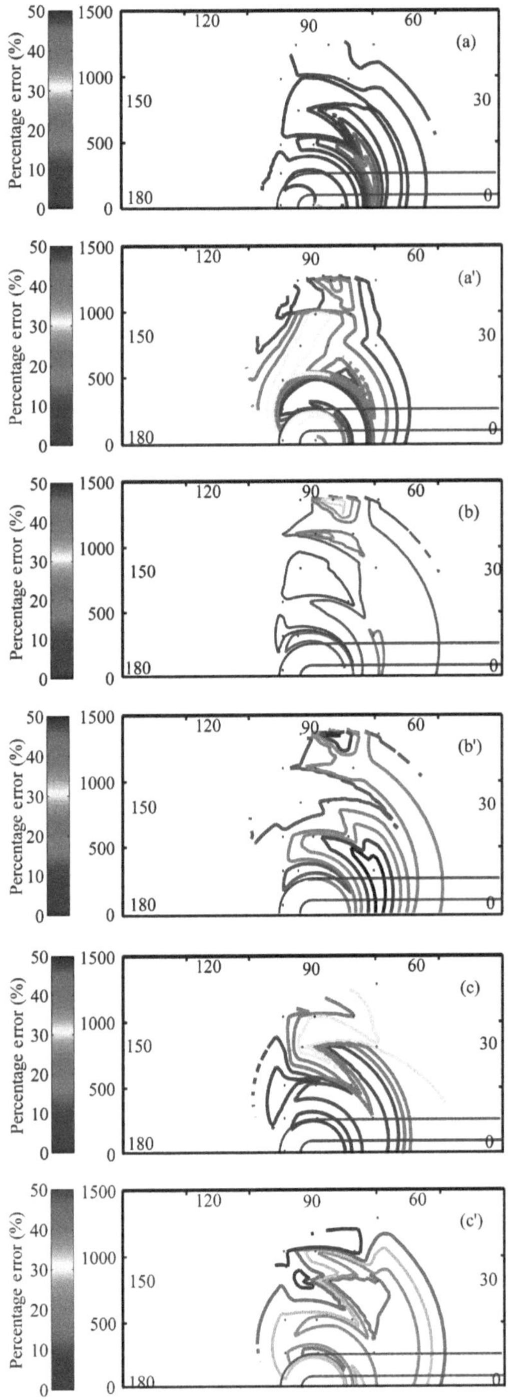

Figure 11.60 Comparison of error contour maps: (a, b, c) ANN3D and (a′, b′, c′) SWE models in three distinct submergence ratios correspond respectively to $h_s = 112$, 62 and 0 mm.

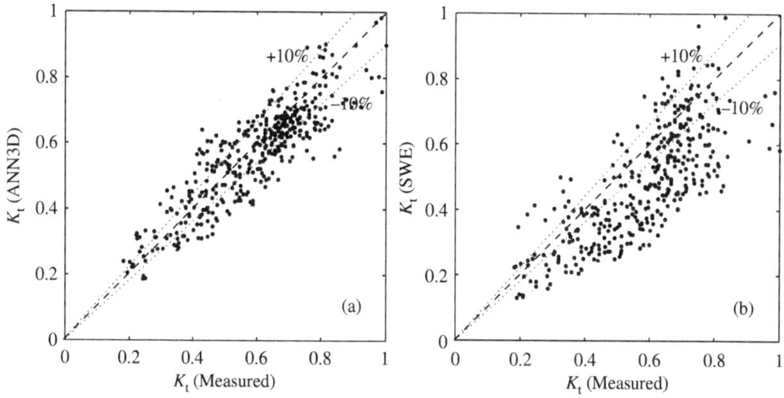

Figure 11.61 Scatter plots for (a) ANN3D and (b) SWE models in whole range of test data.

submerged breakwater for three submergence depths, $h_s = 112$, 62 and 0 mm (respectively, 11.60 a-a′, a-b′ and c-c′). As can be seen, the ANN3D model gives low error (mostly less than 10% up to a maximum of 20%) in most regions behind the breakwater. SWE calculates K_t with an average error of 30%. Higher values of error up to 50% can be seen in some regions. Less accurate results are obtained by SWE when submergence depth decreases.

Finally, Figure 11.61 shows the correlation between K_t predicted by ANN3D and by SWE models and measured values for the whole range of test data using the small scale model. The ANN3D model presents more acceptable results for K_t compared to those from SWE.

11.6.2 Large Scale Model

Experimental data used to evaluate the model are listed in Tables 11.10 and 11.11. The table presents the range of physical parameters for each individual experiment. The submergence depth and total water depth at the toe of the breakwater are 250 mm and 1000 mm, respectively, for these tests.

Figure 11.62 shows contour maps of K_t for the experimental, ANN3D and SWE models respectively from the top to the bottom. The graphs correspond to a specific test ($H_i = 120$ mm and $T = 1.00$ s; Test No.5). As can be seen, the ANN3D model produces K_t with good agreement with experimental data in most regions behind the breakwater.

Figure 11.63 shows the error contour maps for the two models. The ANN3D model produces K_t in relatively good agreement with measured

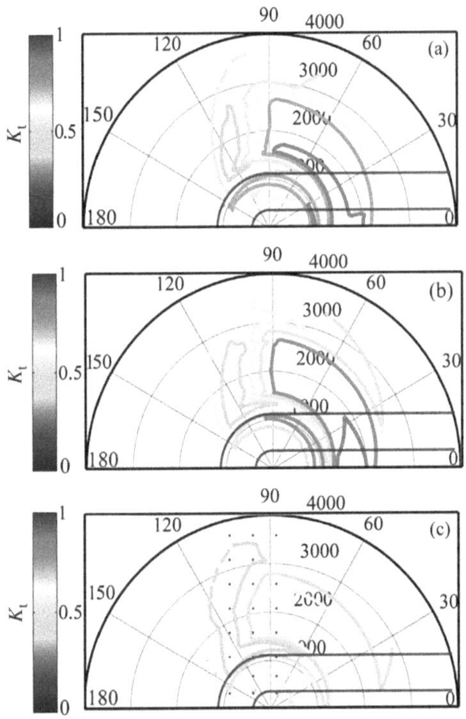

Figure 11.62 Contour maps for K_t; (a) Measured, (b) ANN3D and (c) SWE.

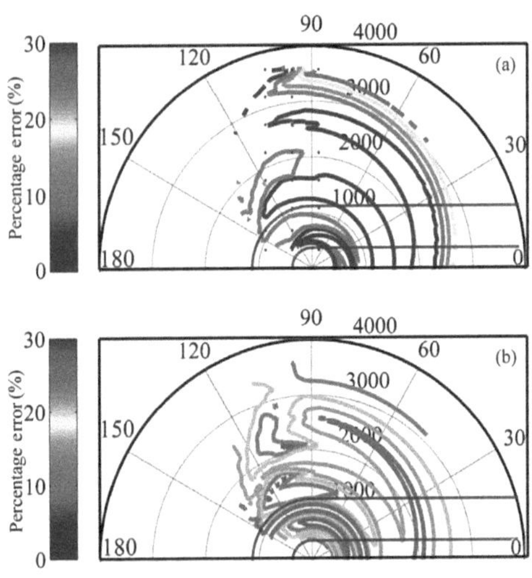

Figure 11.63 Error contour maps: (a) ANN3D and (b) SWE.

values in different areas behind the breakwater (Figure 11.63a) while SWE presents less accurate results in some areas behind the breakwater (Figure 11.63b). Calculated average error by SWE is lower than 20% in most areas but up to a maximum of 30% around the head, while the average calculated error by ANN3D is less than 10% with a maximum of 30% in a very narrow band far behind the breakwater.

Figure 11.64 presents the variation of K_t along the flume in three lines for the two models. A good agreement can be observed for ANN3D predictions with measurements in all lines while SWE performs less well in some cases. In Figure 11.64a (Xgrid = 15; see the bathymetry for details of grid numbers) although SWE produces good results for points far from the breakwater, it presents less accurate results in the gap. In Figure 11.64b, (Xgrid = 25) both models perform very well while in Figure 11.64c (Xgrid = 35) the ANN3D model presents much better results at most points.

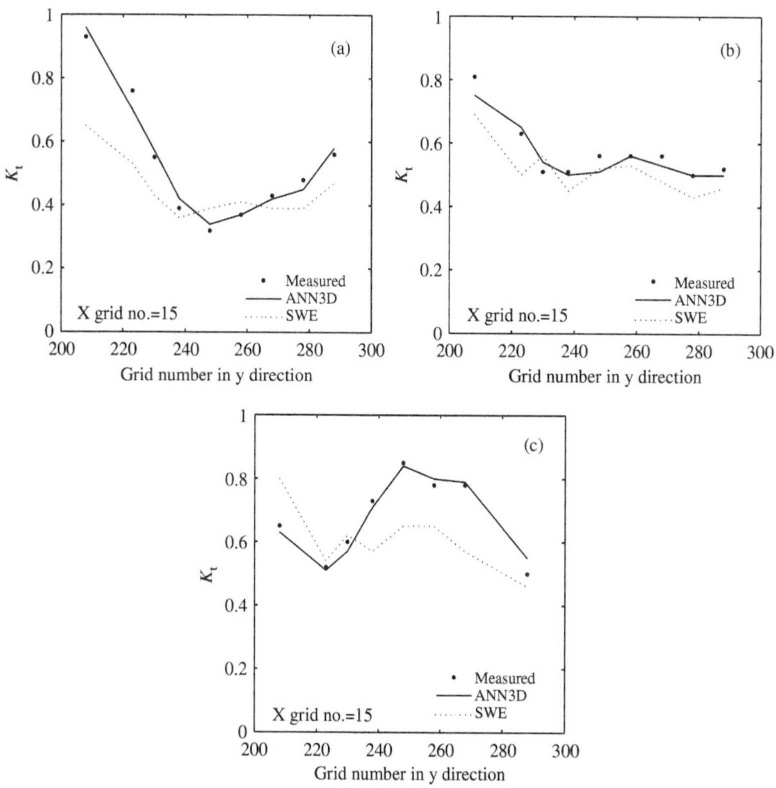

Figure 11.64 2D variation of wave transmission coefficient in three lines along the flume for Measured, ANN3D and SWE results: (a) Xgrid = 15; (b) Xgrid = 25; and (c) Xgrid = 35. See the bathymetry for details of grid numbers.

Figure 11.65 shows the correlation between predicted and measured values for both models for this specific test. Statistical comparison using five indices is also presented in the graphs. The values of the statistical indicators and the graphical comparison both indicate that the ANN3D model is able to calculate wave transmission coefficient with a high accuracy.

The error distribution is shown in Figure 11.66 using box-and-whisker plots. Both graphs confirm the previous results.

Figure 11.67 presents error contour maps for the ANN3D (left column) and SWE (right column) models for six different wave conditions

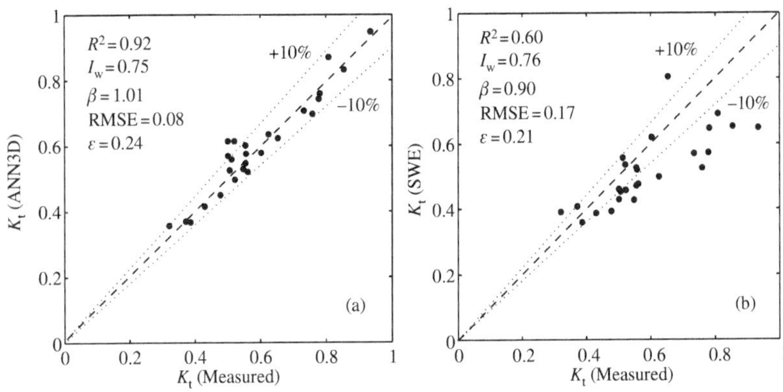

Figure 11.65 Correlation of the predicted K_t (by (a) ANN3D, (b) SWE) and measured values.

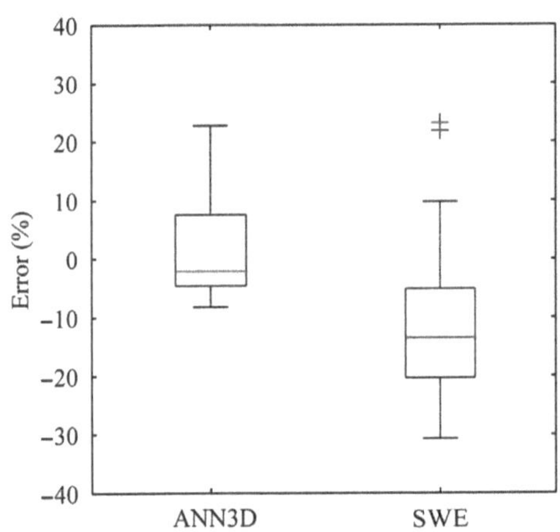

Figure 11.66 Descriptive comparison between predictions from 3D ANN3D and SWE models, by the distribution of percentage error between the estimated and the observed values of K_t using box-and-whisker plots.

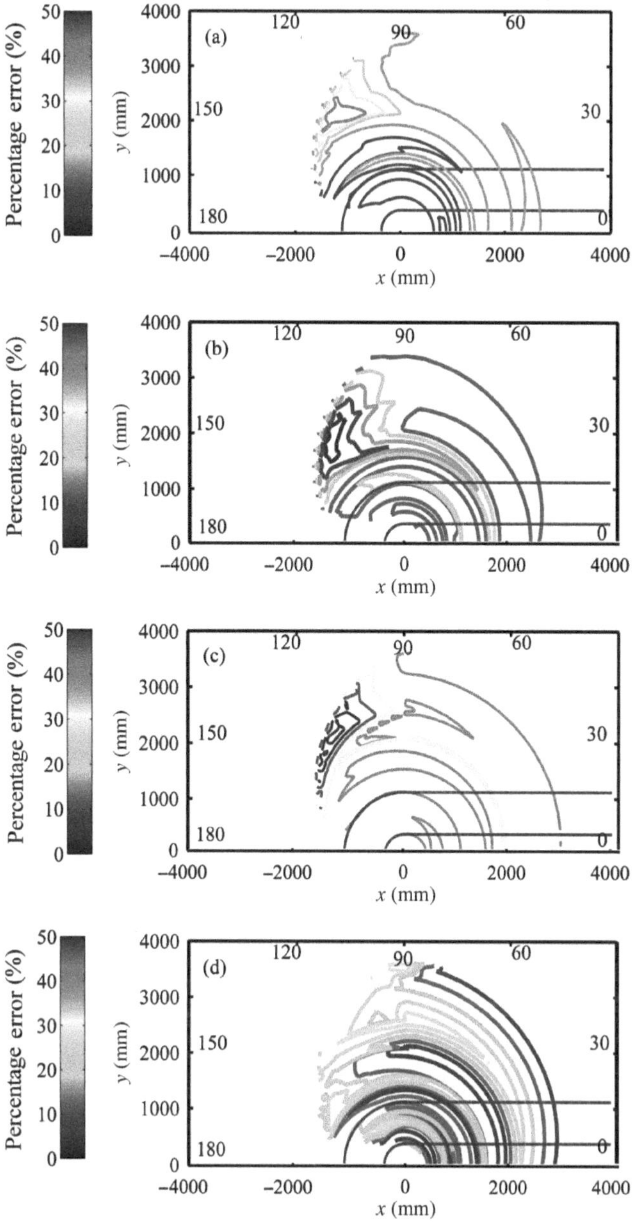

Figure 11.67 Error contour map of ANN3D (a,b,c,d,e,f) and SWE (a′,b′,c′,d′,e′,f′) models for large scale tests (dataset: Set 4; test with: $H_i = 40,60,80,100,120,140$ mm, $T = 1$ s, $h_s = 250$ mm, $h = 1000$ mm)

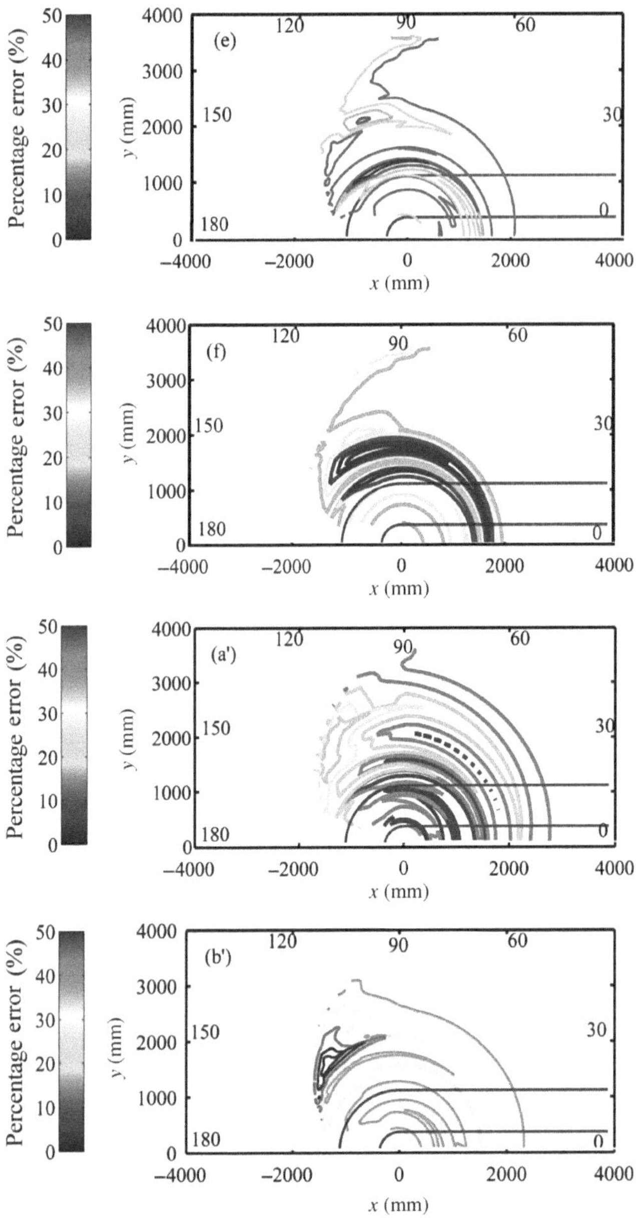

Figure 11.67 (Continued)

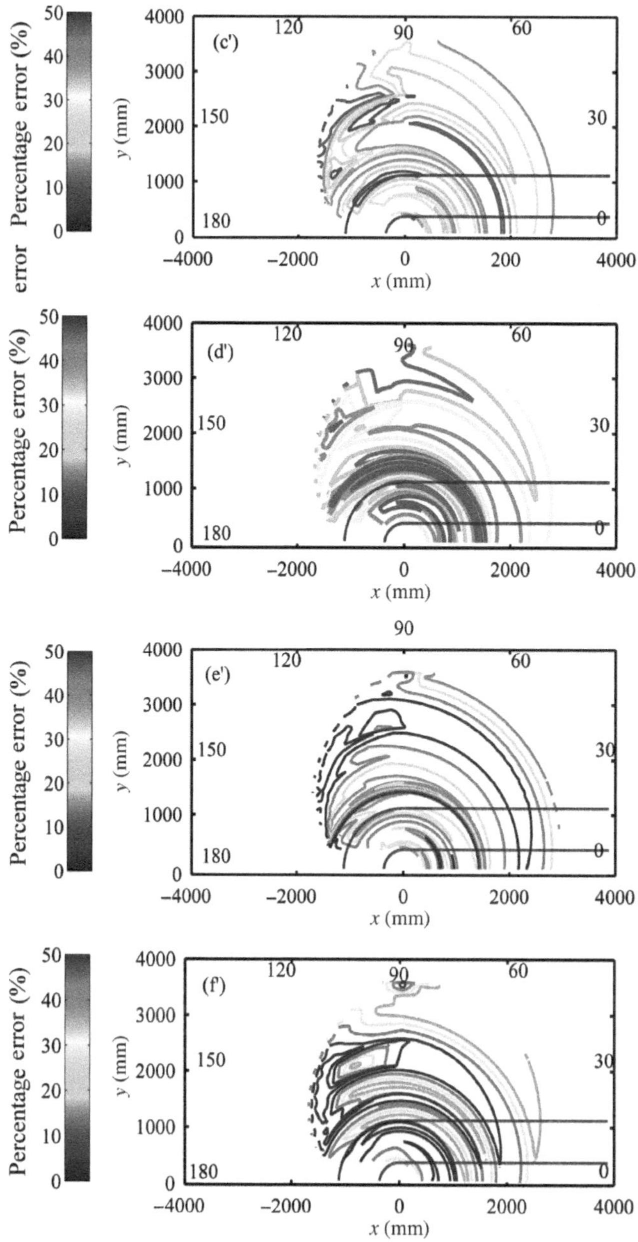

Figure 11.67 (Continued)

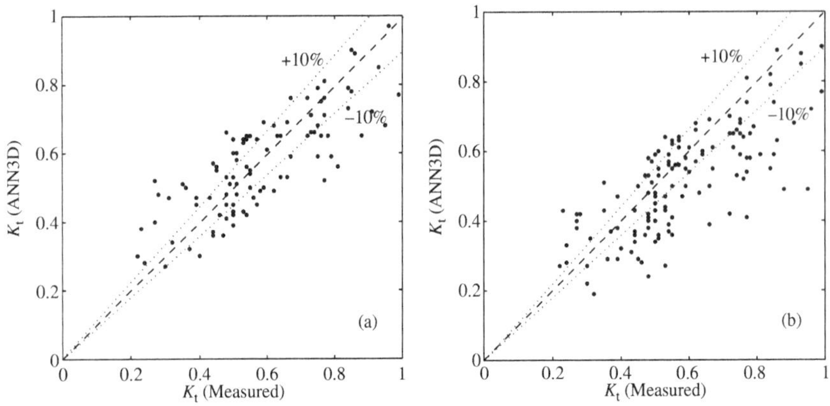

Figure 11.68 Scatter plots of measured computed K_t: (a) ANN3D model and (b) SWE.

(from top graph to the bottom for incident wave heights from 40 mm to 140 mm, respectively). The graphs confirm the more accurate performance of ANN3D compared to SWE in all six categories. Except for one case of $H_i = 40$ mm (Figure 11.67f), the ANN3D model produces results with very good agreement with measured data (error mostly lower than 20%) while SWE produces less accurate results particularly for wave heights higher than 80 mm with an average error of 30% up to a maximum of 50%. For the SWE results, as incident wave height increases from 40 mm to 140 mm, the small area with critical error becomes bigger. The critical error area in the case of the ANN3D model is observed in different locations when the incident wave height changes (around the head for higher wave heights and farther behind the gap for lower wave heights).

Figure 11.68 presents scatter plots for the two models for the whole data set using the large scale breakwater. The ANN3D model presents better correlated results for K_t compared to those from SWE.

11.7 A Simplified 3D Analysis Tool And Preliminary Prediction Scheme For Practical Applications

The effects of submerged breakwaters considering their submergence, distance of structure to shoreline, breakwater crest height and width, the length and number of breakwater segments and their gaps, etc., on wave and current hydrodynamics and morphological seabed and shoreline changes are still developing. Developing and improving the practical knowledge of low-crested/submerged breakwaters and their influence on nearshore wave and current field lead to more accurate, reliable and applicable prediction of sediment transport, morphological changes and beach response in the vicinity of the structure. The functional and practical design of low-crested/submerged breakwaters still needs to be improved, specially with respect to simulation of the wave field around the structure (Sharif Ahmadian and Simons, 2014). An effective design tool needs to calculate both two-dimensional and three-dimensional effects appropriately and accurately to be able to provide reliable results and to be applicable in real practical problems. A three-dimensional numerical model for predicting the spatial wave transmission coefficient around a submerged breakwater is proposed in this example. Two distinct models have been developed based on multilayer perceptron (MLP) and radial basis function (RBF) models. These methods have been applied in this example to design and train ANN models using laboratory experimental data. The collected laboratory data are expressed in terms of both dimensional and nondimensional parameters. Comparisons between the measured data and predictions from the ANN models show that the nondimensional RBF model is best able to predict the wave field around the three-dimensional submerged breakwater. The performance of the model was validated in interpolation, extrapolation and at larger scale. The analyses reveal sufficient agreement with the experimental results. This shows that the model has potential to be applied as a design tool in real applications. Here, first, we intend to have a short review and overview of the topic, justification for the example, and a brief literature review before modeling.

Wave breaking causes several physical processes in the surf zone, including wave-generated turbulence, surge, currents, and sediment transport, which consequently results in shoreline erosion and accretion. Increase of the population in coastal areas as a result of urban development and recreational activities makes this a serious concern. Submerged breakwaters have been proposed and constructed in many coasts around the world to protect the shoreline either by stopping waves reaching the shoreline or at least forcing them to break and to some extent dissipate their energy before reaching the shoreline. Submerged breakwaters are currently being used in coastal areas to protect shorelines from erosion. These structures are widely used due to lower construction costs (compare to the traditional emerged breakwaters) and the significant attenuation of incident wave energy in the sheltered area leeside of the breakwater in an environmentally friendly way (Sharif Ahmadian and Simons, 2014). As mentioned in previous chapters, submerged breakwaters also help to retain a clear view of the sea, preserve alongshore transport and maintain high water quality leeside. Submerged breakwaters should be designed in a proper way to have sufficient impact on decrement of erosion at shoreline, trapping the natural sediments entering that coastal area and restoration of the beach. The shoreline response to the submerged breakwaters is significantly influenced by the wave and current field behind the structure driven by both two-dimensional and three-dimensional coastal processes (Sharif Ahmadian and Simons, 2014). Some examples are permeability through the body and wave overtopping over the crest for two- and three-dimensional cases and diffraction through the heads for three-dimensional cases only.

Two-dimensional wave transmission has been widely studied. The wave transmission coefficient K_t, defined as the ratio of transmitted to incident wave height, is used by engineers and researchers to study the waves transmitted behind the submerged breakwaters. A wide range of two-dimensional design tools has been proposed to calculate K_t behind two-dimensional submerged breakwaters: Buccino and Calabrese, 2007; Calabrese et al., 2002; d'Angremond et al., 1996; Goda and Ahrens, 2008; Panizzo and Briganti, 2007; Seabrook and Hall, 1998; Tajziehchi and Sharif Ahmadian, 2009; van der Meer et al., 2005). However, the only information provided by these 2D models is the average value of K_t rather than the spatial distribution of wave height. This is very important as, obviously, to have a reliable and accurate engineering design tool for the submerged breakwaters we need to have a model with ability to predict the spatial distribution of wave height around the structure. Therefore, prediction of wave diffraction around the head of the structure combined with the wave dissipation by overtopping is needed. This information is required to have a reliable estimation of flow circulation and associated sediment transport (Sharif Ahmadian and Simons, 2012). However, very few studies on the 3D effects (Bellotti, 2004; Caceres et al., 2008; Hur et al., 2012; Johnson et al., 2005) and even fewer on combined diffraction and overtopping effects of similar structures have been conducted in the past (Buccino et al., 2009; Losada et al., 1996; Sharif Ahmadian and Simons, 2012, 2014; Vicinanza et al., 2009). The majority of these studies are restricted to low-crested structures only with a small number considering fully submerged breakwaters. This example intends to model the wave field around submerged breakwaters with particular emphasis on three-dimensional effects.

The spatial distribution of the nearshore wave field around submerged breakwaters is investigated numerically using data-driven algorithms. Such models have proven to be useful in many fields of engineering to solve complicated nonlinear problems where many relationships are involved. ANN models have actually been applied successfully to some coastal engineering problems (Mase et al., 1995; Medina, 1999; van Gent et al., 2007; van Gent and van den Boogard, 1998; Panizzo and Briganti, 2007; Tajziehchi and Sharif Ahmadian, 2009; Sharif Ahmadian and Simons, 2012, 2014).

In the present example, two different ANN models have been designed and trained using new experimental data and the predicted transmission coefficients compared against each other and against actual data measured in facilities of different size.

A brief description of artificial neural network models and the basic concepts, including an explanation of the specific neural network models used in this example, are presented first. A description of the experiments, the methods of data selection, and approaches for data preprocessing to prepare the training data are set out. The different stages of artificial neural network numerical modeling are presented. The capability of the proposed numerical models to simulate wave height transformation over submerged breakwaters is investigated. ANN models are tested in the numerical simulations and the results compared to the laboratory experimental data. A discussion with main conclusions is presented.

Depending on the type of the problem, a proper learning process needs to be used. Here in this example, similar to the majority of engineering problems, we are dealing with a function estimation problem. In applications in the field of coastal engineering, the most popular artificial neural networks are the multilayer perceptron (MLP) and radial basis function (RBF) networks, with the possibility of performing nonlinear approximation. A brief introduction to models with their basic framework employed in this example is set out below.

Based on a supervised learning method for a given training data set including a large number of input-output samples, the algorithm computes weights and biases of the network in order to minimize the mean-square error. To achieve this goal, a back-propagation (BP) algorithm (Rumelhart et al., 1986; Werbos, 1988) is usually employed to train MLP networks. As we deal with a nonlinear regression problem in this example, a batch mode is implemented to adjust the network parameters in BP learning on an epoch by epoch basis, where each epoch consists of the entire set of training input-output pairs. A RBF neural network (Broomhead and Lowe, 1988) also is employed in this example. The entire data set is divided into two main subsets of training and testing. A Gaussian radial basis function is applied in this problem.

In the following section, we describe the preparation process of experimental data used in this example to design and test the models. In the present example, a large number of three-dimensional data with wide ranges of wave height, period, water depth, submergence depth and distance from the beach is used. Laboratory data collected in two wave flumes with different scales of models is used. Impermeable breakwaters are modeled. This can assist us to evaluate the three-dimensional effects appropriately. Therefore, filtration through the structure is not considered here. The data sets used in this example are: Subset No. 1 collected from a 20 m long, 1.2 m wide and 1 m deep small flume, and Subset No. 2 collected from a 20 m long, 2.5 m wide and 1.2 m deep large tank.

Figure 11.69 shows a schematic of the breakwater and defines the breakwater geometry and regions where laboratory data were collected.

Now we need to determine the variables for the model. As stated before, the model should explain the problem in an easy and simple approach using the fewest variables (Stein, 1993). Among a number of physical parameters, the parameters incident wave height H_i, offshore wave length L_o, water depth over the breakwater crest h_s, water depth in the toe of breakwater h, breakwater crest width B and spatial parameters r and θ are selected. This final number of input parameters is selected based on a preliminary architecture of the neural network on the basis of sensitivity to each parameter. Analysis of the dimensional and nondimensional parameters shows the most important physical parameters (as listed in Table 11.15 and Table 11.16 for dimensional and nondimensional parameters, respectively). Tables 11.15 and 11.16 present the range of data used in the neural network models. The tables include minimum, maximum, mean and standard deviation of each individual parameter. Thus, K_t is determined as:

$$K_t = f\left\{\frac{H_i}{L_o}\frac{h_s}{H_i}, \frac{H_i}{h}, \frac{h_s}{h}, \theta, \frac{r}{L_o}\right\} \tag{11.11}$$

Dimensionless parameters should be used. A normalization process is also essential. The number of layers in an MLP model can be variable.

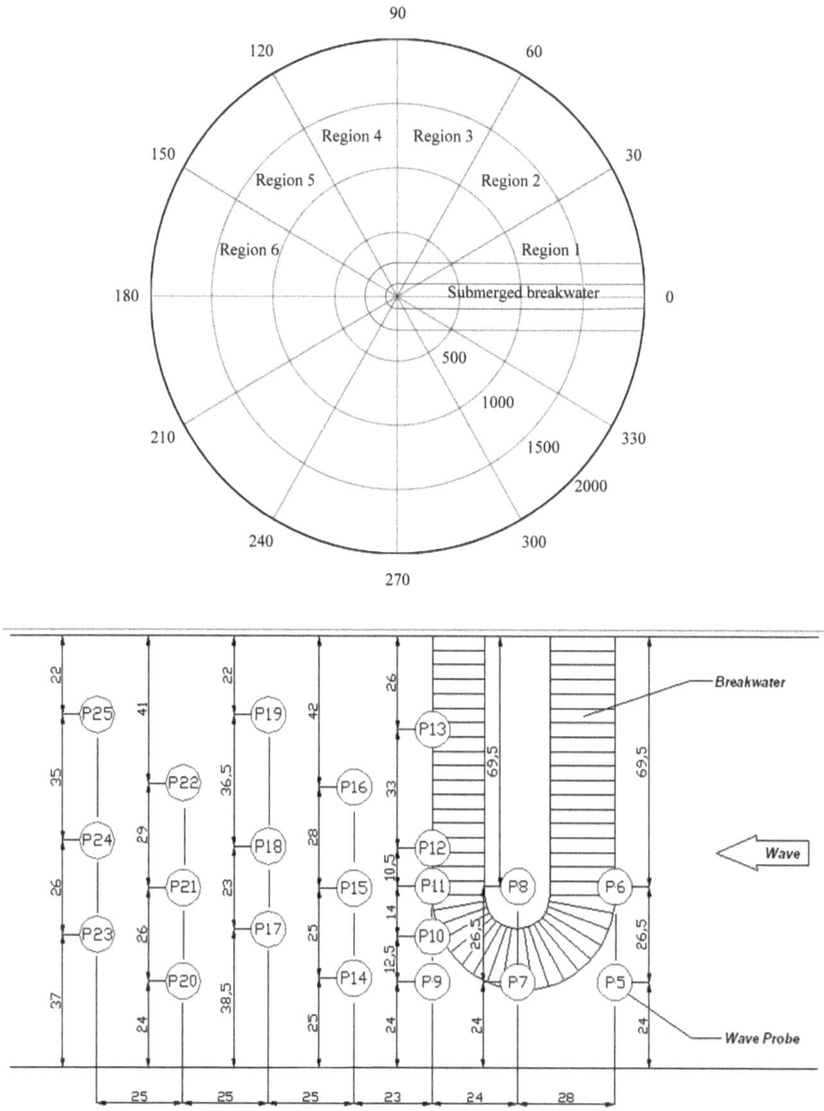

Figure 11.69 Plan of the submerged breakwater and definition of regions around the breakwater and measurement points.

Nevertheless, RBF networks always have three layers. The transfer functions applied in the hidden layer in these two models are different. Both models include biases in their hidden and output layers. However, RBF models usually have more computational units in their hidden layer than

Table 11.15 The range of training and testing data (dimensional datasets: Set 1 and Set 2)

Parameter	Mean	Standard deviation	Min	Max
H_i (mm)	47	21	12	81
L_o (mm)	1812	348	1427	2126
h_s (mm)	42	16	27	62
h (mm)	235	14	222	250
θ (rad)	1.60	0.59	0	3.14
r (mm)	816	363	10	1341

Table 11.16 The range of training and testing data (nondimensional datasets: Set 1 and Set 2)

Parameter	Mean	Standard deviation	Min	Max
H_i/L_o	0.03	0.01	0.01	0.06
h_s/H_i	1.20	0.96	0.33	5.03
H_i/h	0.17	0.06	0.12	0.24
h_s/h	0.20	0.09	0.05	0.37
θ	1.60	0.59	0	3.14
r/L_o	0.47	0.23	0	0.94
K_t	0.55	0.15	0.19	0.89

MLP models. Hence, RBF models usually require less training time compared with MLP models. In our model in this example, the computational units in the different layers were arranged to learn at the same rate by assigning a learning rate parameter. The optimum network can be obtained based on a trial and error scheme, as a predefined rule does not exist to select the number of hidden layers. A cross-validation method (Hagan et al., 1996) is employed in this study to deal with the problem of over-fitting during the training process and to improve the generalization for future applications of the model. The training data used for training were set up to be completely random to improve the learning procedure. The data set described in Table 11.16 is used to train the networks and to investigate the model performance.

The modeling process used for the networks is described here. Higher number of hidden layers might increase the generalization performance of MLP. However, it may increase the computation time. Therefore, in this example, it is assumed that one hidden layer is sufficient. For a more detailed and formal description of MLP neural networks, the reader is referred to Hagan et al. (1996) or Haykin (1999). Once the number of hidden layers in the network is fixed, the size of the hidden layer, the learning rate, the activation function and the minimum gradient should be set. An antisymmetric hyperbolic tangent transfer function is used in the MLP.

We set the training of the RBF model to be terminated once the calculated error reached the goal of 1% or the maximum number of training iterations (500) had been completed. In the case of the MLP network, training is terminated once the number of iterations reached 3000 or when the maximum output error is reduced to below 1% of the target value. The weights that corresponded to the minimum error during the training stage are taken as the network parameters. In the training stage for the MLP model, hidden layers with 10 nodes gave the least errors; hence 10 hidden units were chosen to represent the MLP model's hidden layer. In the case of the RBF model, a 15-noded hidden layer network was chosen.

The predictions of transmission coefficients obtained by MLP and RBF models are plotted against the corresponding observed data in Figure 11.70. The plot shows that they are comparable, with good agreement and some limited exceptions. The RBF model performs particularly

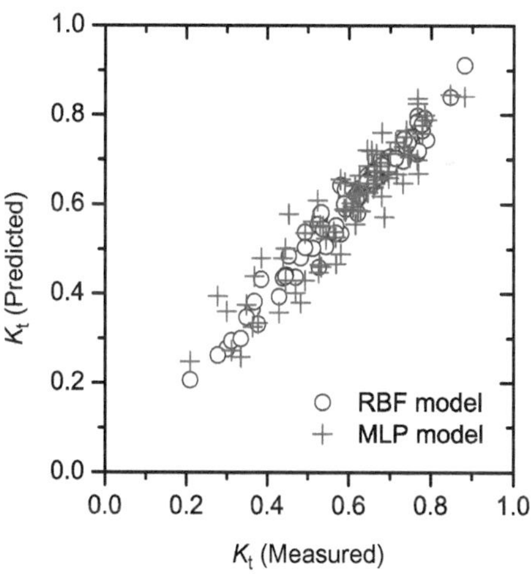

Figure 11.70 Graphical comparison between predictions from MLP and RBF models by the scatter plot of the estimated and the observed values of K_t.

Table 11.17 Statistical comparison between predictions from RBF and MLP models by two indexes of multiple correlation coefficient and root mean squared error

Statistical index	MLP	RBF
R^2	0.86	0.97
RMSE	0.11	0.05

well. The comparison of the two models with the experimental data was investigated by comparing the two statistical parameters, root mean squared error and the squared multiple correlation coefficient R^2 (Table 11.17):

$$RMSE = \left[\frac{\sum_{i=1}^{N} \left(\frac{X_{c_i} - X_{m_i}}{X_{m_i}} \right)^2}{N} \right]^{1/2} \tag{11.12}$$

$$R^2 = \frac{\left(\sum_{i=1}^{N} (X_{m_i} - \overline{X}_m)(X_{c_i} - \overline{X}_c) \right)}{\sum_{i=1}^{N} (X_{m_i} - \overline{X}_m)^2 \sum_{i=1}^{N} (X_{c_i} - \overline{X}_c)^2} \tag{11.13}$$

Both graphical and statistical comparisons indicate a better performance from the RBF model than for the MLP model. Much higher squared multiple correlation coefficient and lower root mean squared error confirmed the better performance of the RBF model.

The error distribution is also descriptively depicted in Figure 11.71, using box-and-whisker plots indicating the degree of dispersion and skewness in the data, as well as identifying outliers. The plot reports the distribution of the error between the estimated and the observed values of K_t. Error box-and-whisker plots in Figure 11.71 illustrate the better performance of the RBF model considering each error data category.

Similar graphical and descriptive comparisons are made between the nondimensional and dimensional models (Figures 11.72 and 11.73) as

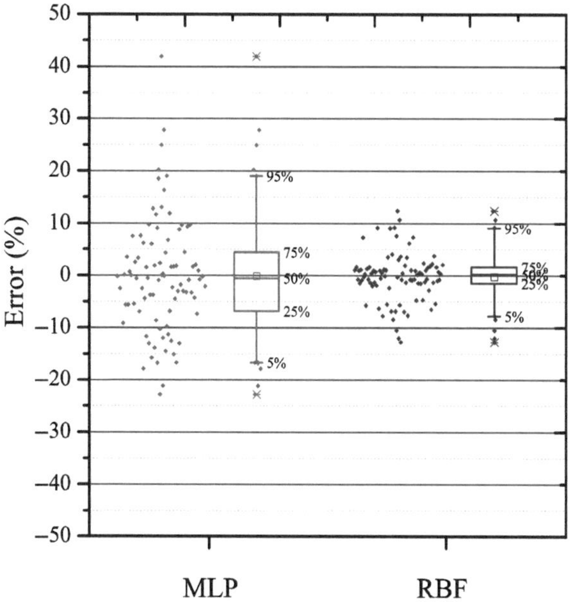

Figure 11.71 Descriptive comparison between predictions from MLP and RBF models, by the distribution of error between the estimated and the observed values of K_t using box-and-whisker plots.

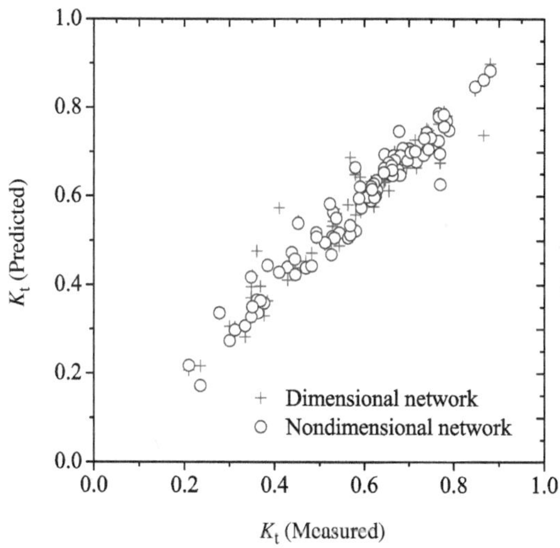

Figure 11.72 RBF model results: dimensional and nondimensional networks.

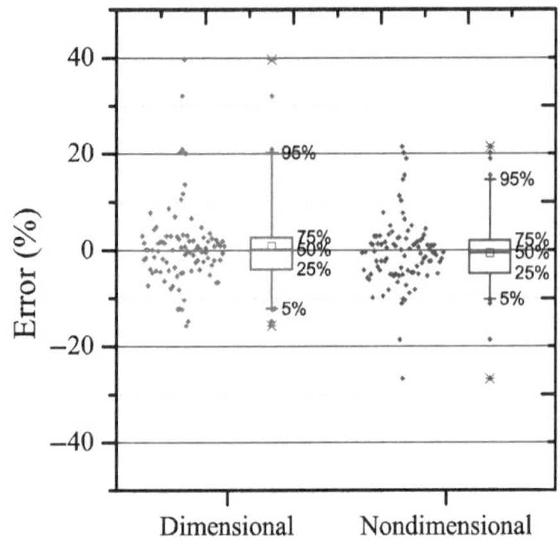

Figure 11.73 RBF model results: dimensional and nondimensional networks.

Table 11.18 Statistical comparison of nondimensional and dimensional RBF models

Statistical index	Dimensional model	Nondimensional model
R^2	0.92	0.95
RMSE	0.08	0.07

well as statistically in Table 11.18. Both models perform well; however, the nondimensional model performs slightly better. This can also be observed from the statistical indexes. Based on all comparisons made in this section, a nondimensional radial basis functional neural network is selected as the best model to simulate wave transmission coefficients around submerged breakwaters. This is validated against experimental data for three different applications and its performance assessed in the following section.

A simplified analysis tool and initial prediction scheme is also provided. The aim of this section is to make the ANN model described throughout this example an available predicting tool, which can be easily implemented and used in the submerged breakwater analysis and design. At this purpose, a K_t calculator based on the ANN model described in this section. The application of the model is recommended for initial analysis and design of the submerged breakwaters. A simplified and step by step procedure is presented here.

Referring to notation and symbols used throughout the example, we will consider a non-dimensional parameters input vector X as:

$$X = [H_i/L_o, h_s/H_i, H_i/h, h_s/h, \theta, r/L_o] \tag{11.14}$$

and which is re-scaled in the range $[-1; 1]$, such as:

$$X_n = (X - X_{min})/(X_{max} - X_{min}) \tag{11.15}$$

Referring to experimental values of Subset 1, vectors X_{min}, X_{max} are characterized by values:

$$X_{min} = [0.0046, 0, 0.0333, 0, -3.1031, 0.0066] \tag{11.16}$$

$$X_{max} = [0.0794, 11.2500, 0.6040, 0.3750, 3.1031, 1.8957] \tag{11.17}$$

The outcomes of the input layer are calculated based on the Euclidean distance between the network input vector and RBF unit's centre vector c. The values of hidden neurons are then calculated by using transfer functions of radial basis functions type, such as:

$$a_1(i,1) = exp(-(dist(c(i,:), a_0)\cdot b)^2); \quad for\ i: 1\ldots 15 \tag{11.18}$$

where

$$dist(p', p) = (\sum (p' - p)^2)^{0.5} \tag{11.19}$$

and

$$a_0(7,1) = 1 \tag{11.20}$$

and

$$b = sqrt(-log(.5))/sp; \tag{11.21}$$

where $sp = 1.2$;

With C being the centre matrix used to pass information from the input layer to the hidden layer, defined as follow. It has to be noted that the seventh column of the matrix is the vector of bias neurons, which are used to adjust the activation value in Eq. (11.18).

$$C = \begin{Bmatrix} 0.0250 & 0.4545 & 0.0701 & 1.0000 & 0.1236 & 0.0508 & 0.6938 \\ 0.1341 & 0.3726 & 0.0419 & 0.6400 & 0.6943 & 0.7569 & 0.6938 \\ 0.2270 & 0.2530 & 0.0894 & 0.6400 & 1.0000 & 0.1376 & 0.6938 \\ 0.0646 & 0.3125 & 0.1285 & 1.0000 & 0.2598 & 0.0358 & 0.6938 \\ 0.1341 & 0.3726 & 0.0419 & 0.6400 & 0.7531 & 0.5659 & 0.6938 \\ 0.2270 & 0.2530 & 0.0894 & 0.6400 & 0.7678 & 0.7099 & 0.6938 \\ 0.2526 & 0.2000 & 0.2337 & 1.0000 & 0.1236 & 0.0828 & 0.6938 \\ 0.3182 & 0.1922 & 0.1361 & 0.6400 & 0.8766 & 0.1998 & 0.6938 \\ 0.7006 & 0.0295 & 0.5835 & 0.3243 & 0.0000 & 0.0930 & 0.6938 \\ 0.5830 & 0.0510 & 0.3130 & 0.3243 & 0.7054 & 1.0000 & 0.6938 \\ 0.4193 & 0.1039 & 0.3014 & 0.6400 & 0.5855 & 0.1944 & 0.6938 \\ 0.0952 & 0.4000 & 0.0876 & 1.0000 & 0.1236 & 0.0828 & 0.6938 \\ 0.3840 & 0.0000 & 0.9945 & 0.0000 & 0.1142 & 0.0541 & 0.6938 \\ 0.3182 & 0.1922 & 0.1361 & 0.6400 & 0.6943 & 0.7569 & 0.6938 \\ 0.2296 & 0.1151 & 0.2664 & 0.6400 & 1.0000 & 0.0613 & 0.6938 \end{Bmatrix} \tag{11.22}$$

The hidden layer is a linear combination of basis functions. As a linear transfer function between the hidden and the output layers has been used, the scaled ANN output K_t can be calculated as:

$$a_2 = w.a_1 + b_2 \tag{11.23}$$

where $b_2 = 0.2742$

With w being the weight vector between the hidden and the output layers, with the bias neuron at the fifteen row of the vector, defined as:

$$w = \begin{Bmatrix} 50.6643 \\ 8.0307 \\ -32.1217 \\ -8.9294 \\ 10.8277 \\ -41.5589 \\ 24.1824 \\ 50.8137 \\ 3.7228 \\ -2.6841 \\ -33.3489 \\ -62.9029 \\ 0.2605 \\ 26.8086 \\ 8.5400 \end{Bmatrix} \tag{11.24}$$

As a last step, we have to re-scale ANN outputs to obtain the estimate of the transmission coefficient K_t:

$$K_t = (Y_{max} - Y_{min}). * a_2 + Y_{min} \tag{11.25}$$

To evaluate the performance of the trained ANN model, it is applied over different conditions. This section considers these analyses. Interpolation predictions are done within the range of input parameters used for training the model as defined in Table 11.17. Detailed analysis of extrapolation predictions is also conducted for input data outside the range of the training data set. A specific application to simulate phenomena at a far larger scale is considered. Finally, a sensitivity analysis is conducted to study the effect of each input parameter on model performance.

Comparison of the proposed model with the experimental data is investigated by comparing the five statistical parameters including the root mean squared error (RMSE) and the squared multiple correlation coefficient R^2 as defined in the previous section, and the three additional parameters, bias or distortion β, Wilmott index I_w (Wilmott, 1981) and error function ε (Haller et al., 2002):

$$I_w = 1 - \frac{\sum_{i=1}^{N} \left(X_{c_i} - \overline{X}_{m_i}\right)^2}{\sum_{i=1}^{N} \left(\left|X_{c_i} - \overline{X}_m\right| + \left|X_{m_i} - \overline{X}_m\right|\right)^2} \tag{11.26}$$

$$\beta = \frac{\sum_{i=1}^{N} \left(\frac{X_{c_i}}{X_{m_i}}\right)}{N} \tag{11.27}$$

$$\varepsilon = \left[\frac{\sum_{i=1}^{N} \left(X_{c_i} - \overline{X}_{m_i}\right)^2}{\sum_{i=1}^{N} X_{m_i}^2}\right]^{1/2} \tag{11.28}$$

where N represents the number of data in each group, X_c is the calculated values, X_m is the measured values and the barred parameters demonstrate the average values of the parameters. Perfect agreement is achieved if R^2 and Wilmott index are 1.0, the error function is zero, the value of distortion approaches 1.0 and RMSE is small. R^2 is considered unsatisfactory as a measure of the goodness of fit for a multivariate regression relationship (Draper, 1984) and thus the validity of the model will be assessed mostly by the other parameters. The value of the five statistical parameters are provided and compared graphically following this section. The proposed ANN model is validated with new experimental data not used during the training process. Accuracy analysis is performed for the three different scales of application referred to previously and explained here.

A statistical comparison is made of wave transmission coefficients around the submerged breakwater for cases involving interpolation. Conditions are as set out in Tables 11.16 and 11.17. Figure 11.74 shows the correlation between predicted and measured values within different regions around the breakwater for a case with input parameters within the range of the training data set. A specific wave period of 1 s and total water depth and submergence depth of 250 mm and 62 mm, respectively, were employed in this test. Incident wave height varied from 22 mm up to 81 mm. The values of the statistical indicators and the graphical comparison both indicate that the ANN model is able to calculate the wave transmission coefficient with a high accuracy. The predicted values gave high R^2 (97%) and Wilmott number I_w (74%) and low values of ε (0.23) and RMSE (0.04) as well as a value of distortion close to 1.0, indicating very good accuracy (Table 11.19).

Figure 11.75 shows an error contour map for the data plotted in Figure 11.74; this indicates the error in different regions around the

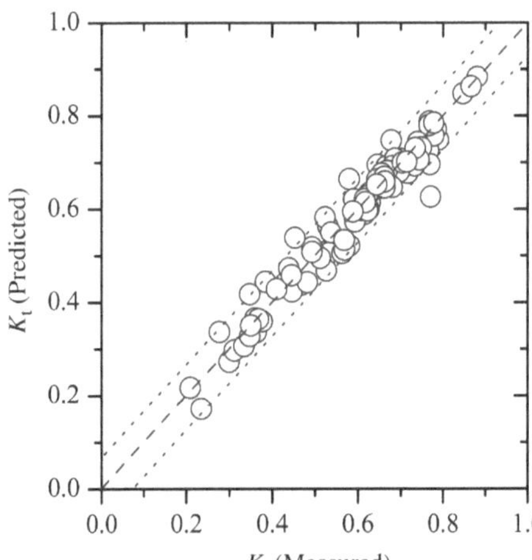

Figure 11.74 Correlation of RBF and experimental values for interpolation cases (dataset: Set 2; test with: H_i = 22,41,60,81 mm, T = 1 s, h_s = 62 mm, h = 250 mm).

Table 11.19 Statistical evaluation of nondimensional RBF model (dataset: Set 2)

Statistical index	I_w	ε	R^2	β	$RMSE$
Nondimensional RBF model	0.74	0.23	0.97	0.99	0.04

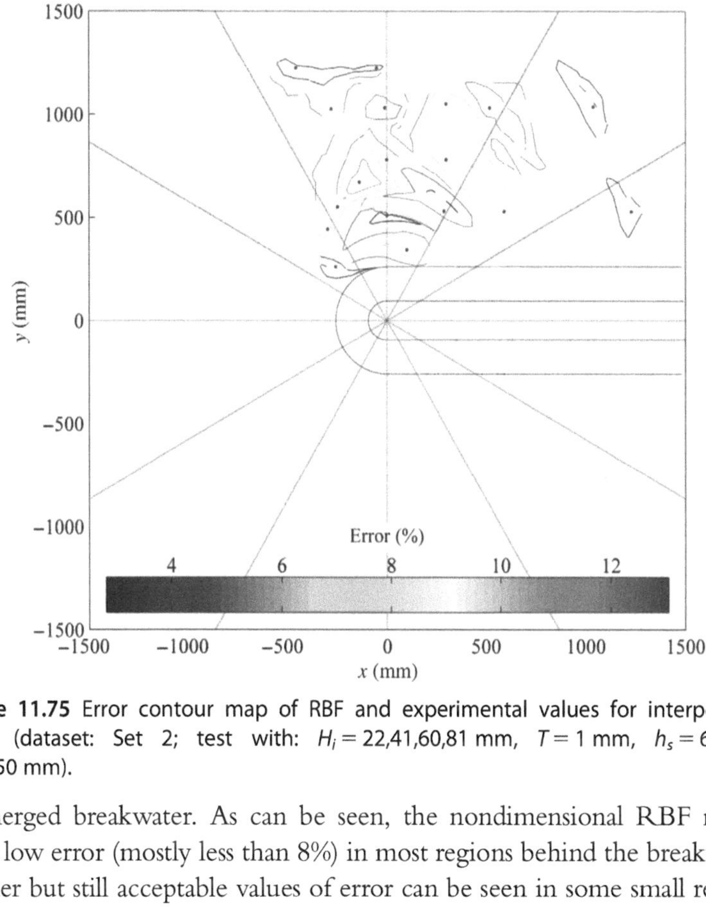

Figure 11.75 Error contour map of RBF and experimental values for interpolation cases (dataset: Set 2; test with: H_i = 22,41,60,81 mm, T = 1 mm, h_s = 62 mm, h = 250 mm).

submerged breakwater. As can be seen, the nondimensional RBF model gives low error (mostly less than 8%) in most regions behind the breakwater. Higher but still acceptable values of error can be seen in some small regions (i.e. far end of regions 3 and 4 and around the head) which are around 12%.

A new data set is employed to test the proposed model in cases involving extrapolation. The parametric range of data is presented in Tables 11.20 and 11.21. Figure 11.76 shows the correlation between predicted and measured values within different regions around the breakwater for a case with input parameters (wave period and water depth over and in front of the structure) within the range of the training data but incident wave heights outside the range of the training data. A specific wave period of 1 s and total water depth and submergence depth equal to 250 mm and 62 mm respectively are employed in this test. Incident wave

Table 11.20 The range of evaluation data in dimensional form (extrapolation; dataset: Set 3)

Parameter	Mean	Standard deviation	Max	Min
H_i (mm)	52	24	10	100
L_o (mm)	2311	808	1427	3380
h_s (mm)	67	36	27	112
h (mm)	258	33	222	300
θ (rad)	1.56	0.56	0	3.14
r (mm)	904	454	10	2099

Table 11.21 The range of nondimensional form of the evaluation data (extrapolation; dataset: Set 3)

Parameter	Mean	Standard deviation	Max	Min
H_i/L_o	0.03	0.02	0.00	0.07
h_s/H_i	1.74	1.68	0.27	11.25
H_i/h	0.25	0.10	0.12	0.37
h_s/h	0.21	0.10	0.03	0.44
θ	1.56	0.56	0.00	3.14
r/L_o	0.44	0.28	0.00	1.47
K_t	0.6	0.16	0.17	0.99

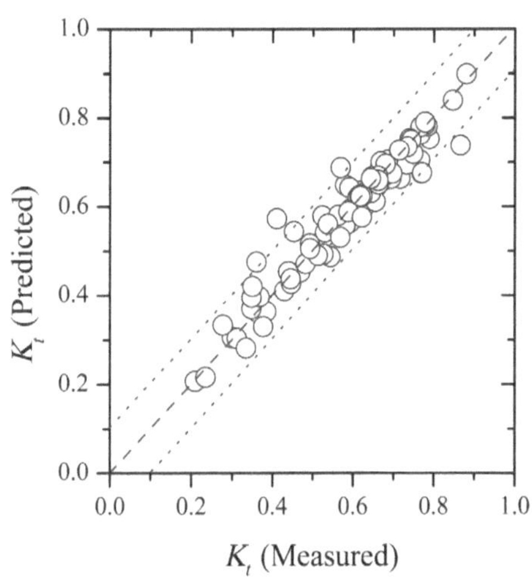

Figure 11.76 Correlation of RBF model results and experiment in the case of extrapolation (dataset: Set 3; test with: $H_i = 90,93,96,100$ mm, $T = 1$ s, $h_s = 62$ mm, $h = 250$ mm).

Table 11.22 Statistical evaluation of nondimensional RBF models (extrapolation; Set 3)

Statistical index	I_w	ε	R^2	β	*RMSE*
RBF model	0.74	0.35	0.90	1.00	0.10

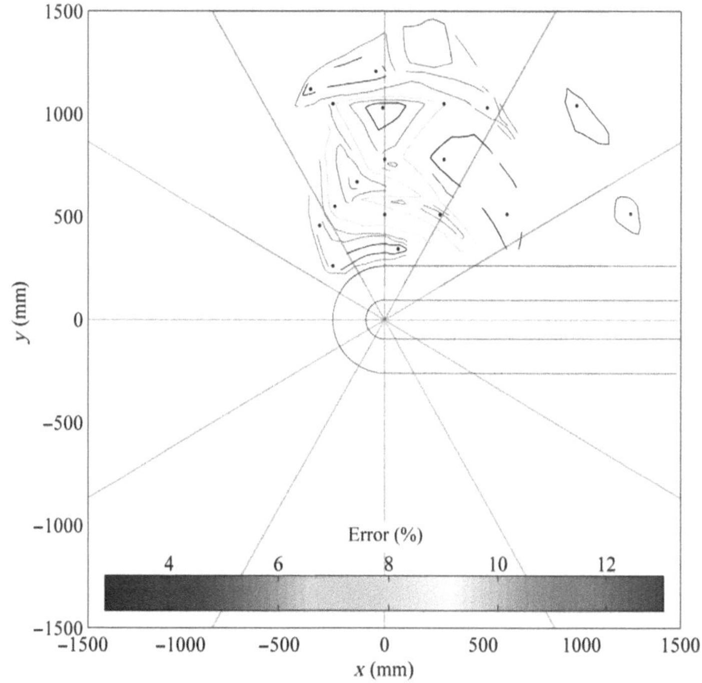

Figure 11.77 Error contour map of RBF model results and experiment for cases of extrapolation (dataset: Set 3; test with: $H_i = 90,93,96,100$ mm, $T = 1$ s, $h_s = 62$ mm, $h = 250$ mm).

height varied from 90 mm up to 100 mm. The values of the statistical indicators and graphical comparison indicate that the ANN model is able to predict wave transmission coefficients with high accuracy. The values calculated by the ANN provide acceptable values of R^2 (greater than 90%) and Wilmott number I_w (greater than 70%) and low value of ε (0.35) and RMSE (0.1) as well as a distortion value of 1.0 that indicates good accuracy (Table 11.22). The graph shows very good agreements in most regions with the higher correlation in regions behind the breakwater, reducing closer to the region around the end of the breakwater.

Figure 11.77 is an error contour map of the same wave transmission calculated by the RBF model as plotted in Figure 11.76. This shows the

Table 11.23 Statistical evaluation of the nondimensional RBF model in larger scale model with a scale factor of 4

Statistical index	I_w	ε	R^2	β	*RMSE*
Upper scale model	0.88	0.07	0.84	1.01	0.11

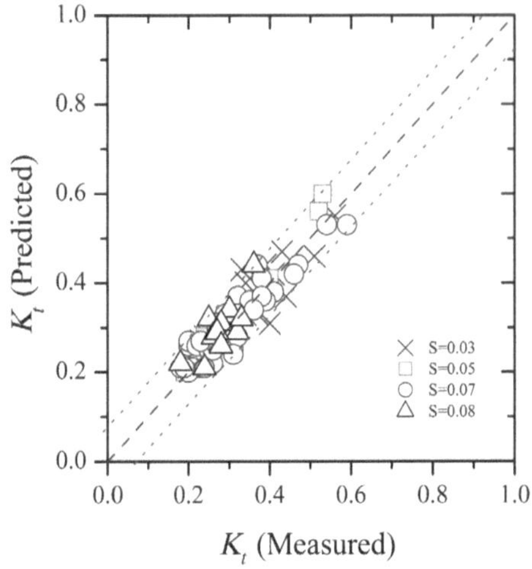

Figure 11.78 Graphical comparison of the nondimensional RBF model tested in larger scale (dataset: Set 2; test with: $H_i = 88,135,180,210,240$ mm, $T = 2$ s, $h_s = 250$ mm, $h = 1000$ mm, and wave steepness = 0.03, 0.05, 0.07 and 0.08).

error in different regions around the submerged breakwater. As can be seen, the RBF model performs with low error (mostly less than 10%) in most regions behind the breakwater. Less accurate results (errors around 15%) are found in some areas in regions 2, 3 and 4 and a narrow band at the far end of the breakwater.

A set of experimental data from the large flume was employed to assess the performance of the ANN model. The evaluation data for these larger scale tests involved a scale factor of 4. The statistical comparison is presented in Table 11.23 and Figure 11.78 shows the correlation between predicted and measured values for these tests. The values of the statistical indicators and the graphical comparison indicate that the ANN model is able to calculate wave transmission coefficients with good accuracy. The values predicted by the ANN model provide fairly good R^2 (84%) and Wilmott number I_w (88%). Low value of ε (0.07) and convincing value of RMSE (0.11) were also obtained. A value of distortion close to 1.0 also points to the good accuracy of the predictions.

Table 11.24 Sensitivity analysis of RBF model

Omitted parameter	H_i/L_o	h_s/H_i	H_i/h	h_s/h	θ	r/L_o
R^2	0.77	0.90	0.87	0.85	0.50	0.76
RMSE	0.14	0.10	0.11	0.11	0.21	0.18

The sensitivity analysis illustrates that the polar coordinate θ is the most significant parameter in the 3D wave field around the breakwater, with wave height and energy varying spatially around the structure. This confirms the importance of the 3D wave height prediction as available 2D models are inefficient in dealing with spatial variation of wave height behind the breakwater.

From Table 11.24, it can be seen that the parameters H_i/L_o and r/L_o have the next lowest correlation coefficients (0.77 and 0.76, respectively) and the highest error ($RMSE = 0.14$ and 0.18); they also have a considerable influence on the results of the networks. It should be noted that the division into training data, validation data and test data for the sensitivity analysis is carried out in the same way as for the construction of the main ANN model.

Comparisons reveal the encouraging performance and reliable predictions from the ANN model. The performance of the ANN model with nondimensional parameters is better than with dimensional parameters in prediction of the wave transmission coefficient around the submerged breakwater. The ANN model based on RBFs presents more acceptable results for K_t compared to those using MLPs. Accuracy analysis illustrates very accurate results for the RBF model when conditions lie within the range used for training the model. When extrapolating wave height outside that range, although results seem less accurate than for interpolation cases, the RBF model still performs very well and presents very acceptable outputs. The RBF model was assessed under larger scale conditions. Outputs under these conditions also showed good agreement. This shows that the performance of the model is not affected significantly by scale changes and the model has the potential to be used in real applications. The sensitivity analysis showed that the nondimensional parameter θ is the most effective among all the parameters that influence the process. H_i/L_o and r/L_o are also found to have a considerable influence on the results. Without these parameters as inputs to the network, the lowest correlation coefficient and the greatest error are obtained. It is concluded that the proposed model offers the potential of a design tool to predict spatial wave transmission coefficients behind submerged breakwaters.

Finally, Figure 11.79 also shows a comparison between the predicted values and measured data in some specific points, which shows a relatively good agreement.

Tables 11.25 and 11.26 also present a selected data set as a sample of data (small scale) used in the modeling in the form of dimensional and nondimensional data with the measured and calculated values of K_t.

Figure 11.80 also shows the correlation between predicted and measured values for these sample tests (small scale). Table 11.27 also shows selected data as a sample of data (large scale) used in the modeling in the form of nondimensional data with the measured and calculated values of K_t. Tables 11.28 and 11.29 also represent the details of the main parameters for the three-dimensional small and large scale tests.

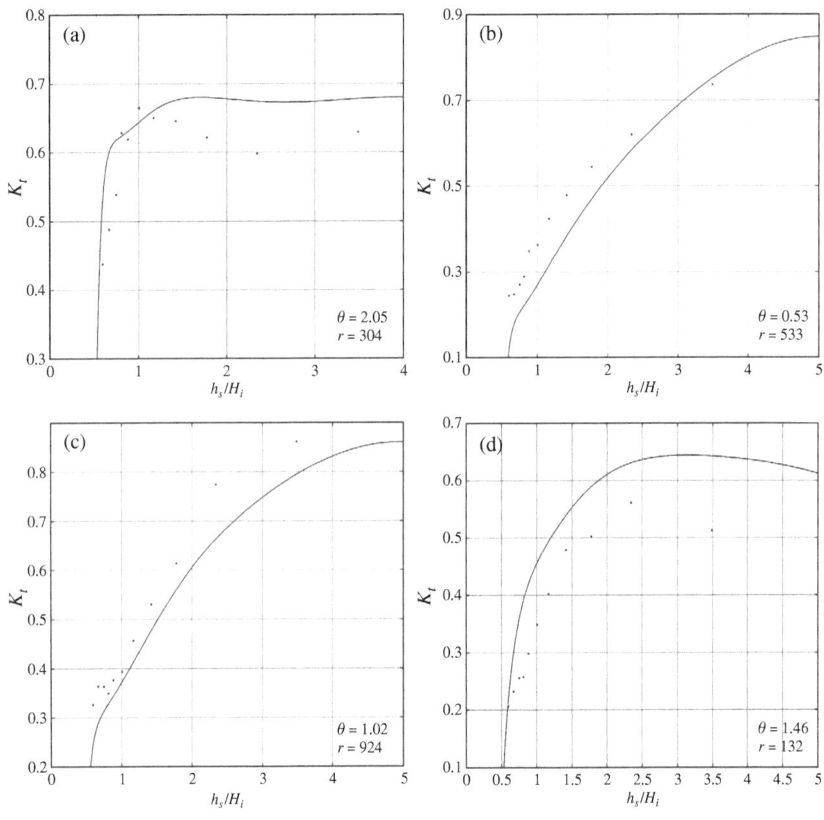

Figure 11.79 Comparisons between the prediction and measured values in some specific points.

Table 11.25 A selected set of data as a sample data (small scale) used in the modeling in dimensional form with the measured values of K_t (All dimensions are in Millimeter. Angles are in Radian)

No	H_l	L_o	h_s	h	theta	r	K_t measured
1	66.11069	976.1623	26.8	214.3	−2.3758	360.6938	0.66
2	65.10727	976.1623	26.8	214.3	−2.3758	360.6938	0.69
3	33.47923	976.1623	0	187.5	−2.39464	367.9674	0.58
4	68.91388	976.1623	0	187.5	−2.39464	367.9674	0.54
5	67.63244	976.1623	0	187.5	−2.39464	367.9674	0.59
6	73.39554	976.1623	0	187.5	−2.39464	367.9674	0.52
7	68.60693	1963.996	0	187.5	−2.39464	367.9674	0.47
8	107.7005	1963.996	0	187.5	−2.39464	367.9674	0.55
9	142.6948	1963.996	0	187.5	−2.39464	367.9674	0.54
10	163.9992	1963.996	0	187.5	−2.39464	367.9674	0.53
11	55.04715	1963.996	26.8	214.3	−2.3758	360.6938	0.62
12	100.5329	1963.996	26.8	214.3	−2.3758	360.6938	0.63
13	135.2266	1963.996	26.8	214.3	−2.3758	360.6938	0.60
14	142.42	1963.996	26.8	214.3	−2.3758	360.6938	0.64
15	33.47923	976.1623	0	187.5	2.356194	367.6955	0.51
16	68.91388	976.1623	0	187.5	2.356194	367.6955	0.34
17	67.63244	976.1623	0	187.5	2.356194	367.6955	0.38
18	73.39554	976.1623	0	187.5	2.356194	367.6955	0.40
19	66.11069	976.1623	26.8	214.3	2.356194	226.2742	0.37
20	65.10727	976.1623	26.8	214.3	2.356194	226.2742	0.47
21	55.04715	1963.996	26.8	214.3	2.356194	226.2742	0.56
22	100.5329	1963.996	26.8	214.3	2.356194	226.2742	0.33
23	135.2266	1963.996	26.8	214.3	2.356194	226.2742	0.23
24	142.42	1963.996	26.8	214.3	2.356194	226.2742	0.22
25	68.60693	1963.996	0	187.5	2.356194	367.6955	0.37
26	107.7005	1963.996	0	187.5	2.356194	367.6955	0.36
27	142.6948	1963.996	0	187.5	2.356194	367.6955	0.24
28	163.9992	1963.996	0	187.5	2.356194	367.6955	0.18
29	142.42	1963.996	26.8	214.3	1.997424	604.1523	0.38
30	135.2266	1963.996	26.8	214.3	1.997424	604.1523	0.41
31	55.04715	1963.996	26.8	214.3	1.997424	604.1523	0.60
32	66.11069	976.1623	26.8	214.3	1.997424	604.1523	0.46
33	65.10727	976.1623	26.8	214.3	1.997424	604.1523	0.55
34	100.5329	1963.996	26.8	214.3	1.997424	604.1523	0.53
35	33.47923	976.1623	0	187.5	1.776788	684.4706	0.45
36	68.91388	976.1623	0	187.5	1.776788	684.4706	0.14
37	67.63244	976.1623	0	187.5	1.776788	684.4706	0.22
38	73.39554	976.1623	0	187.5	1.776788	684.4706	0.19
39	68.60693	1963.996	0	187.5	1.776788	684.4706	0.44

(*Continued*)

Table 11.25 (Continued)

No	H_i	L_o	h_s	h	theta	r	K_t measured
40	107.7005	1963.996	0	187.5	1.776788	684.4706	0.33
41	142.6948	1963.996	0	187.5	1.776788	684.4706	0.28
42	163.9992	1963.996	0	187.5	1.776788	684.4706	0.24
43	66.11069	976.1623	26.8	214.3	1.813533	1081.712	0.35
44	65.10727	976.1623	26.8	214.3	1.813533	1081.712	0.32
45	55.04715	1963.996	26.8	214.3	1.813533	1081.712	0.19
46	100.5329	1963.996	26.8	214.3	1.813533	1081.712	0.17
47	135.2266	1963.996	26.8	214.3	1.813533	1081.712	0.39
48	142.42	1963.996	26.8	214.3	1.813533	1081.712	0.36
49	66.11069	976.1623	26.8	214.3	1.570796	780	0.23
50	66.11069	976.1623	26.8	214.3	1.570796	1050	0.14
51	65.10727	976.1623	26.8	214.3	1.570796	780	0.25
52	65.10727	976.1623	26.8	214.3	1.570796	1050	0.16
53	55.04715	1963.996	26.8	214.3	1.570796	780	0.43
54	55.04715	1963.996	26.8	214.3	1.570796	1050	0.38
55	100.5329	1963.996	26.8	214.3	1.570796	780	0.25
56	100.5329	1963.996	26.8	214.3	1.570796	1050	0.23
57	135.2266	1963.996	26.8	214.3	1.570796	780	0.27
58	135.2266	1963.996	26.8	214.3	1.570796	1050	0.22
59	142.42	1963.996	26.8	214.3	1.570796	780	0.27
60	142.42	1963.996	26.8	214.3	1.570796	1050	0.23
61	33.47923	976.1623	0	187.5	1.570796	510	0.34
62	33.47923	976.1623	0	187.5	1.580505	1030.049	0.37
63	68.91388	976.1623	0	187.5	1.570796	510	0.20
64	68.91388	976.1623	0	187.5	1.580505	1030.049	0.12
65	67.63244	976.1623	0	187.5	1.570796	510	0.20
66	67.63244	976.1623	0	187.5	1.580505	1030.049	0.20
67	73.39554	976.1623	0	187.5	1.570796	510	0.18
68	73.39554	976.1623	0	187.5	1.580505	1030.049	0.16
69	68.60693	1963.996	0	187.5	1.570796	510	0.40
70	68.60693	1963.996	0	187.5	1.580505	1030.049	0.33
71	66.11069	976.1623	26.8	214.3	1.551568	520.0961	0.29
72	65.10727	976.1623	26.8	214.3	1.551568	520.0961	0.32
73	55.04715	1963.996	26.8	214.3	1.551568	520.0961	0.51
74	100.5329	1963.996	26.8	214.3	1.551568	520.0961	0.30
75	135.2266	1963.996	26.8	214.3	1.551568	520.0961	0.29
76	142.42	1963.996	26.8	214.3	1.551568	520.0961	0.32
77	107.7005	1963.996	0	187.5	1.580505	1030.049	0.27
78	107.7005	1963.996	0	187.5	1.570796	510	0.28
79	142.6948	1963.996	0	187.5	1.570796	510	0.26
80	142.6948	1963.996	0	187.5	1.580505	1030.049	0.23
81	163.9992	1963.996	0	187.5	1.570796	510	0.26
82	163.9992	1963.996	0	187.5	1.580505	1030.049	0.20
83	66.11069	976.1623	26.8	214.3	1.292497	1092.016	0.30
84	66.11069	976.1623	26.8	214.3	1.274681	1850.54	0.27
85	65.10727	976.1623	26.8	214.3	1.292497	1092.016	0.31

(Continued)

Table 11.25 (Continued)

No	H_i	L_o	h_s	h	theta	r	K_t measured
86	65.10727	976.1623	26.8	214.3	1.274681	1850.54	0.31
87	55.04715	1963.996	26.8	214.3	1.292497	1092.016	0.39
88	55.04715	1963.996	26.8	214.3	1.274681	1850.54	0.34
89	100.5329	1963.996	26.8	214.3	1.292497	1092.016	0.25
90	100.5329	1963.996	26.8	214.3	1.274681	1850.54	0.26
91	135.2266	1963.996	26.8	214.3	1.292497	1092.016	0.21
92	135.2266	1963.996	26.8	214.3	1.274681	1850.54	0.22
93	142.42	1963.996	26.8	214.3	1.292497	1092.016	0.22
94	142.42	1963.996	26.8	214.3	1.274681	1850.54	0.21
95	33.47923	976.1623	0	187.5	1.266795	1603.527	0.20
96	68.91388	976.1623	0	187.5	1.266795	1603.527	0.26
97	67.63244	976.1623	0	187.5	1.266795	1603.527	0.23
98	73.39554	976.1623	0	187.5	1.266795	1603.527	0.25
99	68.60693	1963.996	0	187.5	1.266795	1603.527	0.20
100	107.7005	1963.996	0	187.5	1.266795	1603.527	0.18
101	142.6948	1963.996	0	187.5	1.266795	1603.527	0.20
102	163.9992	1963.996	0	187.5	1.266795	1603.527	0.19
103	66.11069	976.1623	26.8	214.3	1.203622	835.7033	0.19
104	65.10727	976.1623	26.8	214.3	1.203622	835.7033	0.20
105	55.04715	1963.996	26.8	214.3	1.203622	835.7033	0.33
106	100.5329	1963.996	26.8	214.3	1.203622	835.7033	0.28
107	135.2266	1963.996	26.8	214.3	1.203622	835.7033	0.26
108	142.42	1963.996	26.8	214.3	1.203622	835.7033	0.25
109	66.11069	976.1623	26.8	214.3	1.070129	604.1523	0.29
110	65.10727	976.1623	26.8	214.3	1.070129	604.1523	0.27
111	55.04715	1963.996	26.8	214.3	1.070129	604.1523	0.41
112	100.5329	1963.996	26.8	214.3	1.070129	604.1523	0.32
113	135.2266	1963.996	26.8	214.3	1.070129	604.1523	0.31
114	142.42	1963.996	26.8	214.3	1.070129	604.1523	0.31
115	33.47923	976.1623	0	187.5	1.068706	581.8075	0.25
116	68.91388	976.1623	0	187.5	1.068706	581.8075	0.18
117	67.63244	976.1623	0	187.5	1.068706	581.8075	0.23
118	73.39554	976.1623	0	187.5	1.068706	581.8075	0.20
119	68.60693	1963.996	0	187.5	1.068706	581.8075	0.23
120	107.7005	1963.996	0	187.5	1.068706	581.8075	0.24
121	142.6948	1963.996	0	187.5	1.068706	581.8075	0.24
122	163.9992	1963.996	0	187.5	1.068706	581.8075	0.25
123	33.47923	976.1623	0	187.5	1.103273	1153.82	0.24
124	68.91388	976.1623	0	187.5	1.103273	1153.82	0.26
125	67.63244	976.1623	0	187.5	1.103273	1153.82	0.22
126	73.39554	976.1623	0	187.5	1.103273	1153.82	0.25
127	68.60693	1963.996	0	187.5	1.103273	1153.82	0.18
128	107.7005	1963.996	0	187.5	1.103273	1153.82	0.18
129	142.6948	1963.996	0	187.5	1.103273	1153.82	0.19
130	163.9992	1963.996	0	187.5	1.103273	1153.82	0.20

Table 11.26 Selected data as a sample of data (small scale) used in the modeling in the form of nondimensional data with the calculated values of K_t corresponding to the data in Table 11.25

No	H_i/L_o	h_s/H_i	H_i/h	h_s/h	θ	r/L_o	K_t calculated
1	0.07	0.41	0.31	0.13	−2.38	0.37	0.57
2	0.07	0.41	0.30	0.13	−2.38	0.37	0.57
3	0.03	0.00	0.18	0.00	−2.39	0.38	0.44
4	0.07	0.00	0.37	0.00	−2.39	0.38	0.53
5	0.07	0.00	0.36	0.00	−2.39	0.38	0.53
6	0.08	0.00	0.39	0.00	−2.39	0.38	0.56
7	0.03	0.00	0.37	0.00	−2.39	0.19	0.65
8	0.05	0.00	0.57	0.00	−2.39	0.19	0.67
9	0.07	0.00	0.76	0.00	−2.39	0.19	0.65
10	0.08	0.00	0.87	0.00	−2.39	0.19	0.60
11	0.03	0.49	0.26	0.13	−2.38	0.18	0.75
12	0.05	0.27	0.47	0.13	−2.38	0.18	0.75
13	0.07	0.20	0.63	0.13	−2.38	0.18	0.74
14	0.07	0.19	0.66	0.13	−2.38	0.18	0.73
15	0.03	0.00	0.18	0.00	2.36	0.38	0.46
16	0.07	0.00	0.37	0.00	2.36	0.38	0.41
17	0.07	0.00	0.36	0.00	2.36	0.38	0.41
18	0.08	0.00	0.39	0.00	2.36	0.38	0.41
19	0.07	0.41	0.31	0.13	2.36	0.23	0.44
20	0.07	0.41	0.30	0.13	2.36	0.23	0.44
21	0.03	0.49	0.26	0.13	2.36	0.12	0.55
22	0.05	0.27	0.47	0.13	2.36	0.12	0.47
23	0.07	0.20	0.63	0.13	2.36	0.12	0.42
24	0.07	0.19	0.66	0.13	2.36	0.12	0.41
25	0.03	0.00	0.37	0.00	2.36	0.19	0.47
26	0.05	0.00	0.57	0.00	2.36	0.19	0.44
27	0.07	0.00	0.76	0.00	2.36	0.19	0.39
28	0.08	0.00	0.87	0.00	2.36	0.19	0.36
29	0.07	0.19	0.66	0.13	2.00	0.31	0.37
30	0.07	0.20	0.63	0.13	2.00	0.31	0.38
31	0.03	0.49	0.26	0.13	2.00	0.31	0.53
32	0.07	0.41	0.31	0.13	2.00	0.62	0.42
33	0.07	0.41	0.30	0.13	2.00	0.62	0.42
34	0.05	0.27	0.47	0.13	2.00	0.31	0.44
35	0.03	0.00	0.18	0.00	1.78	0.70	0.37
36	0.07	0.00	0.37	0.00	1.78	0.70	0.31
37	0.07	0.00	0.36	0.00	1.78	0.70	0.31
38	0.08	0.00	0.39	0.00	1.78	0.70	0.31
39	0.03	0.00	0.37	0.00	1.78	0.35	0.37
40	0.05	0.00	0.57	0.00	1.78	0.35	0.32
41	0.07	0.00	0.76	0.00	1.78	0.35	0.29

(Continued)

Table 11.26 (Continued)

No	H_i/L_o	h_s/H_i	H_i/h	h_s/h	θ	r/L_o	K_t calculated
42	0.08	0.00	0.87	0.00	1.78	0.35	0.28
43	0.07	0.41	0.31	0.13	1.81	1.11	0.36
44	0.07	0.41	0.30	0.13	1.81	1.11	0.37
45	0.03	0.49	0.26	0.13	1.81	0.55	0.51
46	0.05	0.27	0.47	0.13	1.81	0.55	0.42
47	0.07	0.20	0.63	0.13	1.81	0.55	0.36
48	0.07	0.19	0.66	0.13	1.81	0.55	0.34
49	0.07	0.41	0.31	0.13	1.57	0.80	0.34
50	0.07	0.41	0.31	0.13	1.57	1.08	0.33
51	0.07	0.41	0.30	0.13	1.57	0.80	0.34
52	0.07	0.41	0.30	0.13	1.57	1.08	0.33
53	0.03	0.49	0.26	0.13	1.57	0.40	0.47
54	0.03	0.49	0.26	0.13	1.57	0.53	0.47
55	0.05	0.27	0.47	0.13	1.57	0.40	0.36
56	0.05	0.27	0.47	0.13	1.57	0.53	0.37
57	0.07	0.20	0.63	0.13	1.57	0.40	0.30
58	0.07	0.20	0.63	0.13	1.57	0.53	0.31
59	0.07	0.19	0.66	0.13	1.57	0.40	0.30
60	0.07	0.19	0.66	0.13	1.57	0.53	0.30
61	0.03	0.00	0.18	0.00	1.57	0.52	0.33
62	0.03	0.00	0.18	0.00	1.58	1.06	0.33
63	0.07	0.00	0.37	0.00	1.57	0.52	0.26
64	0.07	0.00	0.37	0.00	1.58	1.06	0.27
65	0.07	0.00	0.36	0.00	1.57	0.52	0.26
66	0.07	0.00	0.36	0.00	1.58	1.06	0.27
67	0.08	0.00	0.39	0.00	1.57	0.52	0.26
68	0.08	0.00	0.39	0.00	1.58	1.06	0.27
69	0.03	0.00	0.37	0.00	1.57	0.26	0.31
70	0.03	0.00	0.37	0.00	1.58	0.52	0.35
71	0.07	0.41	0.31	0.13	1.55	0.53	0.33
72	0.07	0.41	0.30	0.13	1.55	0.53	0.33
73	0.03	0.49	0.26	0.13	1.55	0.26	0.45
74	0.05	0.27	0.47	0.13	1.55	0.26	0.34
75	0.07	0.20	0.63	0.13	1.55	0.26	0.29
76	0.07	0.19	0.66	0.13	1.55	0.26	0.28
77	0.05	0.00	0.57	0.00	1.58	0.52	0.29
78	0.05	0.00	0.57	0.00	1.57	0.26	0.26
79	0.07	0.00	0.76	0.00	1.57	0.26	0.25
80	0.07	0.00	0.76	0.00	1.58	0.52	0.27
81	0.08	0.00	0.87	0.00	1.57	0.26	0.24
82	0.08	0.00	0.87	0.00	1.58	0.52	0.26
83	0.07	0.41	0.31	0.13	1.29	1.12	0.29
84	0.07	0.41	0.31	0.13	1.27	1.90	0.31
85	0.07	0.41	0.30	0.13	1.29	1.12	0.29

(Continued)

Table 11.26 (Continued)

No	H_i/L_o	h_s/H_i	H_i/h	h_s/h	θ	r/L_o	K_t calculated
86	0.07	0.41	0.30	0.13	1.27	1.90	0.31
87	0.03	0.49	0.26	0.13	1.29	0.56	0.44
88	0.03	0.49	0.26	0.13	1.27	0.94	0.40
89	0.05	0.27	0.47	0.13	1.29	0.56	0.32
90	0.05	0.27	0.47	0.13	1.27	0.94	0.28
91	0.07	0.20	0.63	0.13	1.29	0.56	0.26
92	0.07	0.20	0.63	0.13	1.27	0.94	0.22
93	0.07	0.19	0.66	0.13	1.29	0.56	0.26
94	0.07	0.19	0.66	0.13	1.27	0.94	0.21
95	0.03	0.00	0.18	0.00	1.27	1.64	0.31
96	0.07	0.00	0.37	0.00	1.27	1.64	0.25
97	0.07	0.00	0.36	0.00	1.27	1.64	0.25
98	0.08	0.00	0.39	0.00	1.27	1.64	0.24
99	0.03	0.00	0.37	0.00	1.27	0.82	0.29
100	0.05	0.00	0.57	0.00	1.27	0.82	0.22
101	0.07	0.00	0.76	0.00	1.27	0.82	0.20
102	0.08	0.00	0.87	0.00	1.27	0.82	0.20
103	0.07	0.41	0.31	0.13	1.20	0.86	0.28
104	0.07	0.41	0.30	0.13	1.20	0.86	0.29
105	0.03	0.49	0.26	0.13	1.20	0.43	0.43
106	0.05	0.27	0.47	0.13	1.20	0.43	0.31
107	0.07	0.20	0.63	0.13	1.20	0.43	0.26
108	0.07	0.19	0.66	0.13	1.20	0.43	0.25
109	0.07	0.41	0.31	0.13	1.07	0.62	0.27
110	0.07	0.41	0.30	0.13	1.07	0.62	0.28
111	0.03	0.49	0.26	0.13	1.07	0.31	0.42
112	0.05	0.27	0.47	0.13	1.07	0.31	0.29
113	0.07	0.20	0.63	0.13	1.07	0.31	0.25
114	0.07	0.19	0.66	0.13	1.07	0.31	0.24
115	0.03	0.00	0.18	0.00	1.07	0.60	0.28
116	0.07	0.00	0.37	0.00	1.07	0.60	0.21
117	0.07	0.00	0.36	0.00	1.07	0.60	0.21
118	0.08	0.00	0.39	0.00	1.07	0.60	0.21
119	0.03	0.00	0.37	0.00	1.07	0.30	0.27
120	0.05	0.00	0.57	0.00	1.07	0.30	0.21
121	0.07	0.00	0.76	0.00	1.07	0.30	0.21
122	0.08	0.00	0.87	0.00	1.07	0.30	0.22
123	0.03	0.00	0.18	0.00	1.10	1.18	0.28
124	0.07	0.00	0.37	0.00	1.10	1.18	0.22
125	0.07	0.00	0.36	0.00	1.10	1.18	0.22
126	0.08	0.00	0.39	0.00	1.10	1.18	0.22
127	0.03	0.00	0.37	0.00	1.10	0.59	0.28
128	0.05	0.00	0.57	0.00	1.10	0.59	0.22
129	0.07	0.00	0.76	0.00	1.10	0.59	0.21
130	0.08	0.00	0.87	0.00	1.10	0.59	0.21

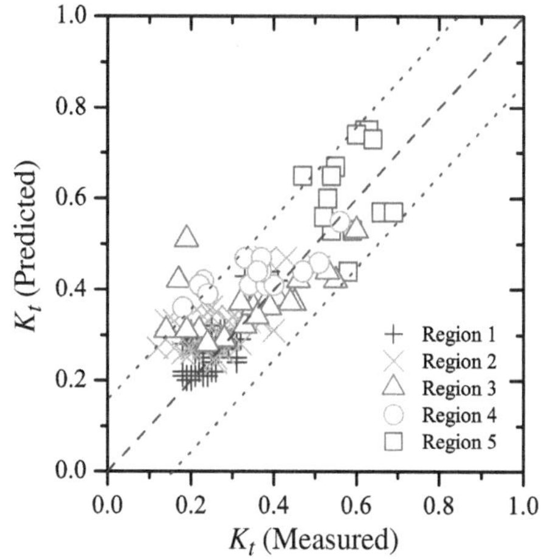

Figure 11.80 Graphical comparison of the RBF model results with measured data for a sample data; small scale (Tables 11.25 and 11.26).

Table 11.27 Selected data as a sample of data (large scale) used in the modeling in the form of nondimensional data with the measured and calculated values of K_t

Test No.	H_i/L_o	h_s/H_i	H_i/h	h_s/h	θ	r/L_o	K_t measured	K_t predicted
1	0.02	5.28	0.05	0.25	0.46	0.05	0.62	0.68
2	0.03	5.05	0.05	0.25	0.32	0.08	0.62	0.65
3	0.04	5.96	0.04	0.25	0.56	0.22	0.58	0.57
4	0.05	5.86	0.04	0.25	0.56	0.26	0.64	0.54
5	0.02	5.28	0.05	0.25	0.79	0.03	0.61	0.65
6	0.03	5.05	0.05	0.25	0.79	0.04	0.77	0.62
7	0.04	5.96	0.04	0.25	0.79	0.27	0.55	0.57
8	0.05	5.86	0.04	0.25	0.79	0.31	0.62	0.53
9	0.02	5.43	0.05	0.25	1.22	0.12	0.65	0.64
10	0.03	5.03	0.05	0.25	1.07	0.17	0.72	0.61
11	0.04	5.96	0.04	0.25	1.11	0.21	0.63	0.55
12	0.05	5.57	0.04	0.25	1.07	0.35	0.63	0.52
13	0.02	5.30	0.05	0.25	1.33	0.09	0.59	0.63
14	0.03	5.03	0.05	0.25	1.42	0.17	0.66	0.60
15	0.04	5.96	0.04	0.25	1.33	0.20	0.67	0.55
16	0.05	5.57	0.04	0.25	1.42	0.36	0.64	0.52
17	0.02	5.28	0.05	0.25	1.57	0.02	0.62	0.61
18	0.03	5.03	0.05	0.25	1.57	0.14	0.64	0.60
19	0.04	5.96	0.04	0.25	1.57	0.19	0.64	0.55
20	0.05	6.12	0.04	0.25	1.57	0.35	0.52	0.52
21	0.01	5.95	0.04	0.25	0.46	0.04	0.71	0.71
22	0.01	5.26	0.05	0.25	0.93	0.05	0.58	0.70
23	0.01	5.26	0.05	0.25	1.11	0.04	0.77	0.69
24	0.01	6.15	0.04	0.25	1.36	0.12	0.65	0.68
25	0.01	5.15	0.05	0.25	1.57	0.01	0.67	0.65

Table 11.28 Test conditions for the collected 3D small scale data set in 1.2 m wide flume

Test No.	T (sec)	H_i (mm)			
		Test series 1	Test series 3	Test series 5	Test series 6
		$h = 300$ (mm)	$h = 250$ (mm)	$h = 220$ (mm)	$h = 190$ (mm)
1	0.80	13	14	14	13
2		20	21	21	19
3		26	28	27	25
4		32	33	32	31
5		36	37	38	37
6		42	43	43	44
7		48	49	47	50
8		54	54	54	56
9		57	57	57	64
10		63	63	62	66
11		60	60	59	65
12	1.00	19	17	19	17
13		27	26	28	27
14		36	34	37	36
15		45	42	46	45
16		53	51	54	55
17		62	60	63	67
18		69	68	72	80
19		75	74	76	88
20		81	80	81	95
21		88	89	90	105
22	1.33	96	100	98	113
23		10	12	12	13
24		16	18	19	20
25		21	23	25	26
26		25	29	31	33
27		30	35	37	38
28		35	41	43	42
29		40	46	49	48
30		44	52	55	52
31		50	57	63	59
32		55	64	70	70
33		60	74	78	83

(*Continued*)

Table 11.28 (Continued)

Test No.	T (sec)	H_i (mm)			
		Test series 1	Test series 3	Test series 5	Test series 6
		$h = 300$ (mm)	$h = 250$ (mm)	$h = 220$ (mm)	$h = 190$ (mm)
34	2.00	22	16	17	28
35		32	24	26	41
36		42	32	34	54
37		51	40	42	64
38		59	48	49	74
39		67	56	56	83
40		74	64	64	90
41		79	72	70	97
42		85	79	76	104
43		90	86	82	109
44		93	92	88	113

Table 11.29 Test conditions for the collected data set in 2.5 m wide flume (Large scale model)

Test No.	T (sec)	a_i [Incident wave amplitude (mm)]									
		Test series:									
		1	2	3	4	5	6	7	8	9	10
1	0.66	15	34	52	70	88	105	123	–	–	–
2	0.71	14	32	50	68	85	101	114	–	–	–
3	0.76	11	26	41	56	70	84	96	–	–	–
4	0.82	12	29	46	62	77	94	108	–	–	–
5	0.92	11	31	50	68	88	104	122	–	–	–
6	1.00	9	31	54	75	95	113	126	148	157	170
7	1.10	9	30	50	69	87	104	117	140	140	169
8	1.23	8	25	41	56	70	81	91	126	139	151
9	1.39	12	29	45	60	75	81	–	–	–	–
10	1.69	15	41	62	73	88	–	–	–	–	–
11	2.00	21	49	72	95	98	113	123	–	–	–

11.8 SUMMARY

The transmitted wave field around submerged breakwaters has been investigated numerically. A wide range of 2D and 3D small scale experiments with different submergence depths and wave conditions were conducted in two wave flumes in the laboratory. A large scale 3D model was also built in the large wave tank for investigation of scale effects. A numerical method for predicting the spatial wave transmission coefficient in the shadow region of 2D/3D submerged breakwaters was proposed. Two distinct models have been developed using machine learning algorithms; these artificial neural networks, based on multilayer perceptron (MLP) and radial basis function (RBF) methods, have been designed and trained against new laboratory experimental data. Each model has been designed in terms of both dimensional and nondimensional parameters. Collected experimental data were employed to train the ANN models to provide a prediction tool for the spatial wave transmission coefficient. Comparisons between the experimental data and predictions from the trained models show that the nondimensional RBF model is best able to predict the wave field around the submerged breakwater. The performance of the model was validated in interpolation, extrapolation and at larger scale, revealing sufficient agreement with the experimental results to suggest it has potential as a design tool in real applications. Numerical modeling of wave deformation using the SWE model has also been carried out. The calibrated SWE model has been applied for comparison with the ANN model results under various conditions and model scales. Comparisons reveal the encouraging performance and reliable predictions from the ANN model. It is concluded that the proposed model offers the potential of a design tool to predict spatial wave transmission coefficients behind submerged breakwaters.

The following conclusions have been arrived at: The spatial variation of wave energy was generally found to depend on whether or not wave breaking occurred over the breakwater crest, with regions of high waves observed at different locations around and behind the breakwater. In cases with waves breaking over the breakwater, the lower wave heights were observed behind the breakwater crown on the shoreward side; for non-breaking wave conditions passing over the submerged breakwater, lower wave heights were observed in the gap between the end of the breakwater and the flume wall. The wave pattern behind the breakwater was strongly dependent on incident wave height and period, with different wave patterns observed for different incident conditions.

The performance of the ANN model with nondimensional parameters was better than with dimensional parameters. The ANN model based on RBF presents more acceptable results for K_t compared to those using MLP. Accuracy analysis illustrates very good results for the RBF model when conditions lie within the range used for training the model. When extrapolating wave height outside that range, although results seem less accurate than for interpolation cases, the RBF model still performs well and presents very acceptable outputs. The RBF model was assessed under larger scale conditions and with data collected in another wave tank with different laboratory facilities. Outputs under these conditions also showed good agreement. This shows that the performance of the model is not affected significantly by scale changes and the model has the potential to be used in real applications.

Analysis of the performance of the 2D model shows that dimensionless parameters ξ_o and H_i/h were the more influential parameters on the wave transmission coefficient calculated by the 2D RBF model. Regarding the 3D RBF model, the investigation illustrated that the dimensionless spatial parameters were the most significant parameter in the 3D wave field around the breakwater, with wave height and energy varying spatially around the structure. This confirms the importance of 3D effects on wave height prediction and highlights the inadequacy of 2D models that are unable to deal with spatial variation of wave height behind the breakwater. The nondimensional hydraulic parameters ξ_o and H_i/h were also found to have a considerable influence on the results. Without these parameters as inputs to the network, the lowest correlation coefficient and the greatest error were obtained.

Comparisons reveal the accurate performance and reliable predictions of the ANN model. The model has proved to be capable of handling wave transmission prediction. The proposed model is a significant advance in that it can be used to predict K_t consistently around submerged breakwaters in a specified range of spatial parameters, whereas existing models in the literature are restricted to only 2D prediction of K_t. The proposed model considers combined transmission and diffraction in a wide range of water depths and wave climates, whereas existing models in the literature are restricted to calculating overtopping and diffraction coefficients separately and in a limited range of input parameters, which makes them less accurate and reliable with lower usefulness and applicability.

A restricted range of input parameters has been found to give more reliable and valid predictions while out of this range the results are still reasonable. Applying the proposed ANN model for predictions, the following guidance can be given as to the most appropriate ANN modeling to predict 2D K_t behind submerged breakwaters. The proposed model is reliable in a specific range of the breakwater crest width particularly for narrow crested and moderately wide crested submerged breakwaters. Analysis reveals that the ANN model gives a relatively high reliability in a wide range of Iribarren numbers. However, outside this interval the model tends to low reliability. The model is reliable in a specific range of nondimensional wave height and submergence ratio. For larger submergence depth the ANN model is unreliable. K_t increases dramatically with submergence ratio from a low value at zero submergence to a highest value at a large submergence ratio. Beyond this point, no significant influence of the submergence ratio is seen. There is a narrow confidence band with high reliability for submergence ratio. Acceptable reliability is still achievable to some extent outside the range of training data considering the breakwater geometry particularly for different values of submergence depth and breakwater crest width.

Applying the proposed ANN model for predictions, the following guidance can be given as to the most appropriate ANN modeling to predict 3D K_t behind submerged breakwaters. The proposed model is reliable in a specific range of the breakwater crest width, particularly for the narrow crested submerged breakwaters. Analysis reveals that the ANN model gives a relatively high reliability in a relatively wide range of Iribarren numbers. Outside this interval the model tends to low reliability. The model is reliable in specific ranges of dimensionless wave height, breakwater crest width and submergence ratio. However, it still works with acceptable reliability outside this range. However, for very large submergence depth the ANN model is unreliable. K_t increases dramatically with submergence ratio from its minimum value at zero submergence to its highest value for a large value of the submergence ratio. Beyond this point, no significant influence of the submergence ratio is seen. The model is also reliable at a specific range of spatial parameters but still works with acceptable reliability outside this range, which suggests that the model is reliable across the entire range of the training data. Further from the breakwater, the model has a rather low reliability.

The SWE model was found to have low accuracy dealing with wave-induced breaking dissipation over the breakwater crest and generally within the breaking zones The SWE model was found to overestimate wave-induced breaking dissipation over the breakwater crest, consequently underestimating K_t values. Particularly as the submergence ratio decreases, a significant increase in error in K_t calculated by model is observed. This is probably due to the overestimated wave energy dissipation in the model which basically increases for steeper waves and shallower depths. It might also be because of the turbulence model adopted in this module, as the turbulence model is not designed to handle turbulence effects properly. The evaluations reveal more consistency between the results of measured data and the proposed ANN model than with SWE. It is evident that the trained ANN model has good potential to be applied in real applications with high reliability.

A simplified 3D analysis tool and initial prediction scheme was also presented. A step by step procedure for practical applications was outlined in this chapter with the intention of providing an initial prediction scheme according to the proposed ANN model described in the book. The main aim is to introduce a simplified tool for preliminary analysis and design purposes which can be used in submerged breakwaters analysis and design. A 3D spatial K_t calculator was described in the chapter, based on the proposed non-dimensional RBF model. As explained in the previous chapters and this chapter as well, the final optimized ANN model has one radial basis functions layer with 15 nodes. The simplified ANN model was proposed and presented in an easy to use and user friendly form and with a step by step procedure including details about the model mathematics.

It is concluded that the applied models offer the potential of a design tool to predict spatial wave transmission coefficients behind submerged breakwaters. The results of this study can be applied not only for wave modeling behind submerged structures but also in flow models around these structures by importing the wave data into the available models and calculating flow patterns behind such structures. A plan for future work on numerical modeling of flow around submerged breakwaters, sediment transport, morphodynamic modeling and beach response is useful to be conducted, to provide a 3D flow, sediment transport and beach response design tool using both traditional and modern numerical models using experimental data collected in the laboratory and artificial data provided by numerical models.

Regarding recommendations for further work, although the model is calibrated and designed using regular wave data, similar prediction can be adopted using significant wave height (H_s) or zero moment wave height (H_{mo}) for incident wave height and the peak wave period, Tp for wave period as inputs. However, some considerations might be required in applying regular wave results to random sea-states. More experimental tests with irregular waves on the smooth submerged breakwaters with the same geometry and water depth are needed to investigate the applicability of the proposed approaches in this book to predict K_t around fully submerged breakwaters under irregular wave conditions. Also experimental tests considering permeable submerged breakwater might be conducted to investigate the effects of permeability on the 3D wave field behind submerged breakwaters. Studying the effects of angle of incident wave attack on the wave pattern behind the submerged breakwaters in very large scale models (in a wave basin) will be useful. Breakwaters with broad crest should be studied in the future. Effects of other physical parameters such as breakwater length, distance to the shore line, and gap length can be tested in the laboratory. Having enough reliable data on irregular waves, permeability, roughness, wave attack angle, gap length, distance to the shoreline, breakwater crest length and width, etc., may be helpful in developing the current ANN model into a more comprehensive tool. Other advanced types of ANN models might be appropriate for this modeling. It is possible to make the model more widely available for future users by creating a stand-alone downloadable program. Field data measurements should also be considered. Physical modeling of segmented submerged breakwaters with different crest and gap lengths in a wave basin should be considered. The current field around the structures and morphological changes behind them as well as shoreline response are also recommended to be investigated further, both experimentally and numerically.

REFERENCES

Ahrens, J., 1987. Characteristics of reef breakwaters. Technical Report CERC-87-17, 45.

Bellotti, G., 2004. A simplified model of rip currents systems around discontinuous submerged barriers. Coast. Eng. 51 (4), 323–335.

Biésel, F., Suquet, F., 1951. Les appareils générateurs de houle en laboratoire. La Houille Blanche 6 (2), 147–165.

Broomhead, D., Lowe, D., 1988. Multivariable functional interpolation and adaptive networks. Complex Syst. 2, 321–355.

Buccino, M., Calabrese, M., 2007. Conceptual approach for prediction of wave transmission at low-crested breakwaters. J. Waterway, Port, Coast. Ocean Eng. 133 (3), 213–224.

Buccino, M., Vicinanza, D., Caceres, I., Calabrese, M., 2009. 3D wave field behind impermeable low crested structures. J. Coast. Res. 56, 477–481.

Caceres, I., Stive, M.J.F., Sanchez-Arcilla, A., Trung, L.H., 2008. Quantification of changes in current intensities induced by wave overtopping around low crested structures. Coast. Eng. 55, 113–124.

Calabrese, M., Vicinanza, V., Buccino, M., 2002. Large scale experiments on the behaviour of low crested and submerged breakwaters in presence of broken waves. Proc. 28th Int. Conf. on Coastal Engineering. ASCE, Cardiff, UK.

d'Angremond, K., van der Meer, J.W., de Jong, R.J., 1996. Wave transmission at low-crested structures. Proceedings of the 25th International Conference on Coastal Engineering. ASCE, Orlando, FL, USA.

Draper, N.R., 1984. The Box-Wetz criterion versus R^2. J. Stat. Soc. 147, 100–103.

Flick, R., Guza, R., 1980. Paddle generated waves in laboratory channels. J. Waterway, Port, Coast. Ocean Div. Amer. Soc. Civil Eng. 106 (WN1), 79–97.

Goda, Y., Ahrens, J.P., 2008. New formulation of wave transmission over and through low-crested structures. Proceedings of the 31st International Conference of Coastal Engineering, 4.

Goda, Y., Okazaki, K., Kagawa, M., 1999. Generation and evolution of harmonic wave components by abrupt depth changes. Proc. Coastal Structure 99. ASCE.

Hagan, M.T., Demuth, H.B., Beale, M.H., 1996. Neural Network Design. PWS Publication.

Haller, M.C., Darlymple, R.A., Svendsen, I.A., 2002. Experimental study of nearshore dynamics on a barred beach with rip channels. J. Geophys. Res. 107 (C6), 14-1–14-21.

Haykin, S., 1999. Neural Networks, a Comprehensive Foundation, second ed. Prentice Hall, New Jersey.

Hur, D.S., Lee, W.D., Cho, W.C., 2012. Three-dimensional flow characteristics around permeable submerged breakwaters with open inlet. Ocean Eng. 44, 100–116.

Johnson, H.K., Karambas, T.V., Avgeris, I., Zanuttigh, B., Gonzalez-Marco, D., Caceres, I., 2005. Modelling of waves and currents around submerged breakwaters. J. Coast. Eng. 52 (10–11), 949–969.

Longuet-Higgins, M., 1977. The mean forces exerted by waves on floating or submerged bodies with applications to sand bars and wave power machines. Proc. The Royal Society of London 352, 463–480.

Losada, I.J., Silva, R., Losada, M.A., 1996. 3-D non-breaking regular wave interaction with submerged breakwaters. Coast. Eng. 28, 229–248.

Losada, I., Patterson, M., Losada, M., 1997. Harmonic generation past a submerged porous step. Coast. Eng. 31 (1–4), 281–304.

Mansard, E., Funke, E., 1980. The measurement of incident and reflected spectra using a least squares method. Int. Conf. on Coastal Engineering (ICCE) Hamburg. ASCE.

Mase, H., Sakamoto, M., Sakai, T., 1995. Neural network for stability analysis of rubble-mound breakwaters. J. Waterway, Port, Coast. Ocean Eng. 121 (6), 294–299.

Massel, S., 1983. Harmonic generation by waves propagating over a submerged step. J. Coast. Eng. 7, 357–380.

Matlab,R2010b. The MathWorks Inc.

Medina, J.R., 1999. Neural network modelling of runup and overtopping. ASCE Proc. Coast. Struct. Santander 1, 421–429.

Panizzo, A., Briganti, R., 2007. Analysis of wave transmission behind low-crested breakwaters using neural networks. Coast. Eng. 54 (9), 643–656.

Rumelhart, D.E., Hinton, G.E., Williams, R.J., 1986. Learning representations by back-propagating errors. Nature 323 (6088), 533–536.

Seabrook, S., 1997. Investigation of the performance of submerged rubblemound breakwaters. Msc. thesis. Queen's University, Kingston, Ontario, Canada, Ph.D. thesis.

Seabrook, S.R., Hall, K.R., 1998. Wave transmission at submerged rubble mound breakwaters. Proceedings of the 26th Int. Conf. on Coastal Engineering. ASCE.

Seelig, W., 1980. Two-dimensional tests of wave transmission and reflection characteristics of laboratory breakwaters. Tech. rep., Technical report, CERC, Fort Belvoir, Report No. 80-1.

Sharif Ahmadian, A., Simons, R.R., 2012. 3-D Wave Field Around Submerged Breakwater. ASCE, New York.. Available from: http://dx.doi.org/10.9753/icce.v33.structures.13

Sharif Ahmadian, A., Simons, R.R., 2014. A 3D numerical model of nearshore wave field behind submerged breakwaters. Coast. Eng. 83, 190–204. Available from: http://dx.doi.org/10.1016/j.coastaleng.2013.10.016.

Stein, R., 1993. Selecting data for neural networks. AI Expert 8 (2), 42–47.

Tajziehchi, M., 2006. Experimental and numerical modelling of wave-induced current and wave transformation in presence of submerged breakwaters. Ph.D. thesis. Civil & Environmental Engineering, Faculty of Engineering, UNSW.

Tajziehchi, M., Sharif Ahmadian, A., 2009. Artificial neural network modeling of wave-induced current over submerged breakwaters, Coastal Dynamics 09, World Scientific, Tokyo, (14), pp. 1–10, doi: 10.1142/9789814282475_0017.

Ursell, F., Dean, R., Yu, Y., 1960. Forced small-amplitude water waves: a comparison of theory and experiment. J. Fluid Mech. 7 (Part 3), 33–52.

van der Meer, J., 1991. Stability and transmission at low-crested structures. Tech. rep., Technical report, Delft Hydraulic, Report No.H 453. 33p.

van der Meer, J.W., Briganti, R., Zanuttigh, B., Wang, B., 2005. Wave transmission and reflection at low-crested structures: design formulae, oblique wave attack and spectral change. J. Coast. Eng. 52 (10–11), 915–929.

van Gent, M.R.A., van den Boogaard, H.F.P., 1998. Neural network modelling of forces on vertical structures, Proc. of the Coastal Engineering Conference, vol. 2. ASCE.

van Gent, M.R.A., van den Boogaard, H.F.P., Pozueta, B., Medina, J.R., 2007. Neural network modelling of wave overtopping at coastal structures. Coast. Eng. 54 (8), 586–593.

Vicinanza, D., Caceres, I., Buccino, M., Gironella, X., Calabrese, M., 2009. Wave disturbance behind low-crested structures: diffraction and overtopping effects. Coast. Eng. 56 (11–12), 1173–1185.

Werbos, P.J., 1988. Backpropagation: past and future. In Neural Networks, 1988, IEEE International Conference on. Neural Networks, vol. 1. New York, pp. 343–353.

Wilmott, C., 1981. On the validation of models. Phys. Geogr. 2, 184–194.

INDEX

Note: Page numbers followed by "*f*," "*t*," and "*b*" refer to figures, tables, and boxes, respectively.

E

MIX
Papier | Fördert gute Waldnutzung
FSC® C083411

Kolibri 360 GmbH
Ferdinand-Jühlke-Straße 7
99095 Erfurt, Deutschland
produktsicherheit@kolibri360.de